Radioisotopes in Biology

Radioisotopes in Biology

A Practical Approach

Edited by

R. J. SLATER

Hatfield Polytechnic
School of Natural Sciences
Herts AL10 9AB, U.K.

at

OXFORD UNIVERSITY PRESS

Oxford New York Tokyo

Oxford University Press
Walton Street, Oxford OX2 6DP

Oxford is a trade mark of Oxford University Press

Published in the United States
by Oxford University Press, New York

British Library Cataloguing in Publication Data
Slater, Robert J.
Radioisotopes in biology.—(Practical approach series).
1. Biology. Research. Use of radioisotopes
I. Title II. Series
574'.072
ISBN 0–19–963080–1
ISBN 0–19–963081–X pbk

Library of Congress Cataloging in Publication Data
Radioisotopes in biology: a practical approach/edited by R. J. Slater.
(Practical approach series)
1. Radioactive tracers in biology. I. Slater, R. J. II. Series.
[DNLM: 1. Radioisotopes. WN 420 R12836]
QH324.3.R33 1990 574'.028—dc20 89–70982
ISBN 0–19–963080–1
ISBN 0–19–963081–X (pbk.)

Set by Footnote Graphics, Warminster, Wiltshire
Printed in Great Britain by
Information Press Ltd, Eynsham, Oxon

Preface

THE value of radioisotopes to the biological sciences is undisputed. Without their availability and use we would be woefully ignorant of a vast range of physiological and biochemical processes including intermediary metabolism, neurochemistry, hormone action, and protein synthesis. The range and detail of applications is vast and beyond the scope of any book. However, there are general principles and techniques common to all or most radiotracer experiments: in the first place, the decisions to use, then choose, a radioisotope; secondly, the establishment of a suitable working environment conducive to safe laboratory practice and the compliance with relevant legislation; thirdly, the principles of safe and efficient experimental design and finally, radiation detection and data analysis. This book attempts to cover those features essential to any radioisotope experiment.

The general principles of experimental design, such as how to go about choosing a particular radionuclide and compound, are discussed in the Introduction. The legal requirements, essentials of laboratory design and safe handling are discussed in Chapter 2 with supplementary summaries of legislation as applied in Europe and the United States of America provided in Chapters 9 and 10. The principles and practice of radiation detection are described in Chapters 3 and 4, covering scintillation counting and autoradiography, respectively. The remaining chapters give background theory, detailed protocols and technical details for some of the most commonly used procedures in current biological research. Finally, the book has Appendices for quick and easy reference to technical and generally applicable procedures. As with all books in this series, the emphasis is on practical considerations rather than theoretical discussion and should be of help to a wide range of radioisotope users from undergraduate project and research students to laboratory supervisors and radiation protection advisors.

Hatfield, U.K. R.J.S.
August 1989

Contents

6. *In vitro* labelling 137

A. Nucleic acids

Martin W. Cunningham, Dennis W. Harris, and Christopher R. Mundy 137

B. Proteins 191

Graham S. Bailey

Appendices

Contributors

G. S. BAILEY
Department of Chemistry and Biological Chemistry, University of Essex, Wivenhoe Park, Colchester, CO4 3SQ, UK

S. A. BOFFEY
Division of Biological Sciences, Hatfield Polytechnic, College Lane, Hatfield, Herts, AL10 9AB, UK

A. BRISTOW
Division of Endocrinology, National Institute for Biological Standards and Control, Blanche Lane, South Mimms, Potters Bar, Herts, EN6 3QG, UK

R. CUMMING
Amersham International Laboratories (Pollards Wood), White Lion Road, Little Chalfont, Bucks, HP7 9LL, UK

M. W. CUNNINGHAM
Amersham International Laboratories, White Lion Road, Little Chalfont, Bucks, HP7 9LL, UK

R. FALLON
Amersham International Laboratories (Pollards Wood), White Lion Road, Little Chalfont, Bucks HP7 9LL, UK

B. FRITH
University of Manchester, Radiological Protection Service, William Kay House, Oxford Road, Manchester, M13 9PL, UK

A. GEARING
Division of Immunobiology, National Institute for Biological Standards and Control, Blanche Lane, South Mimms, Potters Bar, Herts, EN6 3QG, UK

R. GRIFFITHS
Smith, Kline and French Research Ltd, The Frythe, Welwyn, AL6 9AR, UK

D. W. HARRIS
Amersham International Laboratories, White Lion Road, Little Chalfont, Bucks, HP7 9LL, UK

R. A. LASKEY
CRC Molecular Embryology Group, Department of Zoology, Downing Street, Cambridge, CB2 3EJ, UK

C. R. MUNDY
Amersham International Laboratories, White Lion Road, Little Chalfont, Bucks, HP7 9LL, UK

N. S. NELSON
Office of Radiation Program 5, Environmental Protection Agency (ANR 460) 401 M. St. S.W., Washington DC 20460, USA

D. PRIME
University of Manchester, Radiological Protection Service, William Kay House, Oxford Road, Manchester, M13 9PL, UK

G. SIMONNET
Institut National des Sciences et Techniques Nucléaires, Centre d'Etudes Nucléaires de Saclay, Gif-sur-Yvette, F91191, France

R. THORPE
Division of Immunobiology, National Institute for Biological Standards and Control, Blanche Lane, South Mimms, Potters Bar, Herts, EN6 3QG, UK

Abbreviations

ADC	analogue-to-digital converter
ALARA	as low as reasonably achievable
ALI	annual limit of intake
AMP	amplitude
BK	background
BSA	bovine serum albumin
CIAP	calf intestinal alkaline phosphatase
c.p.m.	counts per minute
c.p.s.	counts per second
CSF	cerebrospinal fluid
DG-6-P	deoxyglucose-6-phosphate
DMSO	dimethylsulphoxide
d.p.m.	disintegrations per minute
d.p.s.	disintegrations per second
DTT	dithiothreitol
EBSS	Earle's balanced salt solution
EDTA	ethylene diamine tetracetic acid
ELISA	enzyme linked immunosorbent assay
EM	electron microscopy
E_{max}	maximum energy
E_{mean}	mean energy
F*M*	figure of merit
GI	gastrointestinal
G–M tube	Geiger–Müller ionization monitor
HPLC	high performance liquid chromatography
IDR	Instantaneous dose rate
IRMA	immunoradiometric assay
Kb	kilobase
KTP	cordycepin 5′-triphosphate
LC/MS	liquid chromatography/mass spectroscopy
LM	light microscopy
NMR	nuclear magnetic resonance
NSB	non-specific binding
PBS	phosphate buffered saline
PEG	polyethylene glycol
PF	paraformaldehyde
PPO	2,5 diphenyloxazole
RIA	radioimmunoassay
RPA	radiation protection advisor
RPS	radiation protection supervisor
SDS	sodium dodecyl sulphate
TCA	trichloroacetic acid

Tdt	terminal deoxynucleotidyl transferase
TLD	thermoluminescent detector
UV	ultraviolet
WME	Williams' medium E

1

Introduction

ROBERT J. SLATER

1. The decision to use a radioisotope

On the grounds of safety, whenever an experiment is being considered the need for a radioisotope must be questioned: is a radioisotope absolutely essential to the experiment planned? Or put another way, are non-radioactive methods available that will achieve the same end?

If the answer to these questions is yes, that is there is a non-radioactive technique that is equally effective, then that is the one that must be used. This is a legal requirement. Only if a radioisotope is absolutely essential or if its use results in a significant improvement in benefit is the use of a radioisotope justified. This complies with current thinking and official recommendations regarding radiation hazard, that is the ALARA principle: as low as is reasonably achievable. Although legislatively applied dose limits for radiation workers apply (see Chapter 2), the basic principle of safe laboratory practice is that as little radioisotope should be used as possible.

2. Laboratory facilities

Having decided that a radioisotope is required, it is necessary to satisfy the legal requirements for its use and to establish that the necessary facilities are available. The most commonly used isotopes, ^{3}H and ^{14}C, are of low energy, represent little or no external hazard and are unlikely to require a controlled laboratory. Iodine-125 and ^{32}P, on the other hand, can represent a significant hazard and a controlled facility is required for quantities greater than 270 μCi (10 MBq) and 2.7 mCi (100 MBq) respectively. For a detailed discussion see Chapter 2 and Appendix 2.

3. The choice of labelled compound

Looking at a radioisotope catalogue for the first time can be daunting. A particular compound may appear as several entries with alternative radionuclides (e.g. ^{3}H or ^{14}C), labelling position, specific activity (i.e. Ci or Bq per mol), pack size, formulation (e.g. ethanol or aqueous solutions) and cost. In making a choice the researcher often follows the example of a previously reported experiment. Although this can be of great help, it is not necessarily so. It assumes that the previous worker made the correct choice, it may not be appropriate to the current safety standards, design of laboratory available, or take into account newly available products. Factors that influence choice vary considerably from one experiment to the next and on the facilities available, but a summary of some general features is given below.

3.1 The choice of radionuclide

The vast majority of radioisotope experiments in the biological sciences involve one of the following: ^{3}H, ^{14}C, ^{35}S, ^{32}P, ^{131}I, and ^{125}I. These are presented in order of increasing toxicity (see Chapter 2 and Appendix 2) so, all other factors being equal, use a ^{3}H compound where possible. In some cases the choice will be constrained by the licensing of your laboratory and the advice of your departmental Radiation Protection Supervisor must be sought.

All the radioisotopes listed above are β emitters except for the iodine isotopes which emit electromagnetic radiation (^{131}I emits both β and γ radiation); avoid these if possible because of difficulties in shielding (see Chapter 2). Half-lives are given in Appendix 2; the shorter the half-life the greater the potential specific activity; sometimes, therefore, ^{14}C compounds are inappropriate as they cannot be produced at high specific activity (see Section 3.3).

The energy of the β-radiation emitted increases in the order shown above, ^{3}H being the weakest. This affects the efficiency and method of detection and is discussed in more detail in later chapters. A very important, and sometimes overlooked, consequence is that ^{3}H cannot be detected by an end-window ionization counter (e.g. Geiger counter) often used as a contamination monitor (see Appendix 3). Other consequences of the low β energy of ^{3}H are its relatively low efficiency in scintillation counting and general unsuitability for autoradiography other than for specialist purposes, such as *in situ* autoradiography or as an exercise in patience.

High energy radiation, such as from ^{32}P, is more suitable for autoradiography but because the radiation spreads further there is a corresponding reduction in resolution (see Chapters 4, 6, and 7). A summary of some of the factors involved in choosing a radionuclide is given in *Table 1*. For some applications (e.g. DNA sequencing) the use of ^{35}S derivatives is a useful compromise providing high specific activity and suitable energy for autoradiography.

3.2 Labelling position

Usually there is a choice between different specifically-labelled and uniformly-labelled compounds; the latter generally being cheaper. Few general points can be made about this, as the choice will depend on the precise nature of the experiment; for example, it would be unsuitable to use a compound labelled

Table 1. The relative merits of commonly used β-emitters

	Advantages	Disadvantages
^{3}H	Safety High specific activity possible Wide choice of positions in organic compounds Very high resolution in autoradiography	Low efficiency of detection Isotope exchange with environment Isotope effect
^{14}C	Safety Wide choice of labelling position in organic compounds Good resolution in autoradiography	Low specific activity
^{35}S	High specific activity Good resolution in autoradiography	Short half-life Relatively long biological half-life
^{32}P	Ease of detection High specific activity Short half-life simplifies disposal Cerenkov counting	Short half-life affects cost and experimental design External radiation hazard Poor resolution in autoradiography

in say a carboxyl group ($^{14}COOH$) for a tracer study if the first step was a decarboxylation. It is worth pointing out that a C atom need not necessarily be radioactive to be traceable, it could be tagged with a covalently attached radionuclide such as ^{3}H.

3.3 Specific activity

The specific activity of a radioisotope defines its radioactivity related to the amount of material (e.g. Bq/mol, Ci/mmol or d.p.m./μmol). Suppliers of radioisotope often offer a range of specific activities for a particular compound, the highest often being the most expensive. The advantages of using a very high specific activity compound are as follows:

- products of a reaction using the labelled precursor can be produced at high specific activity (e.g. in the synthesis of high sensitivity nucleic acid probes, described in Chapter 6);
- small chemical quantities of radiolabelled compound can be added such that the equilibrium of metabolic concentrations is not unduly perturbed, this can be important in *in vivo* labelling experiments where the purpose is to observe a physiological process, addition of radiolabelled compound is a balance between adding sufficient label to guarantee statistically meaningful results but insufficient to unbalance endogenous pool sizes;
- calculating the amount of substance required to make up radioactive solutions of known specific activity is simplified as the contribution to concentration made by the stock radiolabelled solution is often negligible.

Sometimes, however, it is not necessary to purchase the highest specific activity available. For example, enzyme assays *in vitro* often require a relatively high substrate concentration thereby necessitating reduction in specific activity by addition of cold carrier (see Chapter 2, Section 8.1).

3.4 The isotope effect

In choosing a radionuclide and compound, consideration must be given to the isotope effect. That is, a radioactive atom may not behave in a manner directly analogous to its non-radioactive isotope. This is particularly true for tritium, being much larger than hydrogen, and can show up, for example, in the rate of reaction involved in forming or breaking a tritium bond. The radioisotope must be positioned on a part of the molecule not involved in the reaction mechanism.

4. Storage and purity

All commercially available radioactive products are accompanied by a specification sheet. This will give details of the formulation and purity of the compound, the quality control data and a recommended means of storage.

Do not store a radioisotope under conditions (e.g. temperature) that differ from that stated. There are several ways in which radiochemicals can decompose during storage. First, there is the entirely unavoidable decomposition caused by radioactive decay

$$\text{e.g. } X\text{–}^{14}CH_3 \rightarrow X\text{–}^{14}NH_2 \rightarrow X\text{–}NH_2,$$

this is referred to as primary (internal) decomposition. Second, there is primary (external) decomposition where emitted radiation is absorbed by other radioactive molecules creating impurities. The extent to which this occurs is dependent on a number of factors: temperature (low temperatures can sometimes increase decomposition by stabilizing transition states), specific activity and energy of the radiation; the lower the energy the more it can be absorbed and the greater the decomposition. High specific activity ^{3}H compounds are therefore highly susceptible to external decomposition. Third, there is secondary decomposition, where labelled molecules interact with excited species such as free radicals produced by the radiation. Fourth, chemical decomposition can occur: radiochemicals are frequently stored or used in minute quantities relative to non-radioactive materials; interaction with, say, glass surfaces which would normally be regarded as of no consequence can result in significant losses. Finally, there is the problem of microbiological spoilage. Every attempt must be made to maintain the radioisotope in a sterile environment (e.g. by retaining the Teflon seal on a storage vial) to prevent uptake or decomposition by micro-organisms.

The following list provides a general guide to storing radioisotopes:

- follow the recommended storage conditions provided by the supplier;
- store radiochemicals at as low a specific activity as possible or appropriate;

- maintain sterility;
- dilute compounds in an appropriate solvent such as that formulated with the isotope and recommended by the supplier;
- include free radical scavengers such as alcohol, sodium formate, glycerol or ascorbic acid, if possible.

5. Experimental design

Before carrying out an experiment with a radioisotope for the first time go through a theoretical exercise to estimate the quantity of radionuclide required and anticipate experimental difficulties. Then do a 'dry run' without radiolabel to identify practical difficulties. Time spent at this stage will considerably increase the chance of success, will reduce unnecessary exposure to radioactivity and will minimize the risk of accident, spillage, and contamination.

For the theoretical exercise consider the following:

- the minimum and maximum quantities of radioactive compound that can be added to the system to allow observation but not disturbance of the process investigated (e.g. substrate concentration for an enzyme assay, endogenous pool sizes for *in vivo* labelling);
- estimate the uptake rate, e.g. of labelled compound by cultured cells;
- estimate the efficiency of the extraction process to recover radioactivity;
- consider the counting efficiency;
- determine the accuracy of counting required and time available for detection (Chapter 4);
- the number of replicates required.

For a dry run, a dye can be used in place of a radiochemical to identify where difficulties may arise in preventing contamination.

Always remember the safety precautions:

- maximize the **distance** between yourself and the source;
- minimize the **time** of exposure, and **maintain shielding** at all times.

2

Radioisotope use

DAVID PRIME and BARRY FRITH

1. Introduction

The subject of this chapter is the safe use of radioisotopes in biological tracer applications. The emphasis of the treatment will be on practical aspects but a certain amount of a more fundamental nature will be included where this is necessary to clarify explanations and terminology. Since biological tracers are used in a wide variety of ways, it is only intended to discuss this topic in general terms paying particular attention to the most commonly used radioisotopes.

2. Radioactivity

2.1 Radioactive decay

The nuclei of atoms consist primarily of combinations of protons and neutrons. The number of protons in a nucleus is known as the atomic number and the sum of the protons and neutrons is known as the mass number. These combinations are only stable in certain ratios (*Figure 1*). Radioactivity is a means by which this ratio in a nucleus can be adjusted in order that a stable configuration may be achieved. This process can take place in one step or in a whole series of steps (as occurs, for example, when uranium decays through a series of radionuclides to a stable isotope of lead). There are a number of different ways in which this readjustment process can occur, several of which are important in biological applications.

The radiations which emanate as a result of the decays discussed below all have characteristic energies or energy spectra. Such energies are most frequently quantified in units of electron volts (eV) particularly in MeV or keV.

2.1.1 Alpha decay

This decay mode is only of importance for nuclides with an atomic number equal to or greater than lead. An alpha particle is the nucleus of an helium-4 atom and therefore consists of two neutrons and two protons. This heavy, slow-moving particle is densely ionizing and is particularly effective at causing biological damage. For example, the decay of radium-226 is shown below:

$^{226}_{88}Ra$	$\rightarrow ^{222}_{86}Rn$	$+ ^{4}_{2}He$
radium-226	radon-222	alpha

The extreme toxicity of alpha-emitting nuclides, due to this effectiveness at causing biological damage, means that they are unsuitable for tracer work.

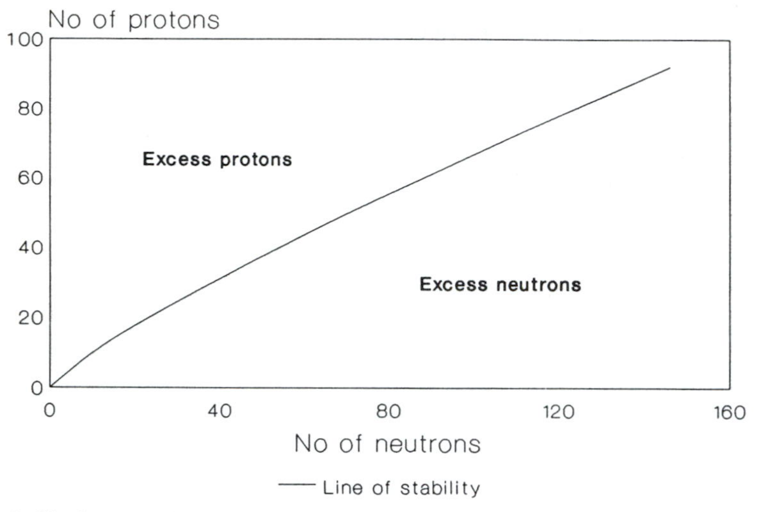

Figure 1. Nuclear stability.

2.1.2 Beta decay

Beta decay is a general term correctly used to describe a series of decays in which either a proton or a neutron may be lost by one of the following processes:

n^0		$\rightarrow$	p^+	$+$ e^-
neutron			proton	electron
p^+		$\rightarrow$	n^0	$+$ e^+
proton			neutron	positron
p^+	$+$ e^-	$\rightarrow$		n^0
proton	electron			neutron

In the first case the net result is that a neutron turns into a proton, whilst in the latter two a proton turns into a neutron. The first process is negatron emission although it is often simply called beta emission. The second process is positron emission and the third is electron capture.

Negatron emission

This decay mode is the most important for those using tracers in the biological sciences. Most of the commonly used radionuclides decay by this mechanism (e.g. ^{3}H, ^{14}C ^{32}P, ^{35}S, ^{45}Ca). For example, the decay of tritium is as follows:

$$\begin{array}{lll} {}^{3}_{1}\text{H} & \rightarrow {}^{3}_{2}\text{He} & + \text{e}^{-} \\ \text{tritium} & \text{helium-3} & \text{beta} \end{array}$$

All forms of beta decay are accompanied by another type of radiation known as a neutrino. The only importance of neutrinos for the user of beta particle emitting nuclides is that they carry away some of the energy available. This distribution gives rise to a spectrum of beta-particle energies a typical example of which is shown in *Figure 2*. Frequently, the nuclide formed as a result of the decay is in an excited state, that is it has some excess energy which it gets rid of by emitting another type of radiation called a gamma ray. Gamma rays are electromagnetic radiation of short wavelength and hence high frequency and can accompany most forms of nuclear decay (see Section 2.2.2 for a discussion of their properties).

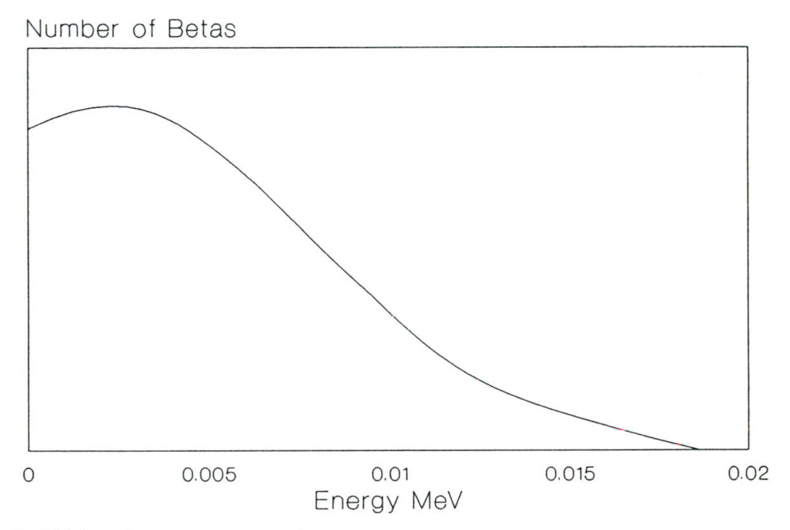

Figure 2. Tritium beta energy spectrum.

Positron emission

The choice of one of the two decay modes which are open to change a proton into a neutron depends on the energy available. If this is greater than 1.02 MeV (twice the rest mass of an electron), then positron emission may occur. This process (see page 8) results in the emission of a positive electron (or positron) from the nucleus of the atom. This particle rapidly encounters an electron and the two particles are destroyed with the creation of two gamma rays each of 0.51 MeV, typical of positron emission. This decay mode is of limited use in the biological sciences and where positron emitting nuclides are used (e.g. ^{22}Na) the counting method is that utilized for other gamma emitters.

Electron capture

If less than 1.02 MeV is available for a decay whereby a proton is transformed into a neutron, the only decay mode possible is that of electron capture. An example of this type of decay is provided by iodine-125:

$$^{125}_{53}I + e^- \rightarrow {}^{125}_{52}Te$$

An electron is captured by the iodine nucleus and combines with a proton to form a neutron. The tellurium atom that is formed has some excess energy which it is able to get rid of in various ways. In 7% of decays, this energy is emitted as a gamma ray and in the remaining 93% the energy is transmitted directly to an electron which is then emitted from the atom. This latter process is known as **internal conversion**.

The capture of an electron is normally from an orbit close to the nucleus, this capture therefore creates a hole which is usually filled by an electron further away from the nucleus. This in turn causes a whole series of further electron transitions as the hole moves away from the nucleus. Each of these electron movements is accompanied by the emission of an X-ray. X-rays are electromagnetic radiations which differ from gamma rays primarily in their mode of formation. With electron capture nuclides only the initial electron transitions are of significance, since it is only these which give rise to X-rays of sufficient energy to be readily detectable.

The products of all this atomic rearrangement are a series of radiations which may well include X-rays, gamma rays, and electrons.

2.1.3 Internal transition

Internal transition is important in certain biological applications. In this type of decay a gamma ray is emitted from a nucleus in an excited state. This is precisely the same as the gamma emission that can accompany other forms of radioactive decay; however, such gamma rays are emitted instantaneously whilst with internal transition the excited state has a significant half-life. This type of excited state is known as a metastable state and its presence in a nuclide is indicated by the symbol m. Technicium-99m is the most frequently used isotope of this class, being particularly important in medical diagnostic work.

$$^{99m}Tc \rightarrow {}^{99}Tc + \text{gamma}$$

2.2 Properties of radiations

Table 1 summarizes the properties of different types of ionizing radiations.

All forms of radiation emitted from a point source follow an inverse square law, that is the intensity of radiation decreases with the square of the distance, or

$$I \propto 1/d^2$$

where I is the intensity of the radiation and d is the distance from the source.

Table 1. Properties of radiations

Radiation	Energy (MeV)	Range (cm)	Shielding
Alpha	4–8	2.5–8 (air)	Paper
Beta	0.1–3	15–1600 (air)	Low atomic number materials
Gamma, X-rays	0.03–3	HVL air 1.3–13 m HVL lead 0.02 mm–1.5 cm	Lead or high density materials

Half-value layer (HVL) is the thickness of material required to reduce the original radiation intensity to one-half.

The above formula strictly only applies to radiation emitted from a point source (i.e. equally in all directions), if the radiation is restricted by shielding into a narrow beam, a process known as collimation, then the distances referred to in *Table 1* apply.

2.2.1 Beta radiation

The nature of beta interactions with matter means that it is possible to absorb this kind of radiation completely. *Table 1* does not include radiation with energy as low as tritium, since this does not constitute an external hazard. Even the most energetic of tritium betas have insufficient energy to penetrate the outer dead layer of skin. *Figure 3* shows the absorption of phosphorus-32 betas and illustrates another important feature of beta interactions with matter: the production of secondary X-rays. It is these X-rays, also called *bremsstrahlung* radiation, which cause the enhanced background illustrated in *Figure 3*. All interactions of beta particles with matter give rise to this type of radiation but the production is greater the higher the energy of the betas and the higher the atomic number of the absorbing material. It is therefore desirable to use a material with a low atomic number, such as plastic or Perspex, for shielding beta emitters.

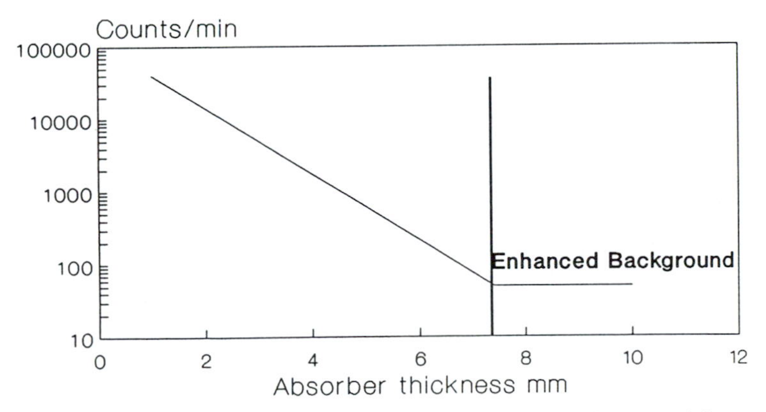

Figure 3. The range of phosphorus-32 beta particles in unit density material.

2.2.2 X- and gamma rays

The interaction of this kind of radiation with matter is such that total absorption does not occur and hence the use of 'half-value layer' as described in *Table 1*. The absorption is dependent on the mass of material through which the radiation passes and hence the denser the material the more efficient it is as an absorber. For this reason, high density materials such as lead, tungsten, and even uranium are used for shielding gamma and X-ray-emitting nuclides. Uranium is itself slightly radioactive, but it still finds shielding applications for high activity sources. *Figure 4* shows half-value layers in lead and water as a function of energy.

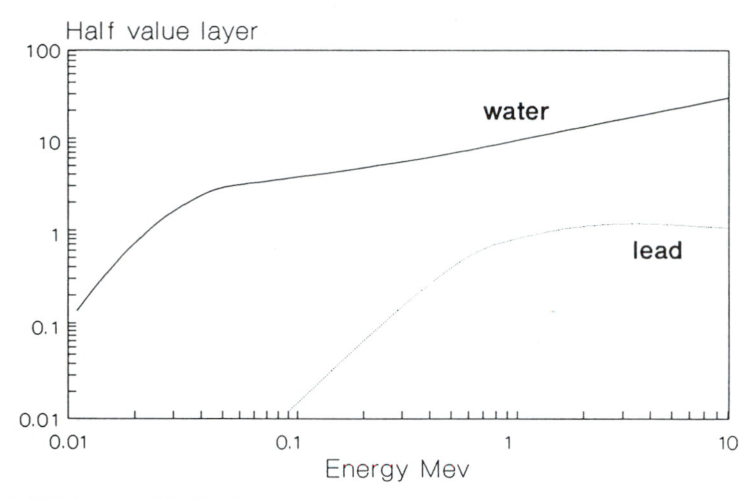

Figure 4. Thickness of half-value layers as a function of energy of radiation for water and lead.

2.3 Radiation dose and dose equivalent

The physical unit of absorbed radiation dose is the gray (Gy), this is defined as the absorption of 1 joule of energy per kilogram of mass. This unit is used for expressing physical measurements of radiation dose and should also be used to describe acute biological effects of radiation. However, it has been found that different radiations have different efficiencies of causing long-term biological effects for the same physical radiation dose. This has led to the introduction of the dose equivalent. For the same dose equivalent one runs the same risk of biological effect irrespective of the type of radiation causing the irradiation. The unit of dose equivalent is the sievert (Sv) and the dose equivalent in sieverts is obtained by multiplying the absorbed dose by a quality factor which reflects the different biological effectivenesses of radiations or,

$$\mathrm{Sv} = \mathrm{Gy} \times Q$$

where Q = 1 for beta particles and gamma and X-rays;
20 for alpha particles; and
20 for neutrons.

2.4 External radiation doses from beta emitters

Apart from tritium, all negative beta emitters pose some degree of external hazard, which becomes greater as the energy of the radiation increases. Various formulae may be used to estimate the dose rate from a source, an example of which is

$$D_\beta = 760A$$

where D_β is the unshielded dose rate in μGy·h^{-1} at a distance of 10 cm from a point source, and A is the source activity in MBq.

It should be noted that this formula is approximate in that it applies strictly only to point sources and ignores absorption of beta particles. The approximations will be minimized when the formula is used for high-energy beta emitters at short distances.

Example: A 10 MBq solution of phosphorus-32 is spilt on to a glove. What will the dose rate be 1 mm from this glove?

$$\text{Dose rate at 10 cm} = 760 \times 10\,\mu\text{Gy·h}^{-1}.$$

Since the quality factor for beta radiation is 1 this will be the same as 7.6 mSv·h^{-1}.

The inverse square rule can then be applied in order to calculate the dose at 1 mm.

$$\begin{aligned}\text{Dose rate at 1 mm} &= 7.6 \times 10^2/0.1^2 \\ &= 7.6 \times 100 \times 100\,\text{mGy·h}^{-1} \\ &= 76\ \text{Gy·h}^{-1}.\end{aligned}$$

The dose rate from a beta source spilt on to skin will be correspondingly larger.

It can be seen from this simple calculation that radiation doses from beta sources can be very high indeed and a considerable local radiation hazard.

2.5 External radiation doses from X-rays and gamma rays

The usual method of calculating dose rates from electromagnetic radiation emitters is to use the specific dose rate constant, Γ. The dose rate from a point source of gamma rays or X-rays is

$$D_{\text{em}} = \Gamma A / d^2$$

where d is the distance in metres, A is the activity of the source in MBq and Γ is the specific dose rate constant, μGy h^{-1} MBq^{-1} at 1 m. *Table 2* lists specific gamma ray constants of some frequently used radioisotopes.

Table 2. Gamma ray doses and annual limits on intake for some commonly used radionuclides

Nuclide	Gamma dose at 1 m × 10^2 (μGy·h^{-1} MBq^{-1})	[a]ALI oral intake (MBq)
Carbon-14	–	90
Caesium-137	8.92	4
Chromium-51	0.5	700
Cobalt-57	2.5	20
Cobalt-60	35.7	1
Iodine-125	1.9	1
Iodine-131	5.7	1
Phosphorus-32	–	10
Sulphur-35	–	80
Technetium-99m	1.6	3000
Tritium	–	3000

[a]Figures based on a whole-body dose equivalent limit of 50 mSv·y^{-1}. ALIs may be reduced to 30% of the figures shown above if a limit of 15 mSV·y^{-1} is adopted (see Section 4).

Example: What is the dose rate from an unshielded solution of 40 MBq caesium-137 at a distance of 25 cm?

From *Table 2* the specific gamma-ray constant for ^{137}Cs at 1 m is 8.92×10^{-2} μGy·h^{-1}MBq^{-1}.

Therefore, the dose rate is

$$= 8.92 \times 10^{-2} \times 40 / 0.25^2$$
$$= 57\ \mu\text{Gy·h}^{-1}.$$

If the energy of caesium-137 gamma rays is 0.662 MeV what thickness of lead shielding would be required to reduce the dose rate to 7.5 μGy·h^{-1}?

From *Figure 4*, the half-value layer for 0.662 MeV radiation is approximately 0.6 cm. If one half-value layer reduces the radiation dose to one-half, two will cause a reduction to one-quarter, three to one-eighth, etc. In general, n half-value layers will reduce the radiation dose by a factor of $1/2^n$.

In this example the attenuation required $= 7.5/57$
$= 0.13$,

therefore, $1/2n = 0.13$ where n is the number of half-value layers required from which,

$n = 2.9$ and

the thickness of lead required $= 2.9 \times 0.6$
$= 1.7$ cm.

This type of shielding calculation is an approximation, but can be used to estimate approximately the shielding required before checking with a dose measuring instrument.

2.6 Internal radiation doses from radionuclides

The calculation of radiation doses from the ingestion of radionuclides can by very complicated. The International Commission on Radiological Protection have published a large amount of information on the subject including computer models for every common radionuclide. For the user of such radionuclides, it is only worthwhile understanding the concept of annual limit on intake (ALI) in order to use this data. The ingestion of one ALI will result in a person receiving a radiation dose of either 50 mSv to the whole body (the stochastic does equivalent limit, see Section 4) or 500 mSv to a tissue group (the non-stochastic limit). A comparison of ALIs will therefore give both the dose received from ingestion or inhalation and an indication of the relative toxicities of radionuclides. *Table 2* lists minimum ALIs for a some commonly used radioisotopes.

Example: A radiation worker spills 20 MBq of iodine-125 in a volatile form. If 5MBq of this has been ingested what is the radiation dose received?

From *Table 2*, the ALI for ^{125}I is 10 MBq and this is a non-stochastic limit to the thyroid. The ingestion will therefore result in a radiation dose of 5/10 of 500 mSv or 250 mSv to the thyroid.

If an incident occurs in which the ingestion of a radioactive material takes place or is suspected, biological monitoring (e.g. urine) should be carried out to check any dose calculations.

3. Biological effects of radiation

The principal reason for precautions and legislation concerning ionizing radiations is because of their ability to cause a variety of biological effects. It is important that a person using radioactive materials should have some idea of the likelihood of occurrence and extent of such effects.

3.1 Tissue level effects

Tissue level effects can be further subdivided into three categories.

3.1.1 Whole-body effects

This sort of effect is of very limited importance as far as the use of radioisotopes in biological sciences is concerned. Below 1 Gy no clinical symptoms of whole-body radiation can be detected, although it is possible to detect cellular chromosome aberrations at lower doses (minimum 10 mGy).

3.1.2 Local effects

The local effect of greatest importance for unsealed isotope users is radiation burns. It has been shown in Section 2.4 that beta emitters can give high local radiation doses and it would be possible to get sufficient radiation dose from the spillage of a stock solution of such an emitter on to unprotected skin to cause a burn. The threshold for burns is 3 Gy; at this dose level, erythema (reddening) can occur which may persist for a few days before disappearing. At higher doses, above 10 Gy, the burn becomes more serious, with the skin breaking down, producing ulceration or blistering. Such burns are slow to heal and in serious cases skin grafting may be necessary.

3.1.3 Effects *in utero*

Figure 5 illustrates the relative importance of the adverse effects of radiation *in utero* at different stages of gestation (2).

The important difference between these effects and those described previously is that these are not generally thought to have a threshold. There cannot be an absolutely safe level of radiation at which such effects do not occur, and the legislation seeks to keep the occurrence of these at an 'acceptable level'.

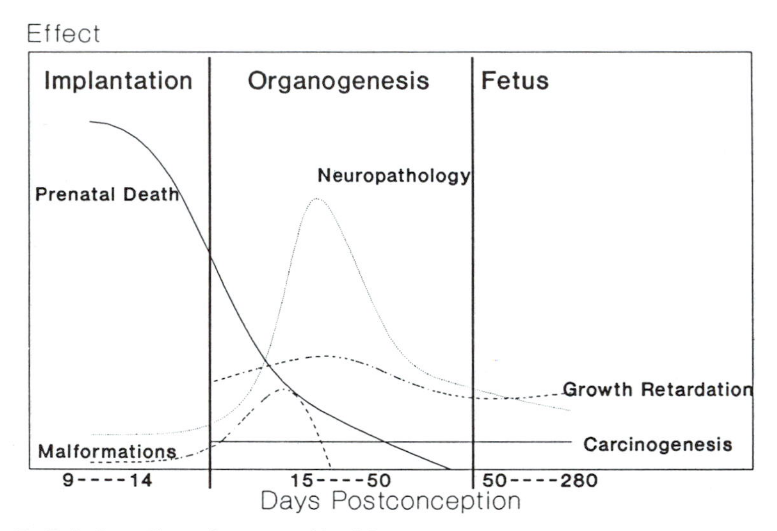

Figure 5. Relative effects *in utero* of ionizing radiations.

3.2 Carcinogenic effects

There is a great deal of evidence for carcinogenic effects of ionizing radiation deriving from groups, including the survivors of the atomic weapons from Hiroshima and Nagasaki, early research workers, uranium and other miners, several types of medical treatments, and also the effects of X-rays. From this type of data estimates can be made of the likely incidence of different forms of cancer for a given radiation exposure. These estimates are continually being refined, with resulting adjustments in the risk estimates. The current estimate (1987) from the International Commission on Radiological Protection is an average risk of fatal cancer of $2.0 \times 10^{-2}\,\mathrm{Sv}^{-1}$ irrespective of age or sex whilst the National Radiological Protection Board (NRPB, Didcot, UK) has suggested a figure of $3.0 \times 10^{-2}\,\mathrm{Sv}^{-1}$. This order of risk rate means that if a population were to receive 100 Sv in total, 2–3 cases of fatal cancer would be expected to occur as a result of that dose. The following factors are important when considering the possibility of radiation induced cancer:

- thyroid and skin are the most likely cancers to be induced but are of low fatality;
- leukaemia, lung and breast cancer are most likely to cause death;
- sensitivity to cancer induction varies with age, the young and the old being at highest risk.

Since women are more prone to breast cancer and certain other radiation-induced cancers than men, they are more likely to suffer from fatal cancer (by a factor of 1.5).

3.3 Genetic effects

No genetic effects of radiation have been demonstrated conclusively in human populations. Extrapolation from animal experiments indicates that such effects do occur but that their incidence is less than that of cancer. The current estimate of risk of significant genetic effects is $0.4 \times 10^{-2}\,\mathrm{Sv}^{-1}$ cases in the first two generations after the irradiation (3). This means that if a population receives a radiation dose of 100 Sv, four significant birth defects would be expected as a consequence in the next two generations.

4. The biological basis of legislation

Two categories of biological effect can be identified from the types described above. Acute effects are all non-stochastic, in other words they occur only above a certain threshold dose and the severity of such effects increases with increasing radiation dose. *In utero*, cancer and genetic effects are all stochastic, or, in other words, there is no threshold and the probability of such an effect occurring increases with increasing radiation dose.

Non-stochastic effects are easy to legislate against. It is simply necessary to ensure that the radiation dose that a person receives is below the threshold at which the particular effect occurs. An example of this is the limit for radiation of the lens of the eye. In this case, the induction of cataract is a chronic non-stochastic effect of radiation and the dose limit is calculated as shown below.

Threshold for cataract induction = 7.5 Sv

Possible exposure time for a radiation worker = 50 years.

Therefore, limit for lens irradiation $= 7.5/50\,Sv{\cdot}y^{-1}$
$= 150\,mSv{\cdot}y^{-1}$.

A radiation dose limit at this level will ensure that even after 50 years' exposure at the limit, no cataract will be induced.

The non-stochastic limit for other types of tissue is 500 $mSv{\cdot}y^{-1}$.

Stochastic effects are more difficult to legislate against. There is no threshold or 'safe' level and limits have been set at what is considered to be an 'acceptable level of risk'. It is thought that the average radiation worker receives a radiation dose of around one-tenth of the limit, and that this dose will impose a risk which is about the same as industries which are considered 'safe'. Recent rethinks about radiation risk rates (see Section 3.2) are likely to result in a reduction in the dose equivalent limit for radiation workers from the present level of $50\,mSv{\cdot}y^{-1}$ to a figure of $15\,mSv{\cdot}y^{-1}$.

In order to protect against *in utero* effects, two limits have been adopted. After the declaration of a pregnancy, the limit to the abdomen of the woman is reduced to one-fifth of the annual dose. Also, women of 'child-bearing capacity' are only allowed to have their radiation exposure at an even rate throughout the year (i.e. a maximum of 13 mSv in any consecutive 3-month interval).

5. Working with unsealed sources

5.1 Getting started

Researchers who propose to undertake work with radioisotopes and labelled compounds must first consider their responsibilities under the law. In the United Kingdom two pieces of legislation are of major importance: The Radioactive Substances Act 1960, often referred to as the Principle Act, and the more recently introduced Ionizing Radiations Regulations 1985. The former deals with the acquiring and disposal of radioactive materials, the latter with its use and manipulation. Similar legislation exists in European countries and the United States of America (see Chapter 10).

There are, however, materials which although radioactive are of such a low radiotoxicity or low radioactivity that exemptions are granted with regard to parts or all of the legal requirements. For example, in Britain, in general, there is no requirement to comply with the Ionizing Radiations Regulations 1985 when the activity concentration of the radioactive material being used is less than $100\,Bq{\cdot}g^{-1}$ unless the material is a significant radiological hazard (see Appendix 2). The majority of tracer work involves the use of material of very much higher activity concentration, and the use of isotopes in the production of labelled compounds requires a level of activity concentration which is higher still. If, in Britain, the researcher decides that the proposed work falls subject to legal controls, and if in doubt advice should be sought from a Radiation Protection Adviser or NRPB, he or she is required to notify the Health and Safety Executive of their intention to start work, giving at least 28 days' notice.

The organization of working procedures, and methods, in the isotope laboratory is based on the fundamental, and legal, requirements to restrict the risks associated with working with isotopes by reducing dose equivalents to 'as low as reasonably achievable', the ALARA principle, while at the same time ensuring that dose equivalent limits are not exceeded. Without prejudice to the ALARA principle, the practical implementation of this basic concept centres around two (for the want of a better phrase) trigger points, which correspond to one-tenth, and three-tenths of the dose equivalent limit.

In essence, the risks to workers who might receive no more than one-tenth the dose equivalent limit from exposure to external radiation or from the ingestion of radioactive material are restricted and controlled by the strict control of the working environment, and those workers who might receive three-tenths of the dose equivalent limit have their exposure controlled by additional personal dose assessment. The machinery for achieving control at these two levels of risk is both administrative and technical. United Kingdom regulations require that a chain of communication be established between the Health and Safety Executive and the controlling authority or employer, through a Radiological Protection Adviser, appointed by the controlling authority, and Radiological Protection Supervisor to the worker. The maintenance of this chain is very important, particularly between the worker and the Radiological Protection Supervisor, who must know what radioactive work is going on in his or her area of responsibility.

Local Rules, which describe how the work will be undertaken in compliance with the regulations, must be written, and must include a hazard assessment and a contingency plan. They form the hinge-pin of the administrative control, but more of this later.

5.2 Designation of working areas

The initial designation of working areas is the first step in the implementation of control procedures. Working areas must be designated as 'supervised' or 'controlled' and a record kept of their location. Whether a working area is designated supervised or controlled depends on the likelihood that a worker who works in the area continually would receive three tenths of the dose equivalent limit. Initially, the Radiological Protection Adviser may use his judgement in designating areas, but often revisions are required once the area is in use since only then can the measurements of dose rate and contamination be made, which are crucial to the decision-making process.

Bearing in mind that the hazard in the open source laboratory comes from both external and internal radiation, there are limiting conditions of dose rate and surface and air contamination above which an area cannot be designated supervised.

5.2.1 Area designation on the basis of external exposure

Definitions: *Instantaneous dose rate* is dose rate in $\mu Sv{\cdot}h^{-1}$ averaged over one minute; that is, the integrated dose over any one minute multiplied by 60 to give μSvh^{-1}.

Time averaged dose rate is the average dose rate in $\mu Sv{\cdot}h^{-1}$ over the working day; that is, the integrated dose over 8 h divided by eight to give $\mu Sv{\cdot}h^{-1}$.

If instantaneous dose rates of more than 7.5 $\mu Sv{\cdot}h^{-1}$ are observed consideration must be given to the need to designate a controlled area. If there are instantaneous dose rates greater than 2 $mSv{\cdot}h^{-1}$ there is no option, the area must be designated a controlled area. However, where intermediate values of instantaneous dose rate are found the area need not be designated a controlled area provided that the exposure is due to a gamma-emitting nuclide only, such that the product of activity (MBq) and gamma energy per nuclear transition (MeV) is less than 50 MBq·MeV.

To illustrate this consider an experiment involving 37 MBq of cobalt-60. (See the decay scheme, *Figure 6*.)

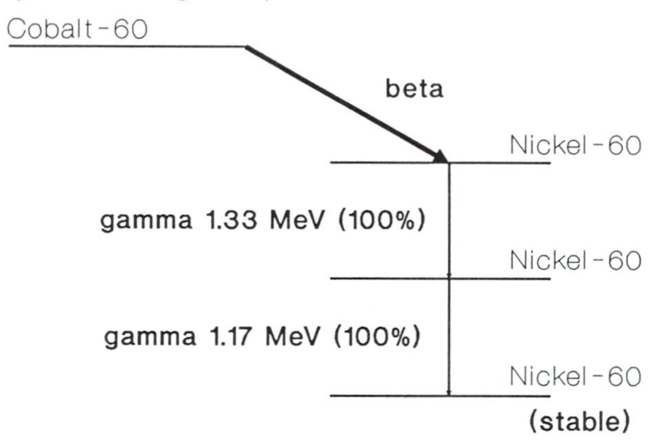

Figure 6. The decay scheme of cobalt-60.

The energy per transition is the sum of the energies of the gamma rays, taking into account their frequency of emission per one hundred nuclear transitions of cobalt-60 nuclei; that is, 1.17 MeV multiplied by 100% plus 1.33 MeV multiplied again by 100%. This yields a total gamma energy per transition of 2.5 MeV. The product is, therefore, 92.5 MBq·MeV.

Similarly, where instantaneous dose rates of between 7.5 $\mu S{\cdot}h^{-1}$ and 2 $mSv{\cdot}h^{-1}$ are found due to a beta-particle emitter of maximum beta energy greater than 0.3 MeV there is no need to designate a controlled area if the total activity is less than 5 MBq. Where the beta energy is less than 0.3 MeV the corresponding activity limit is 50 MBq. It is reasonable for the researcher and the Radiation Protection Supervisor to make decisions up to this point, and to assist a flow diagram is given (see *Figure 7*).

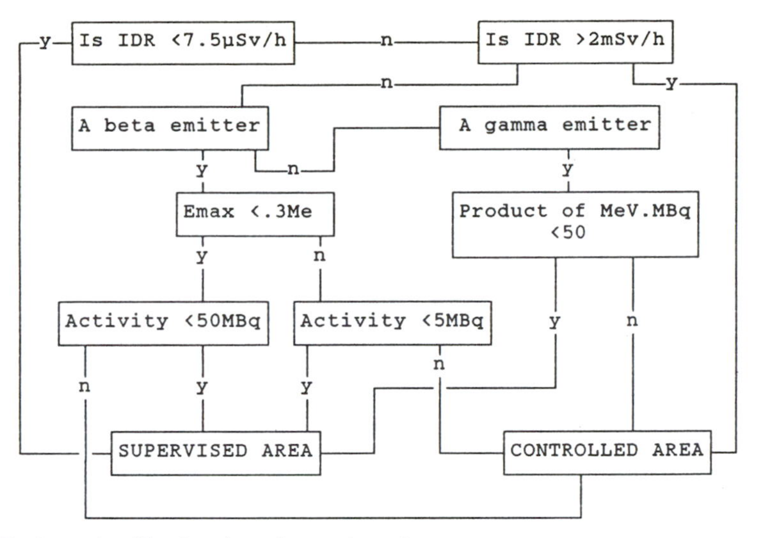

Figure 7. Area classification based on external exposure.
IDR = Instantaneous dose rate.

Other criteria, which are best evaluated by the Radiological Protection Adviser, may also be used to determine the designation of an area.

In addition to the above, there is no need to designate as controlled an area in which the instantaneous dose rate is greater than $7.5\,\mu Sv{\cdot}h^{-1}$ and less than $2\,mSv{\cdot}h^{-1}$ provided that one of the following conditions is satisfied

- the time averaged dose rate does not exceed $7.5\,\mu Sv{\cdot}h^{-1}$;
- only the hands can enter any area in which the instantaneous dose rate exceeds $7.5\,\mu Sv{\cdot}h^{-1}$ and the time averaged dose rate does not exceed $75\,\mu Sv{\cdot}h^{-1}$; or
- the area is one in which
  - (a) the time averaged dose rate does not exceed $240\,\mu Sv{\cdot}h^{-1}$;
  - (b) no person remains in the area for more than one hour in any working period of 8 h; and
  - (c) no person receives more than $60\,\mu Sv{\cdot}h^{-1}$ in any working period of 8 h.

5.2.2 Area designation on the basis of internal radiation

The control of internal radiation is dependent on controlling those parameters which affect the rate at which radioactivity is taken into the body by ingestion, the oral route, and by inhalation. These parameters are judged to be surface contamination, and air contamination respectively. In Britain, Schedule 2 of the Ionizing Radiations Regulations gives derived limits of air and surface contamination above which a controlled area must be declared, unless the total activity in the area is less than ten times the annual limit on intake. It is essential that, in estimating contamination for the purpose of area designation and for later environmental control, that both surface and air contamination are measured.

Just what these area designation criteria mean at a practical level will be discussed later, but the concept of supervised and controlled areas is introduced here because it leads logically on to the classification of the workers.

6. Designation of classified workers

Some workers who work with radioactive materials are designated as *classified radiation workers*. The designation depends upon the likelihood that the worker will receive three-tenths the dose equivalent limit. It would seem to follow, therefore, that non-classified workers are those who spend their time at work in supervised areas, and classified workers are those who work in controlled areas. This is generally what happens, but there may be times when a non-classified worker works in a controlled area for short periods to, say, dispense an isotope from a stock solution, or to feed experimental animals. This is acceptable provided the worker is operating under an 'Approved System of Work' which has been designed to restrict the person's dose to less than three-tenths the dose equivalent limit. In these circumstances, personal monitoring of the non-classified worker is required in order to demonstrate compliance with this requirement.

The classification of workers is an administrative mechanism used to indicate the type, degree, and extent to which monitoring is carried out to assure the required level of risk.

7. Controlling the risks to radiation workers

7.1 The non-classified worker

The reader is reminded that the aim here is to assure the well-being of the non-classified worker by control of the working environment. Regular recorded monitoring of external dose, and surface and air contamination is required to ensure that conditions in the working area comply with those for a supervised area. (See Schedules 2 and 6 of the Regulations.) Dose rates are measured with an instrument which is suitable for the purpose and has been calibrated by an 'Approved Laboratory'. This usually means an ion chamber, proportional counter, scintillation counter, or Gieger–Müller device (of which there are many available), but in the case of tritium contamination monitoring estimates are made by taking swabs of the area (see Appendix 3). The measurement of the activity on the swab by liquid scintillation counting, and the use of a swabbing efficiency factor of ten is used to determine the activity on the area swabbed. A routine for monitoring must be laid down in the Local Rules, and any results which are around one-tenth of the derived limit or above must be recorded.

7.2 The classified worker

In addition to monitoring the working environment in a controlled area, personal monitoring of the worker is required so as to assess the classified worker's individual dose equivalent. Film badges, or thermoluminescent dosimeters, the monitoring of air in the individuals breathing zone, and the assessment of personal contamination by external measurement for skin contamination, and bio-assay for internal contamination are undertaken routinely by an 'Approved Laboratory'. The findings must be recorded and the records kept for 50 years after the last entry. What has been said up to now is intended to make the isotope user aware of the general requirements of the Ionizing Radiations Regulations 1985, and to show something of the framework of controls which enable isotope work to be undertaken within acceptable risk limits. There is, however, the requirement to reduce the risks to 'as low as reasonable achievable'. We must use the experimental and manipulative skills that we have to reduce the already low risk still further.

8. The tracer study

Most isotope experiments follow a systematic procedure; the design of the experiment, the selection of the isotope tracer, the purchase of the isotope, the receiving and opening, the dispensing from the stock solution and dilution to working solution strength, the introduction of the isotope into the experiment followed by sampling and measurement, and finally the decontamination of equipment and monitoring of the working area and personnel. If we work through an imaginary experiment step by step and examine the ways we can reduce the dose, and thence the risk at each stage, we should then have a useful working guide.

8.1 Experimental design

Tracer studies yield information about the movement of stable materials in a system. Reactions and reaction rates are studied by determining how much and how quickly radioactive tracer, and therefore stable material, moves from place to place. Two parameters of the isotope solution or labelled compound solution are very important in achieving good experimental results. They are *activity concentration*, and *specific activity*. Activity concentration is the activity per unit mass or volume of the media containing the activity, and specific activity is the activity per unit mass of the compound being studied, be it compound, ion, or element. Specific activity is a measure of the ratio of active, labelled species to the non-active or stable species which we often refer to as carrier.

The ability to measure the radioactivity in samples taken from the experimental system with a sensitivity and precision, and in some rare cases accuracy, is another important requirement for a successful experiment, but it is also fundamental in considering the appropriate activity concentration, and specific activity of the isotope or labelled compound to be used. The statistics of counting is dealt with elsewhere (see Chapter 3), but in the broadest terms it is necessary to aim for radioactive sources for counting (that is, sources made from aliquots taken from the experiment) which count at a rate significantly greater than the counter background, and generate 10^4 to 10^5 in a reasonably short counting period in order to achieve a counting error (sometimes called a random error, since it is due to the inherent randomness of radioactivity), small by comparison with other systematic experimental errors. Thus, the required specific activity is linked to the radioactivity represented by the above number of counts in the counting period used, divided by the mass of the compound of interest in the aliquot taken. The concentration of the compound of interest may stress the experimental system so as to affect the amount in the aliquot taken, and, unless that is what is being studied, the tracer is added to the experimental system in amounts which do not change the concentration of the compound greatly. Consideration of this problem determines the activity concentration required for the tracer solution. So far we have made the, not unreasonable, assumption that the tracer behaves exactly as the non-active material. In almost all cases the errors attendant on this assumption are negligible, but we should be aware of a mass effect, which is a problem when reaction rates are being studied where the mass of the tracer is very different from the stable species. Problems of significant magnitude are encountered only when elemental hydrogen is being traced using tritium, and since most tracer work involves large molecules the problem seldom worries the biochemist.

There are obvious advantages to be gained, from the safety angle, from using isotope and labelled compound solutions of the lowest activity concentration and specific activity, and adjustments to these parameters should be made as early as practicable in the experiment. The adjustment of activity concentration is straightforward enough, and the reduction of specific activity is achieved by adding carrier in amounts calculated using the formulae

$$W = M \cdot a\,(1/A' - 1/A)$$

where W is the mass of carrier in milligrams to be added to an ampoule of M MBq of a compound of molecular weight a, to change the original specific activity A MBq/mmol to the required specific activity A' MBq/mmol.

8.2 The purchase of isotopes and labelled compounds

Buying isotopes and labelled compounds of the appropriate activity concentration and specific activity has been considered. The choice of isotope must be considered next. Often there are several which might be used. Do have regard for the relative radiotoxicity of the possible candidates before placing your order. How much activity should be purchased? The answer is, not more than is needed to achieve the experimental goal. Do not be tempted to buy more than you need because the cost reduces dramatically with increasing pack size, so that ten units of activity cost little more than twice the cost of one unit. It might make good sense at the purchasing stage of the experiment, but it is contrary to the ALARA principle, and the additional cost of, say, operating a controlled area as opposed to a supervised area, and the additional cost of waste disposal (by no means insignificant), not to mention the increased potential hazard, and the dubious storage properties of the labelled compound in a resealed container, and other factors will probably outweigh any initial advantages.

8.3 Receiving the isotope

The receipt of radioactivity must be recorded, and its temporary storage must be in a suitably secure place which is chosen having regard to the need to control dose rates and contain the activity under likely conditions of use and, to some extent, abuse. The store should be used for active material only and should contain no inflammable or explosive substances. Because of the concentration of radioactive material, the store is likely to be a high dose-rate location, and consideration must be given to the possible need to designate it as a controlled area. If the store is an enclosure other than one which can be entered (for example, a refrigerator into which only the hands may enter) then due regard should be given to this in making your decision. (See Section 4.)

8.4 Isotope manipulations

The manipulation of isotope stock solutions, when dose rates, activity concentrations, and specific activities are at the highest levels which will be experienced throughout the whole experiment, must be recognized as a particularly potentially hazardous operation. It is often wise to manipulate stock solutions in shielded areas constructed, wherever possible in a suitable fume cupboard, so that only the hands can enter, and dose rates outside the shield are less than $7.5\,\mu Sv{\cdot}h^{-1}$. This avoids the need to designate as a controlled area any area other than that within the shield, and, of course, this may not be necessary if the conditions for exposure to hands only are met.

While manipulating the primary container the simple principles of *time*, *distance*, and *shielding* should be used to reduce external dose rates, and it is often useful to have a cold practice run, with a coloured solution perhaps, to identify any possible problems. (See Appendix 1.)

8.5 Design of isotope laboratory

The working area should be designed with the aim of reducing the exposure rates to as low as reasonably achievable in mind, and in particular, the need to control contamination on surfaces by choosing surfacing materials which decontaminate easily. In common with other countries, Britain has an organization (The British Standards Institution) which sets standards and testing procedures designed to assure the performance of products. The British Standard Test, BS4247, enables surfacing materials to be chosen for use in isotope laboratories which are easy to decontaminate. Choosing such surfacing materials and making them as continuous as possible and without difficult-to-clean corners should be given a high priority. The better materials are not obvious. Stainless steel is extensively used in radioactive areas, but it can be difficult to decontaminate, especially when radio-halides are used. Examples of materials which are very easily decontaminated are: unfilled polypropylene, Perspex, and paints of the epoxy or polyurethane type. It is always best to ask for certification of the surfacing material involved when laboratory refurbishment takes place or a new piece of equipment is purchased. Strippable coatings and the use of temporary surface coverings can be advantageous, but their use can greatly increase the volume of radioactive waste produced. Air conditioning which gives 5–30 air changes per hour is a desirable feature, and a fume-cupboard with a design specification which will assure the containment of any radioactive gas generated within it is essential. A draught of 0.5 m per sec with the sash at a normal working position is regarded as satisfactory in Britain, but the overriding priority is that the fume-cupboard should 'contain' radioactivity generated in gaseous or volatile form within it. The laboratory sink, if it is designated for the disposal of radioactive waste, should be given special attention. The joint between bench and sink should be properly sealed to prevent the ingress of radioactivity; there should be no large trap which might collect and accumulate radioactive items or solids; the waste should find its way to the main drain by the most direct and preferable discrete route and be carried by high integrity pipe work which is marked at intervals with appropriate warning labels. Where the area is a controlled area there must be provision for washing, (a wash basin with elbow-or foot-operated taps is required), and somewhere to leave lab coats in the area. To help prevent accidental entry to controlled areas it should require a positive action to enter; for example, the lifting of a physical barrier, the opening of a door, or the stepping over of a shoe-change barrier. With attention to these details an ordinary good well appointed laboratory is usually sufficient for most levels of radio tracer work.

9. Personal protection in the isotope laboratory

In all laboratories some form of protective clothing is required, a laboratory coat, of suitable design (e.g. wrap-over design without open neck, elasticated and cuffs and no pockets), and safety spectacles being a bare minimum. The use of latex or vinyl gloves is a useful first device in the prevention of skin contamination. They must be used in a manner which does not permit their being removed and put back on with the original, possibly contaminated, outside in. Reusable ambidextrous gloves should be marked to prevent this, or used only once. Over shoes should be worn if not wearing them would result in the spread of contamination to other areas.

10. Routine area monitoring

Monitoring the radiation and contamination in the working environment and of the individual on a routine detailed in the local rules is required. Most of the effort will, in normal circumstances, be aimed at the environment. The relevant derived limits are, in the case of external dose, and air contamination, derived in a fairly straightforward way. On the one hand, the annual limit is divided by the number of working hours in the year, and on the other, with some knowledge of a worker's breathing rate we can calculate the limiting air concentration which would lead to an intake of an annual limit on intake. Here the simplicity ends, the author cannot find sufficient evidence to dispel his fear that confusion exists in the minds of whoever derived the limits for surface contamination.

10.1 External dose measurement

The measurement of dose rate requires the use of a measuring device which is sensitive to the type of radiation being given of by the isotope used and, which will give a reading which reflects the true radiation dose (see Appendix 3). The simplest monitor is a Geiger–Müller (G–M) counter with the output fed to a rate meter. The visual indication is of a count rate. The relationship between the count rate and true dose rate is given by calibrating the monitor in conditions of accurately known dose rate. Errors occur in measurement when the energy of the radiation given off by the isotope used differs from that of the source used in calibration. For example, a simple G–M device which has been calibrated using a cobalt-60 source, will probably show a reading much higher than the true value when used to measure a dose rate produced by 100 keV X-rays. This response energy dependence is all but eliminated with more elaborate (compensated) G–M devices, and in monitors which employ ion chambers as their detector. If a single isotope is used, expense can be avoided by choosing a simple G–M monitor and having it calibrated against standard sources of the same, or similar emitter. The multi-isotope user will require the more sophisticated, and more expensive, monitor of the compensated G–M or ion chamber type. In either case it is important to be able to detect with confidence dose rates of $7.5\,\mu Sv{\cdot}h^{-1}$ and $2\,mSv{\cdot}h^{-1}$ for routine monitoring, since these are important limiting values in the overall scheme of control.

10.2 The measurement of air contamination

The measurement of activity concentration in air is undertaken by direct reading instrumentation in only a very few circumstances. For example, tritium as a gas, or radon/thoron gases are sometimes measured by passing air through a chamber containing two electrodes. The resulting ionization current being a measure of the activity concentration. More often than not, a known volume of air is drawn through a trap, which may be, for example, a glass-fibre filter paper or an activated charcoal absorber, and the trap analysed later in the radiochemistry laboratory.

10.3 The measurement of surface contamination

Surface contamination measurements are best used as nothing more than an indicator of the overall control and housekeeping in a laboratory. The general level of surface contamination in the open source laboratory should be many orders of magnitude less than the derived limit shown in Schedule 2 (see Appendix 2 for an abbreviated list) of the regulations. Changes in the normal or usual level of contamination should be investigated. They usually indicate a malpractice or the need for a change in working conditions. Two methods are used to measure surface contamination. The direct method involves observing the response of a suitable detector when held close to the surface, and comparing it with the response of the instrument held in a similar position close to a 'standard of surface contamination'. This method is suitable when the ambient radiation level is low, and when the isotope being used is one which may be detected by a detector outside the source. Tritium contamination is not usually measured by the direct method. The indirect method involves the swabbing of the area with the subsequent measurement of the activity removed.

Monitoring results should be recorded in the manner previously discussed. (See Section 7.1.)

10.4 The estimation of individual dose

In Britain the monitoring of individual dose involves a responsibility under the regulations to keep records of individuals doses for 50 years after the last entry. Laboratories who undertake measurements of activity in samples taken to assess personal dose must be approved by the Health and Safety Executive, as must any laboratory offering a calibration service for measuring instruments.

11. Transporting radioactive materials

A particularly hazard prone event in the use of active material is its transport from one place to another. Hand-held containers slip from wet fingers, glass vials break, and contamination results. There is a need to control radiation dose and contain the radioactive material during its movement by hand within, say, the building. Hard beta emitters such as ^{32}P, which should be in closed containers, are conveniently shielded in transit by using a Lucite or Perspex box. There are, however, more stringent controls when wheeled transport is involved. Compliance with International Atomic Energy Agency's Regulations for the Safe Transport of Radioactive Materials, which is embodied in the Department of Transport's Code of Practice and the International Air Transport Authority's Dangerous Goods Regulations, is required. Here again, control of radiation dose, containment, and freedom from external contamination of the package must be assured.

There are, however, certain exemptions which apply to 'professional users', to permit the transport of small sources in private vehicles, and certain low levels of activity and activity concentration are exempt from parts of the regulations and may be shipped as 'exempt packages' with very much less fuss than there would otherwise be. Radioactive materials whose radioactivity is less than the relevant exemption limit may be shipped as an exempt package provided that the dose rate at the surface of the package is less than $5\,\mu Sv \cdot h^{-1}$, there is no leakage of radioactivity from the package, and the package contains a warning, which is visible to anyone opening the package, that the package contains radioactive material. The commonly used nuclides ^{3}H, ^{14}C, ^{32}P, in solution may be shipped as exempt radioactive material, for example, if these conditions are met, and the activity is less than 37 GBq, 370 MBq, and 111 MBq, respectively.

12. Waste disposal

The disposal of radioactive waste must be authorized under the Principle Act. Individual authorizations, specific to users' particular requirements, are issued which allow disposal of low-level wastes to the drain, to the ordinary trash, to tip burial, and by incineration. Other more active, or more hazardous, waste is disposed via the National Disposal Service. For whichever route is used, careful recording of disposals in terms of activity and name of isotope is required. Radioactive waste with an activity concentration less than $4\,Bq \cdot g^{-1}$ of tritium and carbon-14, and $0.4\,Bq \cdot g^{-1}$ for all other nuclides may be disposed of without authorization.

Organ liquid scintillation cocktails deserve a special mention. Where activity concentrations exceed those above an Authorization to Dispose is required. Incineration is the usual disposal route, with a condition that the incinerator ash is treated as radioactive (except in the case of waste containing no nuclides other than ^{3}H and ^{14}C which are believed to be removed in the effluent gases), and disposed via a solid waste route if appropriate. Alternatively, with the agreement of the local water authority, some cocktails may be disposed as aqueous waste down a designated sink, when the authorized limits for aqueous waste will apply.

References

1. Ballance, P. E., Day, L. E., and Morgan, J. (1987). *Phosphorus-32 Practical Radiation Protection*. Occupational Hygiene Monograph No. 16. Science Reviews Ltd., Northwood, Middlesex, UK.
2. Mettler, F. A. and Moseley, R. D. (1985). *Medical Effects of Ionizing Radiation*. Grune & Stratton, Orlando, Florida.
3. ICRP (1977). *Recommendations of the International Commission on Radiological Protection, Publication 26*. Pergamon Press, Oxford, UK.
4. *Code of Practice for the Carriage of Radioactive Materials by Road*. Her Majesty's Stationery Office.

3

The scintillation counter

GÉRARD SIMONNET

1. Introduction

All experiments based on the use of radionuclides depend at some time or another on the detection of radioactivity. Therefore, the quality and reliability of the experimental results and subsequent interpretation depends on the accuracy and dependability of the detection system.

This is achieved by:

- an appropriate choice of detector;
- accurate calibration of the detector;
- suitable preparation of the radioactive sample to be measured.

The choice of detector, and procedures used, depends on the nature and energy of radiation produced by the experimental samples. Radionuclides used in biology are obviously isotopes of the elements forming live matter, i.e. H, C, S, P, I, Na. Radioactive isotopes of these elements are: ^{3}H, ^{14}C, ^{35}S, ^{32}P, ^{33}P, ^{125}I, ^{131}I, and ^{24}Na.

When a radionuclide disintegrates it usually emits a charged particle β^-, β^+, or α and some electromagnetic radiation, γ or X-rays. Radioactivity measurement, is most readily achieved through detection of electromagnetic radiation X or γ. In this case, a gamma counter is used. This is appropriate for ^{125}I and ^{131}I. However, most radionuclides used in biology, i.e. ^{3}H, ^{14}C, ^{35}S, ^{32}P, and ^{33}P emit only β^- particles and no γ radiation. Detection of these radionuclides, is achieved through counting β^- particles. This needs a different device, the liquid scintillation counter.

Before describing the use of these counters in detail two points, common to both β^- and γ measurements (and any other measure of radioactivity), will be discussed.

1.1 Counting efficiency

The radioactivity (A) of a sample is expressed in terms of the number of nuclei which disintegrate per unit time:

disintegration per second d.p.s. (1 d.p.s. = 1 Bq)
disintegration per minute d.p.m.

The measured radioactivity (n) of a sample corresponds to the number of disintegrations counted and is expressed in counts per unit time:

counts per second c.p.s.
counts per minute c.p.m.

The measured radioactivity (n) is proportional to the real radioactivity (A) and they are connected by the expression:

$$A = n/E \text{ where } E \leqslant 1.$$

E is the counting efficiency, a parameter determined by the nature of the sample and detector.

$$\text{Thus: } E = \frac{n}{A} \text{ or } \frac{\text{c.p.m.}}{\text{d.p.m.}} \text{ (sometimes expressed in per cent).}$$

Measurement of counting efficiency (E) is a critical factor in radiation detection and will be discussed in detail later (Section 2.8.3).

1.2 Errors and precision of measurements

1.2.1 Single measurements

If the radioactivity of a sample is measured say, a hundred times, with any given counter, it is observed that the results:

- are not all the same;
- are arranged between limits;
- are centred around a mean value ($\bar{N}$) corresponding to the most probable value.

These observations lead to two conclusions:

- the estimated count is aleatory;
- the true rate of decay cannot be measured, it is only possible to determine an interval around the mean value $\bar{N}$ in which the true value may be assigned.

It is possible to estimate the true value of activity, and uncertainty with which it is associated, from one measurement by applying the mathematical principles of Poisson and Gauss. When applying the Gauss law to a value N_m, it is possible to measure the standard deviation σ:

$$\sigma = \sqrt{N_m}.$$

In the interval:

(a) $N_m - \sigma$ to $N_m + \sigma$, there is a 68% chance that the true value is situated within this limit,

(b) $N_m - 2\sigma$ to $N_m + 2\sigma$, there is a 95% chance that the true value is situated within this limit.

The result of the N_m measure can be expressed by:

$$N_m \pm 2\sqrt{N_m}$$

where the value $2\sqrt{N_m}$ shows the uncertainty associated with the measurement quoted as δN, as for example, as follows:

$$N_m = 1\ 600$$

then $$\delta N = 2\sqrt{N_m} = 80.$$

Result: 1600 ± 80 for a 95% probability.

The relative accuracy (P) on the N_m measurement is given by the quotient:

$$P = \frac{2\sqrt{N_m}}{N_m} = \frac{2}{\sqrt{N_m}}.$$

The accuracy can also be expressed in per cent:

$$P\% = \frac{2 \times 100}{\sqrt{N_m}}.$$

The data shown in *Table 1* illustrates that the relative accuracy of a measurement improves at higher count rates. Alternatively, an improvement in the precision of a measurement can be obtained by merely increasing the counting time. However, it is inappropriate to work to a precision beyond that of the experiment carried out, as it would lead to unnecessary use of the counter. For example, if the error on an experiment is about 15%, it is not necessary to work to a 1% error for the radioactivity measurement. Conversely, if an experiment has an error ranging from 2 to 5%, it would be unwise to taint it with a 10% additional error from the radioactivity measurement.

When results are expressed in count rate (n, i.e. in counts per unit time (e.g. c.p.m. then $n = N/t$. In this case the calculation for accuracy must be done not only on the counting rate (n) but also on the total number of measured pulses (N). However, we can use the following relationship for this calculation:

$$\Delta n = 2\sqrt{\frac{n}{t}}$$

where n is the count rate, Δn is the uncertainty associated with the count rate, and t is the counting time.

For example, consider the following data:

counting rate	640 c.p.m.
counting time	10 min

The error can be calculated in two ways using the equations described above.

Table 1. Uncertainty and precision associated to different counting values

Measured values	Uncertainty $N = 2\sqrt{N}$	Precision $P\%$
100	20	20%
1000	62	6.2%
10000	200	2%
40000	400	1%

1st method

Calculate the total number of pulses (N) measured during the time t:

$$N = n \times t = 640 \times 10 = 6400$$

$$\Delta N = 2\sqrt{N} = 2 \times 80 = 160$$

$$\frac{\Delta N}{N}\% = \frac{160}{6400} \times 100 = 2.5\%.$$

2nd method

Use the relationship:

$$\Delta n = 2\sqrt{\frac{n}{t}}$$

$$\Delta n = 2\sqrt{\frac{640}{10}} = 2 \times 8 = 16$$

$$\frac{\Delta n}{n}\% = \frac{16}{640} \times 100 = 2.5\%.$$

The result is $640\,\text{c.p.m.} \pm 16\,\text{c.p.m.}$

$640\,\text{c.p.m.} \pm 2.5\%.$

Notice that the answer is the same with both methods; that is,

$$\frac{\Delta N}{N} \text{ is equal to } \frac{\Delta n}{n}.$$

1.2.2 Combinations of measurements

Usually a result (R) is obtained from the combination of several independent measurements (for example, the net activity of a sample is equal to the difference between the gross activity and the background count). The possible combinations of two measurements (A and B) are subtraction, addition, division or multiplication and the uncertainty associated with the result is given by the following general formulae:

$$\left.\begin{array}{l} R = A + B \\ R = A - B \end{array}\right\} \sigma(R^2) = \sigma\,(A)^2 + \sigma\,(B)^2$$

$$\left.\begin{array}{l} R = A\,/\,B \\ B = A\,.\,B \end{array}\right\} \left(\frac{\sigma\,(R)}{R}\right)^2 = \left(\frac{\sigma\,(A)}{A}\right)^2 + \left(\frac{\sigma\,(B)}{B}\right)^2.$$

The use of these formulae can be illustrated by using the substraction of background as an example.

Suppose a gross count A (c.p.m.) is obtained during a time t_A and a background count B (c.p.m.) obtained during a time t_B. Then:

the net activity is $R = A - B$
and the uncertainty on R is $2\,\sigma(R) = \Delta R$

$$\frac{\Delta R}{R}\,\% = \frac{200\sqrt{\left(\frac{\sqrt{N_{(A)}}}{T_{(A)}}\right)^2 + \left(\frac{\sqrt{N_{(B)}}}{T_{(B)}}\right)^2}}{A - B}$$

where $N = \text{c.p.m.} \times t$.

Consider a numerical example:

$$A = 95 \text{ c.p.m.} \qquad t_{(A)} = 5 \text{ min}$$
$$B = 37 \text{ c.p.m} \qquad t_{(B)} = 10 \text{ min}$$
$$R \text{ (c.p.m.)} = 95 - 37 = 58 \text{ c.p.m.}$$

$$\frac{\Delta R}{R}\,\% = \frac{200\sqrt{\left(\frac{\sqrt{95 \times 5}}{5}\right)^2 + \left(\frac{\sqrt{37 \times 10}}{10}\right)^2}}{95 - 37} = 16.43\%.$$

The difference in count rates can therefore be expressed as:

58 c.p.m. $\pm$ 16.43% or 58 $\pm$ 10 c.p.m.

1.3 Preset time and preset count

The previous section has illustrated that the precision with which the rate of decay can be estimated varies with the level of measured activity. When comparing various samples it is advisable that all the samples are measured with the same accuracy, irrespective of their level of radioactivity. This means that samples should be measured not for the same time but for various times; longer as the activity is weaker.

All scintillation counters include commands which allow for the measurement of a series of samples at the same accuracy. The user can then fix the number of impulses measured, or error in per cent. Each sample is therefore measured for a time required to reach the determined accuracy. If a series of samples includes weak radioactive samples (for example, experimental controls), it is possible by means of a command on the counter referred to as 'Low Count Reject', to count some samples for a relatively short time and thus avoid long and unnecessary use of the counter.

Therefore the three parameters: preset time, preset count, low count reject, may be combined and it is the first parameter to be reached that determines when counting stops. Use of these three parameters allows for:

- an optimal accuracy to be set for radiation detection; and,
- a rational utilization of counting time, particularly important when a counter is used by several people.

2. Gamma counters

An understanding of the mechanism of radiation detection requires some knowledge of the physical nature of the radiation(s) being detected and of the mechanism by which ionizing radiations interact with matter. In this section the physical nature of gamma radiations and their interaction with matter will be described prior to a discussion on the utilization of gamma counters.

2.1 γ and X electromagnetic radiation

2.1.1 Gamma radiations

Following the emission of β^+ or β^- charged particles the new nucleus that has just been formed is in an excited state (see Chapter 2). This nucleus will remain for a short time only in that excited state before it returns to its fundamental state either directly, or through intermediary stages, there by releasing energy (E) through electromagnetic radiation called *gamma radiation* or *gamma photons*.

The gamma radiation(s) emitted by a given radionuclide have a distinctive energy spectrum with a maximum energy (E_{max}) that does not go beyond a few MeV. In the case of ^{131}I a peak at 364 keV is observed (*Figure 1*).

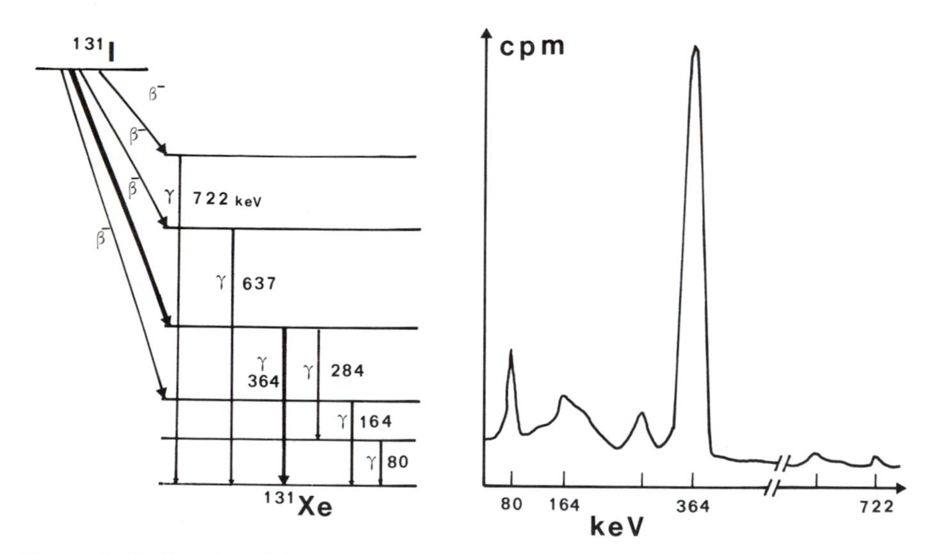

Figure 1. Radioactive disintegration diagram and γ spectrum of ^{131}I.

2.1.2 X-rays

In the case of certain radionuclides the excess energy of the nucleus (ΔE), instead of being released under the form of γ radiation, is transferred to an electron, usually of the K shell. This electron is ejected from the atom with an energy (E_c) equal to the difference:

$$E_c = E - Wl \quad \text{where } Wl \text{ is the link-energy.}$$

These electrons are called *electrons of internal conversion*. The vacancy on the K shell is filled up by a more peripheral electron; a transition that will lead to the emission of an X_K radiation. The phenomenon of internal conversion with X_K emission is observable with ^{125}I.

The γ and X radiations are the same in physical nature (electromagnetic); the difference is in their origin: γ-rays originate from the nucleus whereas X-rays have an extranuclear origin. γ and X radiations are detected by the same means: γ counters. All that has to be done is to adapt the energy calibration.

2.2 Interaction of γ and X radiations with matter

As γ or X-rays pass through matter some of their energy is absorbed, both by electrons and nuclei of the absorber. The interaction with electrons involves two mechanisms: the photoelectric effect and the Compton effect. Interaction with a nucleus creates an electron pair: that is, the materialization effect. The relative importance of these three modes of interaction depends on the radiation energy:

- for weak energies (< 200 keV), the photoelectric effect prevails over the Compton effect; the materialization effect is nil;
- for intermediate energies (500–1000 keV) the Compton effect becomes predominant;
- for high energies (> 1 MeV), the materialization is the predominant effect.

None of the radionuclides commonly used in biology emits gamma radiation of high enough energy to allow a materialization effect; this phenomenon will not be discussed further.

2.2.1 Photoelectric effect

The γ or X radiation of energy $h\nu$ interacts with an electron with link-energy to the atom We. If $h\nu > \text{We}$ the electron may be released and, moreover a kinetic energy E_c (given by $E_c = h\nu - \text{We}$) is conveyed to it. In this process, all the γ or X energy is transferred during a single interaction.

2.2.2 Compton effect

During an interaction by the Compton effect, only a fraction of the energy ($h\nu$) of a γ or X-ray is transferred to an electron. Following this interaction there will be a diffused photon of energy ($h\nu'$), where $h\nu' < h\nu$. The resulting electron is called a *Compton electron*, that has a kinetic energy E_c where:

$$E_c = h\nu - h\nu'.$$

2.2.3 Interaction of photoelectrons and Compton electrons with matter: ionization and excitation

As previously discussed, the γ or X radiation transfers some energy to electrons that, in their turn, interact with the medium giving up their kinetic energy. It mainly involves the peripheral electrons. If the quantity of transferred energy is sufficient, *ionization* occurs. If the interaction is insufficient, there will only be *excitation*, i.e. the electron goes from an initial state S_0 to an excited state S_1. The atom or molecule is then in an excited state.

During their passage through matter, electrons will undergo a great number of interactions and progressively lose their energy. The rate of energy loss is such that an electron with an energy of 500 keV will have transferred all its kinetic energy after passing through 2 mm of a material with a density of 1 g/cm^3.

2.2.4 Fluorescence

Molecules that absorb kinetic energy from a photoelectron or Compton electron are in an excited state (S_1) that quickly returns to the initial states S_0 with the concomitant emission of energy. In scintillation counters a material is chosen that emits energy as fluorescence; the material involved being referred to as a scintillator.

2.3 The principle of radiation detection

The physical principle behind detection is the interaction of radiation with matter. As described above, γ or X radiations lead to the formation of ionized and excited atoms or molecules in the absorbing material. It is this ionization and excitation that is the basis for two main types of detector:

- gas ionization counters (that is Geiger–Müller ionization chamber and proportional counters);
- and scintillation counters respectively.

The remainder of this chapter is concerned exclusively with scintillation counters.

2.3.1 The scintillators

A scintillator is a material that must have:

- fluorescence properties;
- a transparency to emitted light; and
- a high absorption of the radiation to be detected.

The detection of γ and X-rays involves scintillators of sodium iodide crystal activated by thallium. The radiation interacts with sodium iodide but, photons are emitted by the thallium. The wavelength of light emitted ranges between 350 and 500 nanometres and shows a maximum of 410 nanometres.

Sodium iodide is hygroscopic and it must be perfectly protected from extraneous light. Therefore the crystal is wrapped in an aluminium cover inwardly coated with magnesium oxide or aluminium oxide, to reflect the light emitted by the crystal. One of the surfaces is covered in a thin glass through which the emitted light passes.

The shape of scintillators is cylindrical, with a diameter that may be 32, 51, or 76 mm. The choice in size is a function of the energy of radiation to be detected; the higher the energy the greater the size of scintillator required.

Table 2. Diameter of the scintillators required for detection of γ and X-rays

Radionuclides	**Energy γ or X (MeV)**	**Diameter (mm)**
^{22}Na	γ 0.51 (β^+) γ 1.28	51–57
^{24}Na	γ 1.37 γ 2.75	76
^{42}K	γ 1.54	51–76
^{131}I	γ 0.364	51–76
^{125}I	γ–X 0.030	32
^{51}Cr	γ 0.320	32
^{57}Co	γ 0.122	32

The scintillator NaI may be used not only to measure the activity of a source but also to do a spectrophotometric analysis. The latter aspect will not be discussed further as it is not relevant to biology and in any case is now achieved by means of Ge(Li) coaxial detectors that allow a much better resolution.

The scintillator has a cylindrical shape and is bored in its centre, through part of its height, with a cylindrical well also protected with the aluminium wrapper. The sample to be measured is placed in this well. This placing of the sample, with respect to the detector, allows for a higher counting efficiency than if the sample were simply settled above the detector (*Figure 2*).

For reasons of counting geometry, the height of the radioactive sample in the test-tube must be low with regard to the well depth, and should be kept constant for a series of samples.

The device, including the photomultiplier (PM) tube associated with it, is located within lead shielding as protection from external radiation. The

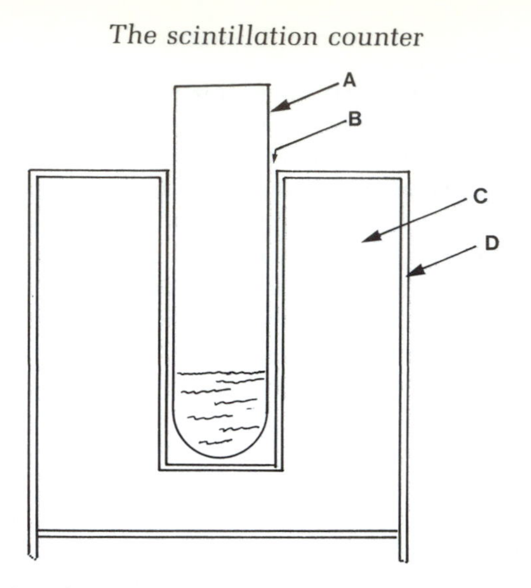

Figure 2. Cross-section of a well NaI scintillator crystal. A—tube containing the sample to be counted; B—well; C—NaI (Tl) crystal; D—aluminium wrapper.

weight of shielding may be as high as 400 kg for crystals with a diameter equal to 71 mm.

2.3.2 Detection of photons emitted by the crystal

Interaction of γ or X-rays with the crystal causes the crystal to fluoresce and emit photons. The level of fluorescence is measured in two stages. First, the emitted photons are detected by a photocathode that changes photons into electrons; a current is obtained that is very weak and must be amplified. Second, a photomultiplier tube amplifies the electric current and emits electrical pulses.

The duration of the whole phenomenon is very short and the photomultiplier tube registers a very fast flash of light (scintillation). The resulting electric pulses are counted and analysed as follows.

The number of pulses

For each γ or X-ray that interacts with the crystal, an electric pulse is generated. If every atom that disintegrated emitted a detectable radiation, it would be possible to directly evaluate the activity of the radioactive source. In fact, the relationship between the number of disintegrations and the number of provided electric pulses is less than 1; however, its value for a given radionuclide under any given set of conditions is steady, therefore the activity of various samples can be compared.

The height of pulses

Each electrical pulse emitted from the photomultiplier tube has a height, measured in volts, that is proportional to the intensity of the light-flash (scintillation). The light-flash intensity is itself proportional to the energy transferred to the medium by γ or X radiation, therefore pulse height analysis can be used to determine the energy of the detected radiation (*Figure 3*). This is discussed further in Section 2.4.3.

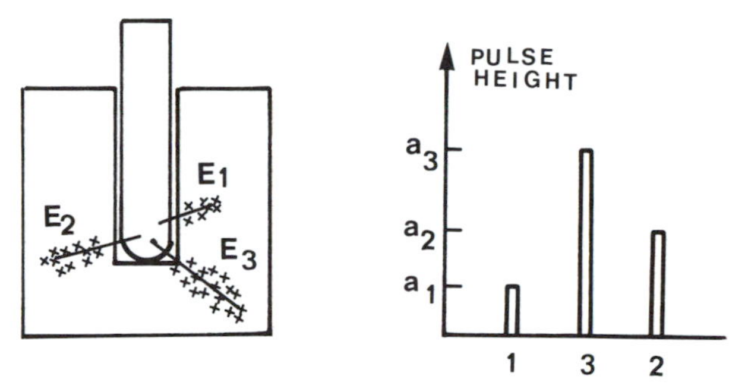

Figure 3. Quantitative and qualitative information obtained from detection by scintillation. Three γ radiations of energy $E_1 < E_2 < E_3$ lead to the formation of electric pulses of amplitude $a_1 < a_2 < a_3$, respectively.

To summarize, the electronic signal issued from the photomultiplier supplies two types of information:

quantitative: that is the number of measured pulses related to the radioactivity of a source; and

qualitative: where the height value of electrical pulses relates to the energy of the detected radiation.

2.4 Auxiliary electronic instrumentation associated with the detector

2.4.1 The preamplifier

Despite the amplification factor provided by the photomultiplier tube, the signal is still very weak and could not travel across the coaxial cable that connects the photomultiplier tube to the rest of the counter. A preamplifier, placed immediately next to the photomultiplier tube slightly increases the signal and also serves as an impedance matcher.

2.4.2 The amplifier

The amplifier is an electronic device that increases the height of all the electric pulses. This amplification is essential for pulse height analysis.

2.4.3 Pulse height analyser and multichannel analyser

i. Pulse height analyser

For counting and spectrum analysis, it is necessary that the electric pulses are analysed with respect to pulse height. The device that carries out this function is the pulse height analyser. With this instrument it is possible to demarcate a height interval located between a *lower level* and an *upper level*. This interval is called a counting *window* or *channel*. It is therefore possible to select only those pulses with a height ranging between two limits, i.e. upper to the lower level, and lower to the upper level (*Figure 4*)

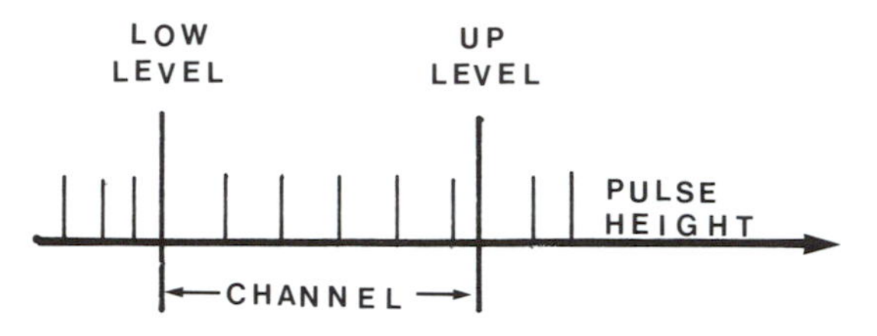

Figure 4. Principle of a monochannel amplitude discriminator. Only pulses recorded are those with amplitudes between the upper and lower channel levels.

The device that allows for these setting of a counting channel is called a *single channel analyser*. It is a device that in principle is similar to the wavelength calibration of a spectrophotometer. All counters have a pulse height analyser with two or three such channels working in parallel.

ii. Multichannel analyser

A multichannel analyser is a pulse height analyser built up from a large number of very narrow channels, that is from 512 to 8192 channels. Multichannel analysers are made of an analogue-to-digital converter (ADC) and a memory as in a microcomputer. The analogue-to-digital converter converts the electronic signal, i.e. the counting pulse, into a numeric signal, the value of which is proportional to the pulse height. The numeric signal is stored in a memory, and each memory cell corresponds to a channel of the analyser. The number of cells or channels is, according to the device, 512, 1024, 2048, 4096, or 8192. The operation of the multichannel analyser can be illustrated as follows.

If the pulse height analyser operates within the range 0–2 MeV energy and comprises 1024 channels, each channel will store pulses with a height corresponding to an interval of $2\,\text{MeV}/1024 \simeq 2\,\text{keV}$. The lower address channel (channel No. 1) will receive pulses from 0 to 2 keV and so on to the upper address channel (or channel No. 1024) that will receive pulses from 1998 to 2000 keV.

Each time a pulse is recorded by a channel, its content is incremented by one (+1) it is therefore possible to obtain, on a data output from the counter, a histogram representing the content of each channel as a function of the channel number (height).

2.5 Numeric processing and data output

All counters built in recent years either contain a microprocessor or are connected to a microcomputer. Their role is to manage the running of the counter and to perform otherwise time-consuming data manipulations.

The microprocessor calculation programs are not modifiable by the operator, who simply has the choice between various pre-established programs. If the operator wishes to carry out other calculations, the counter can be coupled to a microcomputer (1) through a connection using the standard RS-232 interface.

The calculations directly provided by the counter are:

- counting rate: counts per minute;
- background subtraction;
- statistical precision: $P_s = 2/\sqrt{N}$, where P_s is statistical precision and N is number of pulses measured.

Also, most counters will automatically analyse data from radioimmunology and radioimmunometric assays; that is:

- plotting of calibration curves by means of different mathematical methods;
- calculation of concentration in the unknown samples;
- quality control.

2.6 The different types of counters

Counters may be classified into two categories depending on whether they comprise one or several detectors.

2.6.1 Monodetector counters

Counters having a sole detector are either old devices, or instruments where the scintillant crystal has a wide diameter, 51 or 76 mm with which it is possible to detect γ radiations of high energy. A sample changer associated with the detector allows for an automatic measurement of the samples.

2.6.2 Multidetector counters

The γ multidetector counters have been especially developed for radioactivity measurements of radioimmunoassay. In these assays the radionuclide used is ^{125}I, this isotope emits X and γ radiations of weak energy, about 30 keV. To detect radiation of this energy a NaI crystal of small size (32 mm) is adequate. These detectors may also be used for detecting ^{57}Co (122 keV) and ^{51}Cr (320 keV). In the same instrument the number of detectors is from 4 to 24.

Depending on the model, the counter may or may not include a sample changer. The advantage of multidetector counters is the high number of samples counted in a short time, a particularly important point for a laboratory that carries out a great number of assays.

The multidetector counters need specific calibration, discussed below.

2.7 Counter calibration

This section is concerned with the calibration of newly acquired counters, periodic, routine calibration, and calibration prior to the counting of a particular radionuclide. All are necessary to control background, resolution, and counting efficiency.

2.7.1 Calibration sources

Calibration and counting of controls is done with standard sources supplied by radiation metrology laboratories and commercially by firms that sell radioactive products. Radionuclides usually available as a standard source are, for example, ^{22}Na, ^{60}Co, ^{51}Cr, ^{137}Cs. For radionuclides of short half-life such as ^{125}I ($t½ = 60$ days) or ^{131}I ($t½ = 8$ days) it is more convenient to use *calibration simulators*. These contain one or more radionuclides that provide a spectrum that is close or identical to the radionuclide to be simulated. For example the calibration simulator for ^{125}I is made of a mixture of ^{129}I ($t½ = 1.6 \times 10^7$ years) and americium-241 (^{241}Am, $t½ = 433$ years). The simulated source may reproduce the peaks in a spectrum for a given radionuclide well but the relative proportions of the peaks may be different.

For γ counters containing a well-shaped crystal, calibration sources are the shape of a filled tube of 12 mm diameter and height of approximately 100 mm, and may be preserved for several years, depending on half-life.

2.7.2 Background counting and figure of merit

In the absence of a radioactive sample, the detector gives a background count that must be subtracted from all sample readings (*Figure 5*).

Background arises from:

- cosmic and telluric radiations;
- radioactivity from materials making up and surrounding the counter;
- noise from the PM tube.

In addition, the background is higher with increasing size of the NaI crystal. For example, the background from a 75-mm crystal is double that from a 50-mm crystal.

Superimposing on the same graph the background and signal from the detection of ^{131}I shows clearly that if the counting is carried out in a narrow energy area located under the main peak (that is 364 keV for ^{131}I) the best ratio between signal and background is obtained. This ratio is called the figure of merit: *FM*(2).

$$FM = \frac{(\text{net counting} \cdot \text{c.p.m.})}{\text{background} \cdot \text{c.p.m.}}$$

The figure of merit value must be maximal and is the reason why γ or X

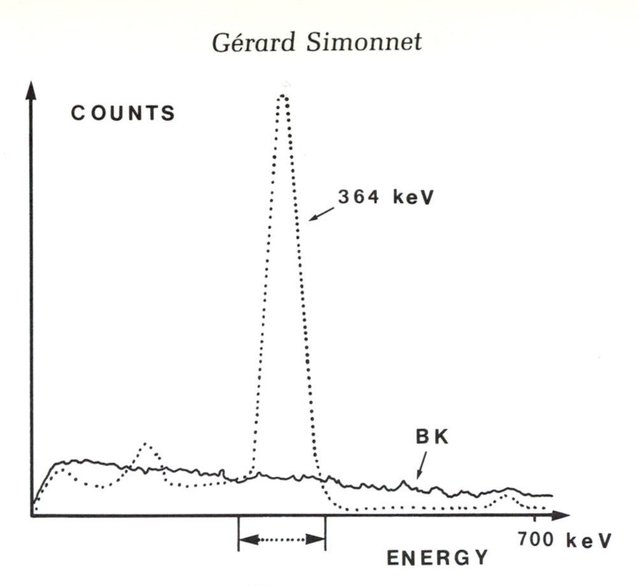

Figure 5. The background (BK) and ^{131}I spectra between zero and 700 keV.

counting is done by positioning a window on the main peak of emission and not on the whole γ or X emission of a particular radionuclide.

2.7.3 Resolution of a detector

In spite of the fact that γ emission is monoenergetic, analysis by means of a well NaI detector makes the spectrum appear like a bell-shaped curve. This loss of resolution is due to the crystal, the photomultiplier tube, the optical connection between the two, and the electronic set-up.

From the obtained spectrum with the γ counter it is possible to determine what is called the *resolution* of the detector; it is usually determined from a spectrum obtained with a ^{137}Cs source. The employed peak is the one corresponding to the γ emission of 662 keV (*Figure 6*). The resolution is calculated by dividing the width at half-height of the peak, by the peak abscissa as described in *Protocol 1*.

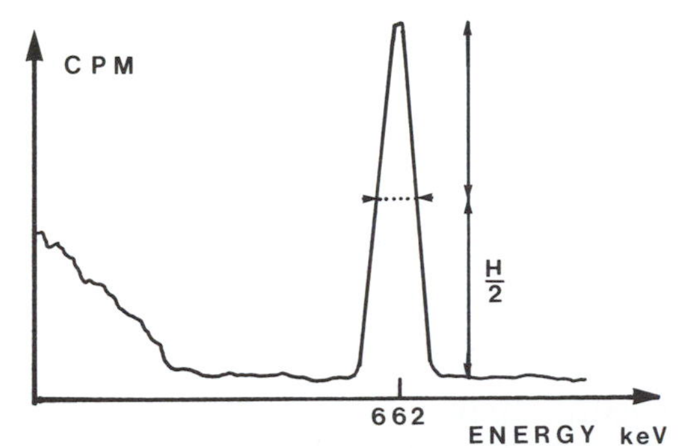

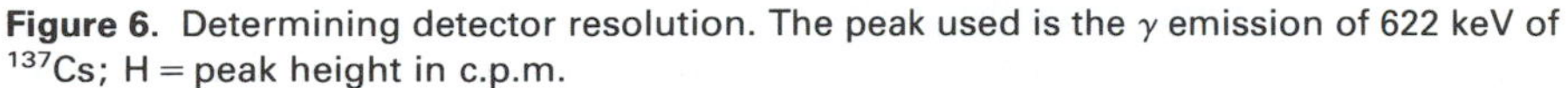
Figure 6. Determining detector resolution. The peak used is the γ emission of 622 keV of ^{137}Cs; H = peak height in c.p.m.

Protocol 1. Determination of the detector resolution

Counter with manual pulse height analyser (without multichannel analyser):

1. Place the ^{137}Cs source in the detector.
2. Set the energy scale on 0 to 1 MeV.
3. By means of the threshold potentiometer for window width set the window width at 50 keV.
4. Set the threshold potentiometer window successively at 0, 50, 100, 150, . . ., 700, 750 keV.
5. For each value count the source for 1 min. Between 500 and 700 units, an important peak appears corresponding to the emission at 662 keV.
6. Plot on standard millimetric graph paper the counting values (c.p.m.) as a function of the potentiometer units (or keV). On the graph, measure:
   (a) the width at half-height of the peak, and
   (b) the abscissa of the peak summit.
7. Calculate the resolution, $R\%$

$$R\% = \frac{\text{width at half-height}}{\text{peak abscissa}} \times 100.$$

Counter with multichannel pulse height analyser:

1. As steps 1 and 2 above.
2. Set the counting time selector to 5 or 10 min.
3. Obtain a graphical output of the spectrum on the printer associated with the counter.
4. On the graph, measure, as steps 6 and 7 above.

The resolution value is usually about 10–15% for well NaI detectors. It is important that the resolution is checked periodically as an increase would mean a deterioration of the detector and a consequent invalidation of all pre-established calibrations. If this occurs, the manufacturer of the counter should be consulted.

2.8 Automatic and manual calibrations of counting channels

The calibration of a counter consists of setting a counting channel appropriate to the γ peak to be measured.

2.8.1 Automatic calibration

All counters include the facility of automatic calibration: all that is required is to input, by means of the keyboard, the symbol of the radionuclide to obtain the appropriate calibration. With γ counters containing a 51- or 76-mm diameter crystal automatic calibration is usually available for: ^{125}I, ^{57}Co, ^{51}Cr, ^{131}I, ^{137}Cs, ^{42}K, ^{60}Co, ^{24}Na, and ^{59}Fe. For counters with a 32-mm diameter crystal that are specific to radioimmunology, automatic calibration is for ^{125}I and ^{57}Co. The procedure for automatic calibration is described in *Protocol 2*.

The automatic calibrations will be correct only if the gain and the value of high voltage applied to the photomultipliers are suitable. If these values are too high, the peaks will be shifted towards the upper part of the counting windows; and inversely if the values are too weak.

Protocol 2. Automatic calibration control

Where the d.p.m. of a standard calibration sample or the counting efficiency is known:

1. Set one of the windows appropriate to the standard; e.g. at 662 keV for ^{137}Cs.
2. Count the calibration source for 2 min.
3. Remove the calibration source.
4. Take a background count.
5. Subtract the background from the measured activity.
6. Compare the measured activity with the precedent value. If the variation between values is above statistical fluctuations, modify the high-voltage applied to the photomultiplier tube (steps 7–9). If this process cannot be done by the operator, contact the manufacturer.
7. Count the standard calibration source.
8. While the counting is on, shift very slowly towards the bottom, then towards the top, the value of high voltage. Observe the variation in count rate.
9. Record the value of high voltage that provides the maximum count-rate, it will correspond to the peak centring in the counting window. Henceforth, if for example the standard is ^{137}Cs and the pêak of 662 keV is perfectly centred in the counting window, 'peak ^{137}Cs', then all the pre-selected windows on the counter, on the same scale of energy, will be correct.

2.8.2 Manual calibrations

Manual calibration is necessary if the counter has not automatic calibration for the radionuclide used. Manual calibration involves defining the *width* and the *position* of a counting channel to detect a γ or X-ray peak. These values may be either determined graphically from the spectrum obtained from the counter, or calculated from the value of the resolution of the counter and from the energy of the peak detected. The method is described in *Protocol 3*.

Protocol 3. Manual calibration

Determination from the graph

1. Place a sample of the appropriate radionuclide in the counter. Obtain a spectrum (see Section 2.7.3).
2. From the spectrum, for the main γ or X-rays peak, read the value of lower and upper energy levels.
3. Set threshold and windows to these values for counting experimental samples.

Determination from the resolution

If the resolution of the detector and the energy of the peak to be detected are known, the instrument can be calibrated.

Example:

Resolution: 12%.

Peak detection 364 keV of ^{131}I.

The width of the peak at half-height is equal to the resolution multiplied by the peak abscissa divided by 100:

$$\text{therefore width at half-height} = \frac{12 \times 364}{100} 43.6 \text{ keV.}$$

The width of the counting channel set at twice the width of the peak at half-height:

e.g. channel width = 43.6 × 2 = 87 keV.

The lower channel level set at the energy value of the peak, less the width at half-height:

e.g. channel level = 364 − 43 = 321 keV.

Therefore the appropriate calibration for the counting of the 364 keV peak of ^{131}I is:

channel width	87 keV
channel level	321 keV.

2.8.3 Calculation of counting efficiency

When the calibration is established, measure the activity of the calibration source and the background in the channel, then calculate the counting efficiency (E):

$$E\% = \frac{\text{c.p.m. (source)} - \text{c.p.m. (background)}}{\text{d.p.m. (source)}} \times 100.$$

As shown above the counting efficiency of a detector may be determined by means of a calibration source, but it is also possible to use a sample which is not calibrated (3). In that case, the method is based on the detection of a 30 keV peak that comes from the γ or X_K (see Section 2.1.2), and, from the peak referred to as the *summation* in energy, 60 keV from a γ plus a X_K or from two X_K peaks. The summation peak appears when a ^{125}I atom emits two radiations in series that are detected not as two radiations of, say, 30 keV each, but as a single radiation of $30\,\text{keV} \times 2 = 60\,\text{keV}$ (*Figure 7*).

The real activity in d.p.m. is given by the formula:

$$\text{d.p.m.} = \frac{[(\text{c.p.m. } A) + (2 \times \text{c.p.m. } B)]^2}{4.028 \times \text{c.p.m. } B}$$

where : c.p.m. A
c.p.m. B
4.028 = activity in c.p.m. of the peak at 30 keV,
= activity of the peak at 60 keV,
constant

$$E = \frac{\text{c.p.m. } A + \text{c.p.m. } B}{\text{d.p.m.}} \times 100\%.$$

2.9 Calibration checking for multidetector gamma counters

As discussed in Section 2.6.2, multidetector counters can comprise between 4 and 24 detectors. Each of these detectors will not have the same counting efficiency for any given radionuclide, it is necessary to determine the counting efficiency for each one of them.

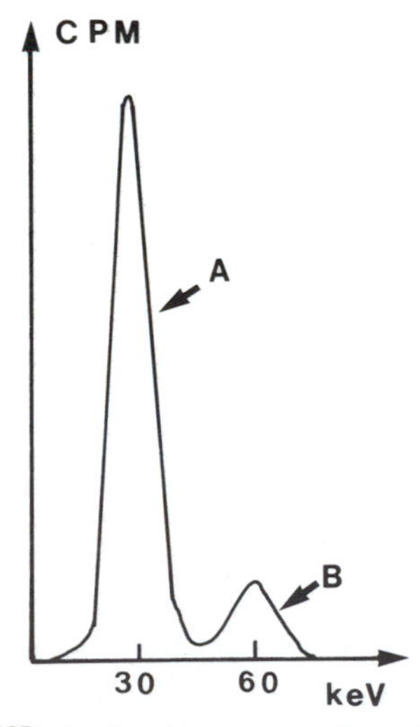

Figure 7. γ and X spectra of ^{125}I obtained by means of a well NaI crystal. A—peak due to γ or to X_K; B—summation peak $\gamma + X_K$ or $X_K + X_K$.

The efficiencies are expressed as a relative value, the value 1 being assigned to the detector that provides the highest counting efficiency. It is therefore not necessary to use a calibrated source; any source of the radionuclide of sufficient activity may be used. During this calibration process, the background for each detector is determined at the same time.

The principle behind the protocol is as follows: the same source is measured successively in each of the (N) detectors of the counter. During each counting, the background in the remaining ($N-1$) detectors is registered.

After the (N) readings have been taken, the relatively efficiency of each detector is calculated.

All these values of efficiency and background are stored in the counter's computer, so that correction factors will automatically be applied to readings. A steady control of these parameters is necessary, its frequency should be monthly if the counter automatically calibrates the counting channels and daily if not.

Conclusion

To summarize, the following parameters must be determined for a detector:

- the resolution;
- the background;
- the counting efficiency.

These values must periodically be checked to detect any fluctuations in the detector and associated electronics. Moreover, if at any given time a counting anomaly is suspected, these calibrations can be carried out and the anomaly corrected.

The counting of γ and X-rays does not involve very great complexity for calibration, the handling of counters, or for the preparation of samples. In view of this, to make the operator's work easier, apparatus built since around 1985 incorporate a great increase in automation, controlled by an automatic data processing unit, including counter calibration. However, this does not relieve the operator from all responsibility and the controls described in the previous sections are required.

3. Liquid scintillation counters

3.1 Principle

As discussed in Section 1, the majority of radionuclides used in biology: ^{3}H, ^{14}C, ^{35}S, ^{32}P, ^{45}Ca, are pure β^- emitters; that is, there is no γ emission associated with their decay. From the point of view of recording levels of radioactivity this is a handicap because the detection of β^- particles presents more difficulties than detection of X or γ radiations; β^- radiation of energy equal to a X or a γ radiation has a much lower penetration. For example, β^-

particles of 50 keV are completely absorbed by 50 μm of water whereas an X or γ-ray of the same energy (50 keV) is only diminished by 1.5%. It is because β^- particles have a very low penetration that their detection is achieved by introducing the radioactive source into the very substance of the detector; in this case a liquid with fluorescent properties.

3.1.1 β^- spectra

Not all β^- particles emitted by a given radionuclide have the same energy; they are distributed between the value zero and a maximal energy value, E_{max}, characteristic for every radionuclide (4). The E_{max} values are as follows: ^{3}H 18.6 keV, ^{14}C 155 keV, ^{35}S 165 keV, and ^{32}P 1708 keV. Between zero and E_{max}, the spectrum has a maximum (E_{medium}) located at about a third of the E_{max} value (*Figure 8*).

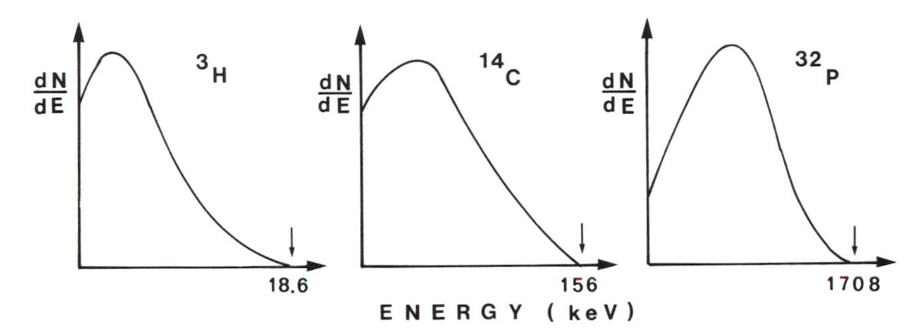

Figure 8. β^- spectra of ^{3}H, ^{14}C, and ^{32}P; N = c.p.m., E = energy (keV).

3.1.2 The interaction of β^- particles in the scintillator

When β^- particles pass through matter they yield their kinetic energy to the medium. This produces three types of phenomena: ionization excitation and heat. As previously described (Section 2.2.3) in scintillation-type detectors, it is the excitation that accounts for detection. When a β^- particle excites fluorescent molecules constituting the scintillator, the latter return to ground state with the emission of fluorescent photons that can be detected by means of a photocathode:

$$\beta^- + S \rightarrow S^*$$
$$S^* \rightarrow S + h\nu$$
$$h\nu \rightsquigarrow \text{photocathode}$$

S = molecules of the scintillator,
S^* = excited state,
$h\nu$ = photons of fluorescence.

3.1.3 Scintillation fluid

To allow for the introduction of the radioactive sample into the substance of the detector the scintillator has to be a liquid. Compounds answering to the double criterion fluorescence and liquid are essentially the aromatic organic compounds of the benzene family. However, benzene itself is never used as a scintillator because of its high toxicity; instead, methyled aromatic solvents are used such as toluene, xylene, cumene, or pseudo-cumene. The wavelength for fluorescence of these compounds is located in the ultraviolet at 270 nm. However, photons with this wavelength cannot go through glass or plastic, the constituent materials of vials in which radioactive samples and scintillator are placed. Therefore it is necessary to use a solvent such as xylene in which other fluorescent compounds are dissolved, these being called *solutes*. Their role is to shift the fluorescent wavelength of the solvent (270 nm) into the visible part of the spectrum at about 450 nm. Usually, two solutes are used: a 'primary' and 'secondary', their concentration in the solvent being about 4 g litre^{-1} and 0.25 g litre^{-1} respectively.

The energy transfers are as follows:

$$\beta^- + T \rightarrow T^*$$
$$T^* + S_1 \rightarrow S_1^* + T$$
$$S_1^* + S_2 \rightarrow S_2^* + S_1$$
$$S_2^* \rightarrow S_2 + h\nu \text{ (wavelength} = 450\,\text{nm)}$$

where: T = solvent,
S_1 = primary solute,
S_2 = secondary solute,
$*$ represents the excited state.

Figure 9 shows these stages of energy through the fluorescence spectra of the solvent and the two solutes. A scintillation cocktail made of solvent (T) and S_2 would be inefficient as the energy transfer between those two compounds is impossible, a S_1 intermediate solute is absolutely necessary. In a later section (3.6.1), the chemical nature of solutes (that are usually of the phenyl-oxazole type) will be discussed.

The number of scintillations is proportional to the number of particles emitted in a scintillating medium (*Figure 10*), but also there is a direct relationship between the energy of a β^- particle and the number of photons produced, or the intensity of emitted light. As will be discussed later, this property is exploited by liquid scintillation counters.

To summarize, from the fluorescence yielded in a scintillating liquid, we can obtain two kinds of information:

quantitative: the number of scintillations is proportional to the sample radio-activity;

qualitative: the light intensity of a scintillation is proportional to the energy of the absorbed β^- particle.

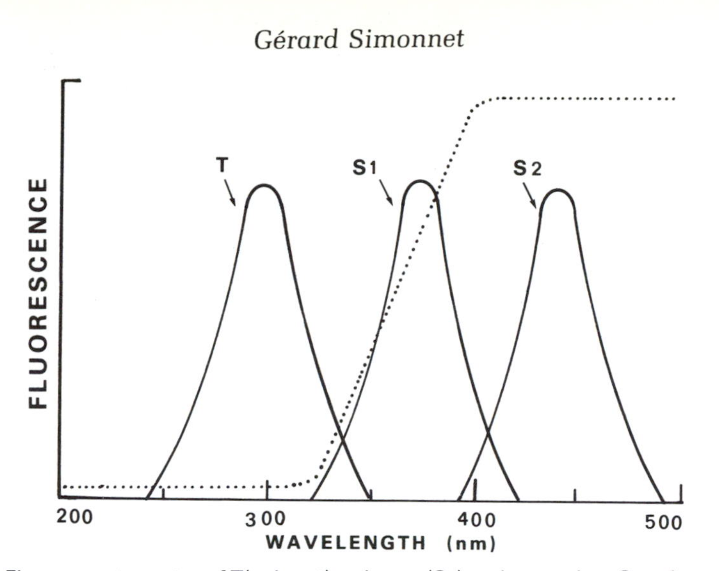

Figure 9. Fluorescent spectra of T(solvent), primary (S_1) and secondary S_2 solutes. The S_2 solute has a spectrum with wavelengths that are entirely transmitted by glass (---glass transmission curve).

3.2 Electronic devices used in liquid scintillation counters

3.2.1 Photocathode and photomultiplier tube

Scintillations emitted from the vial containing the radioactive source and the scintillant liquid are counted by means of a photocathode. The various stages in the process for a β^- of 50 keV are as follows:

$$\begin{array}{c} 1\ \beta^-\ \text{particle of } 50\,\text{keV} \\ \downarrow \\ 1\ \text{scintillation of } 200\ \text{photons} \\ \downarrow \\ 20\ \text{photoelectrons (photocathode)} \\ \downarrow \\ 1\ \text{electric pulse, amplitude } (A). \end{array}$$

The electric current from the photocathode is very weak, therefore it is necessary to multiply the electrons in the photomultiplier tube and to pre-amplify the signal before analysis. The photocathode and the photomultiplier tube because of thermic 'bubbling' spontaneously release electrons that give rise to electric pulses of about 10^5 c.p.m. This phenomenon is called *dark current*. The dark current pulses and those coming from β^- particles have closely related amplitudes, it is impossible to discriminate them on the basis of an amplitude difference.

The method used to get rid of the dark current is based on the following fact: when a β^- particle is emitted in the scintillator a scintillation occurs that is made up of a number of fluorescent photons going off in all directions (the

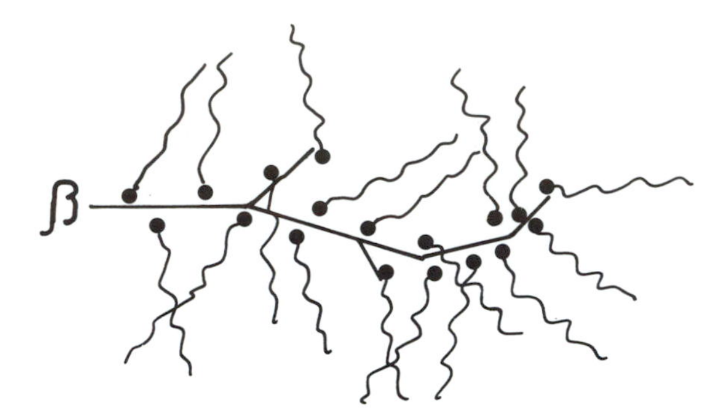

Figure 10. Trajectory of a β^- particle of energy 5 keV. The distance run is about 1.2 μm. Along the trajectory about 20 photons are emitted forming a luminous point or scintillation.

four pi steradians of space). Photons may thus be detected at the same time by *two* photomultiplier tubes diametrally opposed to the vial (see *Figure 11*). The two photomultiplier tubes will then *simultaneously* yield an electric pulse. Thus, there are two pulses in coincidence. For the dark current, pulses are not formed simultaneously. A coincidence circuit can therefore cancel the pulses coming to it with a time difference of around 15 nsec or above; however, it will retain the pulses emitted from a radioactive disintegration. The probability of dark current pulses being in coincidence is of the order of one or two pulses per minute.

3.2.2 Summation circuit

Following the detection of a scintillation, each photomultiplier tube provides a pulse with an amplitude proportional to the number of photons collected on each photocathode. To obtain a signal proportional to the energy of the detected β^- particle, the two pulses are summated, providing a pulse with amplitude equal to the sum of the two original amplitudes (see *Figure 11*). The electronic device that carries out this process is a *summation circuit*.

3.2.3 Amplification

After summation, electric pulses that are in coincidence undergo an amplification. Depending on the make of instrument, it may be a linear (Packard, Kontron, LKB) or logarithmic (Intertechnic) amplification. The logarithmic approach may also be obtained by means of logarithmic amplification photomultiplier tubes as used by Beckman Instruments. The shape of the spectra showing the distribution of electric pulses obtained with these two types of amplification are symmetrical (see *Figure 12*). In LKB instruments, after a linear amplification the display spectrum is logarithmic.

The shape of spectrum for a given radionuclide is dependent on the type of counter and will be the result of various processes: detection (the scintillator

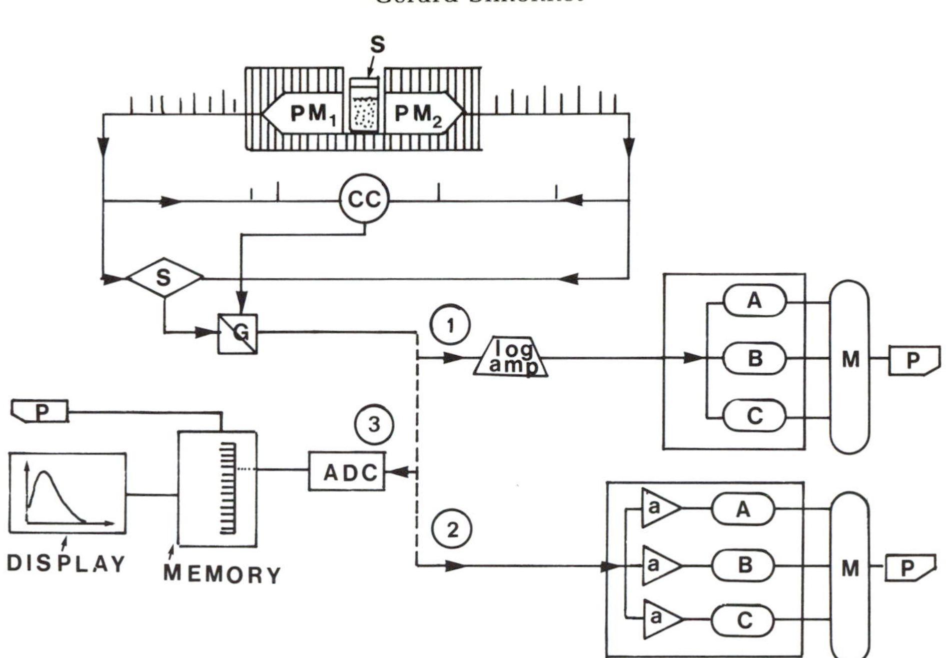

Figure 11. Summation circuit of a liquid scintillation counter. The vial containing the scintillation fluid and the radioactive source (S) is placed between two photomultipliers PM_1 and PM_2. The coincidence circuit (CC) allows the elimination of dark current. The pulses of each PM are summated (S). Then pulses are sent through a gate, G, in the pulse height analyser which can be (1) a logarithmic or (2) a linear amplifier. The pulse height analyser can also be of the multichannel type (3). A, B, C refer to counting channels, (a) is a linear amplifier, (M) a microcomputer, (P) a printer, and (ADC) an analogue-digital-converter.

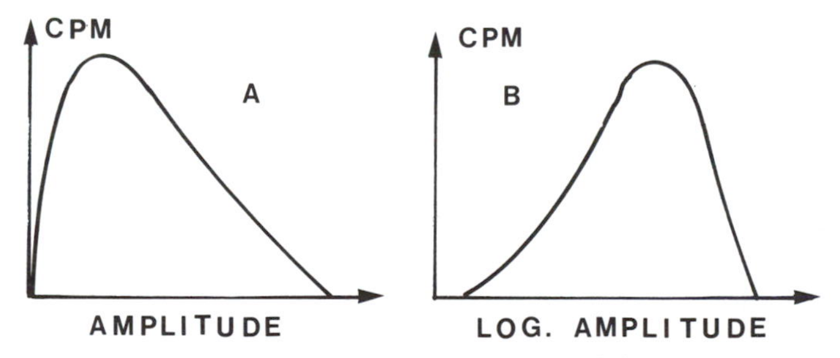

Figure 12. Electric pulse height spectrum obtained from: (A) linear, and (B) logarithmic amplifier.

and optics of the chamber) and the treatment of the electronic signal (coincidence, summation). Because of this deformation between the β^- spectrum and the counting pulses, liquid scintillation counters cannot be qualified as β^- spectrometers. However, radioactivity measurements are often based on the spectrum analysis of the electric pulses.

3.2.4 Pulse height analyser

A pulse height analyser sorts electric pulses according to their amplitude (see Section 2.4.3). What is required of liquid scintillation counters is the facility to:

- select the range of amplitude corresponding to pulses originating from a radioactive disintegration and eliminate background pulses;
- correct for quenching (see Section 3.4);
- measure the activity of samples containing two different radionuclides.

The selected pulses are sent to a scaler for output by the counter.

3.3 Calibration of counters

The calibration that an operator will have to perform on a liquid scintillation counter is solely concerned with the pulse height analyser. For each radionuclide, a specific channel has to be established (see *Figure 13*).

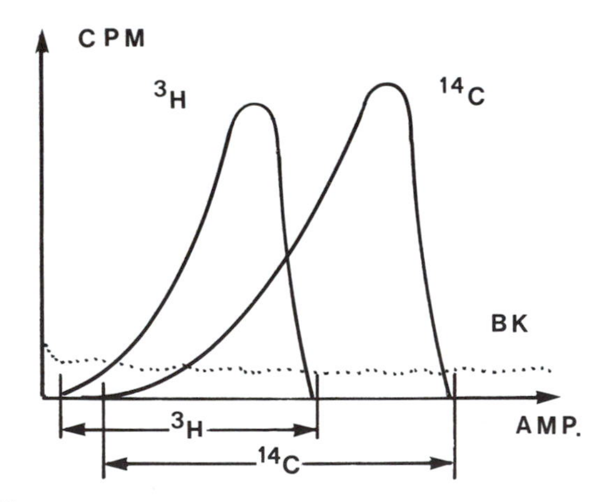

Figure 13. ^{3}H, ^{14}C, and background (BK) spectra. (AMP) = log of amplitude.

3.3.1 Calibrations

All counters have preset calibrations for at least ^{3}H, ^{14}C, and ^{32}P and no other action is required. For other nuclides, manual calibration as described in *Protocol 4* will be necessary.

Protocol 4. Calibration of liquid scintillation counters

Counters with logarithmic amplifier and no multichannel analyser

1. Put a sample of the radionuclide (possibly a standard), into the counter.

2. Set the lower level (threshold) on zero and the upper level (window) on maximum.
3. Start counting the radionuclide.
4. Progressively raise the lower level until counting stops. Note the level value; it will be the upper level value of the channel.
5. Put the lower level back to zero.
6. Progressively reduce the upper level until counting stops. Note the value of this level; it will be the lower level value of the channel.
7. Set the levels to the given values.

Counters with a multichannel analyser

1. Start counting an appropriate sample (possibly a standard).
2. Register the spectrum.
3. Visualize the spectrum on the video screen or printer.
4. Note the abscissa values of spectra end-points and adjust the counter from these values.

3.3.2 Performance data of a counter

The characteristics that must be determined are: the counting efficiency (E) of any measured radionuclides, the background value (BK) and the figure of merit (FM).

After measuring the standards and background activity, calculate the counting efficiency:

$$E = \frac{\text{standard c.p.m.} - BK \text{ c.p.m.}}{\text{standard d.p.m.}}.$$

In the case of ^3H standards, note the date of their preparation and if necessary, calculate the remaining activity (d.p.m.), knowing that the half-life is of 12.26 years (see *Table 3*).

Table 3. Radioactive decay of tritium ($t½ = 12.26$ years)

	Proportion remaining after period shown (months)					
Years	**0**	**2**	**4**	**6**	**8**	**10**
0		0.99	0.98	0.97	0.96	0.95
1	0.95	0.94	0.93	0.92	0.91	0.90
2	0.89	0.88	0.88	0.87	0.86	0.85
3	0.84	0.84	0.83	0.82	0.81	0.80
4	0.80	0.79	0.78	0.78	0.77	0.76
5	0.75	0.75	0.74	0.73	0.73	0.72

The figure of merit (5):

$$FM = \frac{E^2}{BK} \qquad BK = \text{background given in c.p.m.}$$

The values of efficiency may be slightly inferior to those given by the manufacturer, it's usually due to a difference in the nature of standard samples. The background itself may also be different as it depends on the environment of the counter. When monitoring a counter, what's most important isn't the absolute value of a characteristic such as the figure of merit, but good stability of the value with time.

3.4 Quenching

The term quenching indicates a phenomenon that often appears during liquid scintillation counting. Quenching entails an underevaluation of the activity of the measured samples. Moreover, samples of the same radionuclide measured under apparently the same conditions (i.e. same scintillant and channel) but with variable levels of quenching, will not be detected with the same counting efficiency (E). It is in the scintillation fluid that quenching occurs.

3.4.1 Mechanisms

Recall the sequence of reactions that take place in the scintillator (see Sector 3.1.3). This chain of reactions proceeds satisfactorily unless the medium contains no other chemical compounds other than those forming the scintillation fluid. In practice, however, various compounds are always introduced with the radioactive sample; for example, water, alcohol, different organic matters, dyes, etc. ...

All these compounds may interfere in several ways within two categories of quenching:

- chemical quenching: produced by compounds acting on the energy transfers between T and S_1 or S_1 and S_2, they capture energy and give it back, but as heat (internal conversion) not fluorescence;
- colour quenching: produced by dyes that absorb within the violet–blue spectrum, and constitute a kind of absorbent filter for the photons between their emission point and the internal wall of vials.

NB. Various compounds (powders, filters, fibres, etc.) may absorb a part of the kinetic energy of the β^- particle before its interaction on the solvent; this is not considered as a quenching it is a *self-absorption* phenomenon. In no case can the loss in counting efficiency by self-absorption be corrected by an automatic quenching correction method but an internal standard can be used in certain cases (see Section 3.4.3).

When we have to compare quenched samples, two cases may occur:

- the quenching rate is the same in all samples. The counting efficiency is unknown and is the same for all samples.

In this case, it is possible to compare the relative activities (c.p.m.) between samples.

- the quenching rate is variable from one sample to another. Therefore for each sample the efficiency is different.
  In this case, after measuring the c.p.m. we must, for each sample:
  (a) determine its counting efficiency (E);
  (b) then calculate its real activity (d.p.m.): d.p.m. = c.p.m./E.
  These operations constitute what is called quench correction.

3.4.2 The influence of quenching on spectra

Compared to an unquenched sample the spectrum of a quenched sample is shifted towards weak amplitudes and has its area diminished. This area is proportional to the measured activity (c.p.m.) (see *Figure 14*). These two modifications can be explained as follows. When a β^- particle gives up its energy in unquenched scintillation liquid, the number of photons arriving on the photocathode is, for example, N, giving an electric pulse of amplitude A.

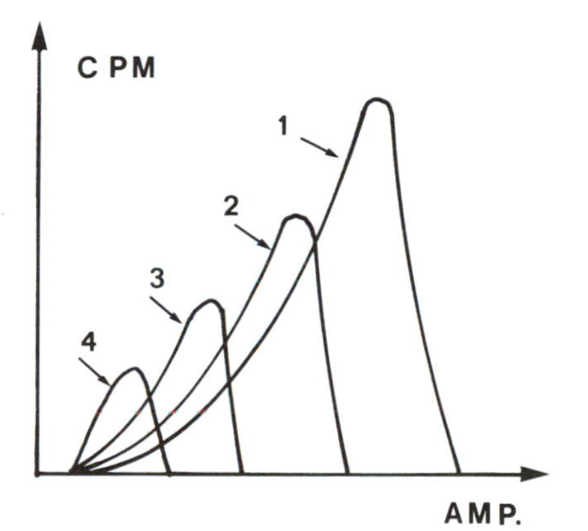

Figure 14. The shift in spectrum observed from a ^{14}C sample that is (1) unquenched, or increasingly quenched (2), (3), and (4).

The same β^- particle in a quenched medium will produce less photons (N' where $N' < N$), and an electric pulse of lower amplitude A'. Also it may happen that the number of emitted photons has become insufficient to lead to a pulse event, in which case the counting rate is diminished.

3.4.3 Quenching corrections

The evaluation of the shift in spectrum is the basis for most quenching correction procedures. Three methods will be described:

(a) internal standard;

(b) channels ratio and sample spectrum analysis;

(c) external standardization.

Quench correction by internal standard

The principle is as follows: after measuring the activity (A c.p.m.) of a sample, a known quantity (N d.p.m.) is added of the same radionuclide in the same physico-chemical form. The added N d.p.m. constitutes an internal standard. The sample is measured once more, its activity is then A' c.p.m. Where A' c.p.m. > A c.p.m. The difference (A' c.p.m. − A c.p.m.) is due to the N d.p.m. added to the sample. The counting efficiency (E) of the internal standard is therefore:

$$E = \frac{A' \text{ c.p.m.} - A \text{ c.p.m.}}{N \text{ d.p.m.}}.$$

It is assumed that the counting efficiency of the sample is the same as for the internal standard, therefore the real activity of the sample is given by:

$$\text{d.p.m. (sample)} = \frac{A \text{ c.p.m.}}{E}.$$

Here is a numerical example:

Sample activity (less background) = 3525 c.p.m.
Internal standard added = 10000 d.p.m.
Activity of sample including internal standard = 6847 c.p.m.
Activity measurement in c.p.m. of internal standard:

$$6847 \text{ c.p.m.} - 3535 \text{ c.p.m.} = 3322 \text{ c.p.m.}$$

Therefore, detection efficiency of internal standard:

$$\frac{3322 \text{ c.p.m.}}{10000 \text{ d.p.m.}} = 0.33.$$

Therefore, real activity (d.p.m.) of the sample:

$$\frac{3525 \text{ c.p.m.}}{0.33} = 10681 \text{ d.p.m.}$$

The internal standard method is the most reliable method. However, it is time-consuming as it involves: (a) two measurements; (b) internal standard addition with the precision problem of pipetting; and (c) calculations; and the whole process cannot be automated. This method is only advisable therefore when there is a small number of samples to be corrected; with one exception: the counting of filter papers. As discussed earlier, automatic quench correction is not feasible in this case. The only way to estimate the counting efficiency of such samples is to remove the filters from the scintillant after the first count, dry the filter, add a standard, dry the filter and recount.

Quench correction by channels ratio method and analysis of sample spectra

As shown in *Figure 14*, counting spectra are shifted towards weak amplitudes through the quenching effect. The more serious the quenching the further the shift and the lower the counting efficiency. In this manner, the spectrum position along the abscissa gives information about the level of quenching and the counting efficiency. If we were to set a relationship of the type:

efficiency = *f* (spectrum position)

we would then be able to determine the counting efficiency of the sample and then its real activity (d.p.m.).

Before trying to establish this relationship it is necessary to quantify the spectrum position; there are two methods available:

Channels ratio

This method consits of measuring the number of pulses in two channels: one covering the whole area of the unquenched spectrum of the radionuclide, and the other covering only the lower part of the spectrum (see *Figure 15*). Alternatively, two overlapping channels can be set up covering the lower and upper regions of the spectrum.

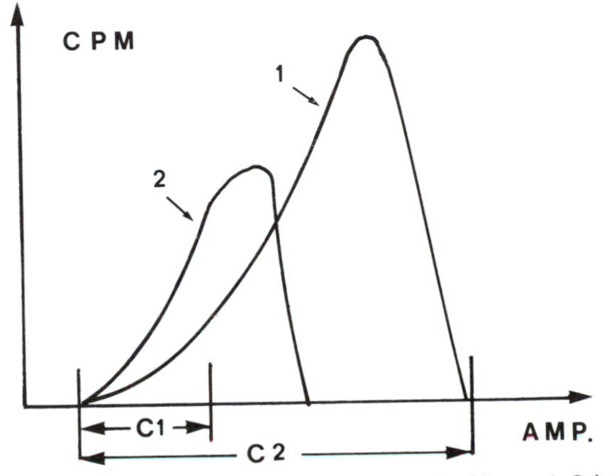

Figure 15. Relative positions of two counting channels (C_1 and C_2). The ratio value (C_2/C_1,*R*) of the measured activities in the channels, lowers with increasing quench (1) unquenched; (2) quenched.

The sample c.p.m. in both channels is determined and the ratio (*R*) of these two values calculated. Counting is first done on samples with different levels of quenching such that the channel ratios (*R*) will vary. Do this with 8–10 samples ('quenching standards') containing a known constant amount of radioactivity (d.p.m.) but with variable quenching (see Section 3.4.6). After counting each vial calculate the channels ratio (*R*) and the counting efficiency (*E*) in the wider channel. Plot *R* against *E* on a graph. This is the quench correction curve, that links efficiency (*E*) to the *R* parameter (*Figure 16*).

The counting efficiency (*E*) of an unknown sample is read from the quench correction curve using its *R* value. Calculate the true radioactivity (d.p.m.) by dividing the relative activity (c.p.m.) measured in the wider channel by the efficiency (*E*).

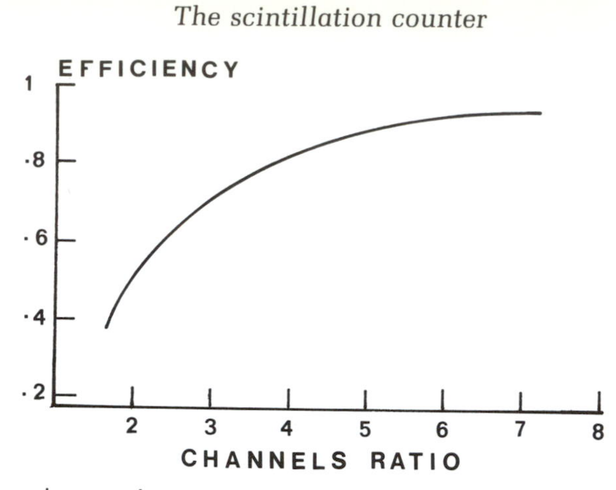

Figure 16. Quench correction curve. A graph that links the counting efficiency (*E*) to the quenching parameter (channels ratio or spectrum analysis).

This correction method is easy to manage, it can be automated as modern counters will store the quench correction curve in the microprocessor, but the method lacks precision in the case of samples with low activity.

Spectrum analysis

As far as counters including a multichannel analyser and specific software are concerned, the analysis of sample spectra provides a numeric parameter that is a function of the shape and position of the spectrum and varies with quenching. As for the channels ratio method, the parameter is linked to the counting efficiency. However, the precision obtained is better than for the channels ratio, because it is the total spectrum that is used for determination.

Counters exploiting such a device include: LKB, with the 'Spectral Quench Parameter of Isotope' SQP(I), and Packard, with the 'Spectral Index Sample' SIS, that indicate the channel number in which the centre of the spectrum is counted. To use this method follow the manufacturers' instructions.

Quench correction by external standardization

To cope with the deficiency in precision of the channels ratio or spectrum analysis correction method, the external standardization method has been developed. This method is also based on measurements made in two channels or by spectrum analysis but the method is based on the activity of a γ source set outside the vial. Hence the term external standardization. The source is shifted automatically from a lead container to the counting chamber where it irradiates the scintillation fluid producing a high count rate of more than 2×10^5 c.p.m., thereby providing good statistical precision.

The nature of the external source depends on the counter; it can be ^{137}Cs, ^{133}Ba, or ^{226}Ra, with activity ranging between 185 and 740 kBq (5 and 20 μCi) Because the energy of the γ radiation and the chemical composition of the scintillation fluid (mainly C and H) the gamma interaction is done through the

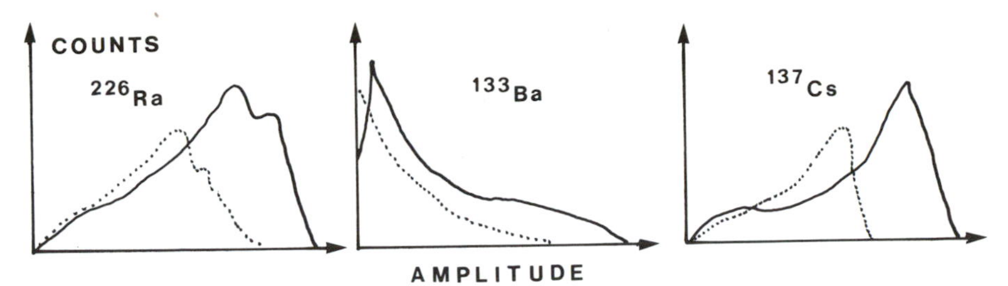

Figure 17. Spectra of external sources of: ^{133}Ba, ^{137}Cs, and ^{226}Ra in unquenched (—) quenched (---). Scintillation figure.

Compton effect; the energy spectrum of the Compton electrons is similar to the β^- spectrum of ^{32}P (see *Figure 17*).

The external standard spectra will, under a quenching effect, undergo the same modifications as those of ^{3}H or of ^{14}C contained in the vial; there will be as shown in *Figure 17* a shifting of the spectrum towards weak amplitudes and a reduction in area (reduced count rate).

The shift in spectrum is evaluated either by analysing the ratio of the measured activity in two channels, or better, by spectrum analysis (only available on counters that include a multichannel analyser and the necessary software).

External standardization with channels ratio

In the channels ratio method, the calibration is set by the manufacturer and the ratio value is directly provided by the counter (see *Figure 18*a). In fact, the ratio value, named external standardization, results from a calculation of the ratio of two measurements: first, in the presence of the external source, during which the external source activity plus that due to the sample contained in the vial are detected; and second, in the absence of the external source where the measured activity is only due to the sample. This is done in two different channels.

To subtract the sample contribution in the channels from external standardization, calculate the difference between the 'first' and the 'second' measure. The quotient between these two differences gives the ratio value of external standardization and is usually calculated automatically by the instrument.

Spectrum analysis

The methods for external standard spectrum analysis differ from one manufacturer to another; below is a description of those used by Beckman, Packard, LKB, and Kontron.

Beckman

The spectrum of the external source (^{137}Cs) has an 'inflexion point' towards its right end. The abscissa value, on a logarithmic scale, of the inflexion point of the unquenched external standard spectrum is assigned the value zero. For quenched samples, the instrument automatically determines an abscissa value separate from the point of inflexion with zero. This value number is called the Horrocks, after the physicist who perfected this method (see *Figure 18*b).

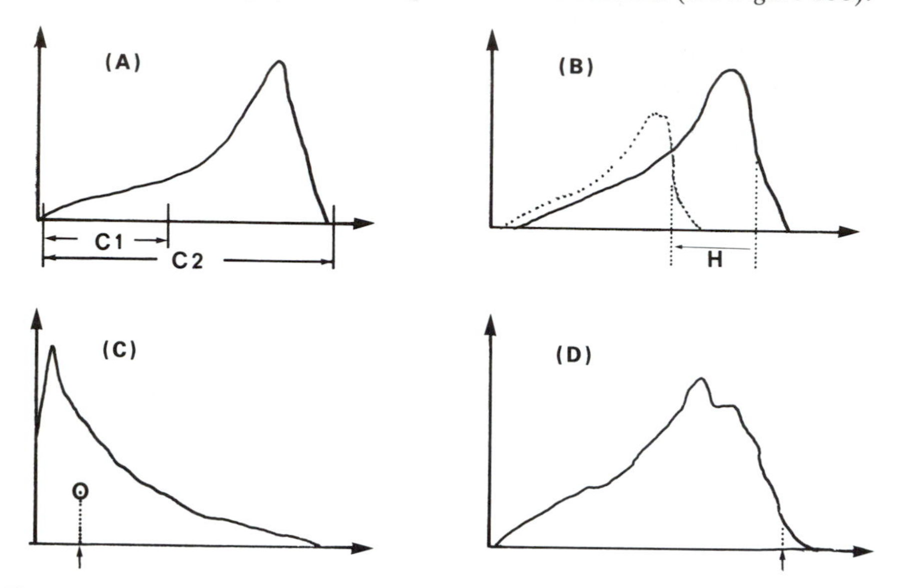

Figure 18. Methods for evaluating a spectral shift of the external source. (A) channel ratio; (B) number H, Beckman counters; (C) Centre of gravity, Packard counters; (D) spectra end-point, LKB and Kontron counters.

Packard

After registering the spectrum of the external source (^{133}Ba), the curve is mathematically transformed and then a numerical value is determined that represents the mean point of the area under the new curve tSIE (Transformed Spectral Index of the External Standard).

LKB and Kontron

These devices use a method that consists of determining an abscissa value [the Standard Quench Parameter, SQP(E)] that divides the external standard into two parts consisting of 95% and 5% (Kontron) or 99% and 1% (LKB).

In this approach, the lower region of the spectrum is not used or eliminated. This means that interference events from the sides of plastic vials, volume effects, colour quenching, or effects of vial wall thickness are reduced. Nevertheless, some interference may still exist and it is up to the operator to check for this when preparing the samples by comparison of different quench correction curves.

To carry out quench correction the Standard Quench Parameter, which is a function of the quenching, has to be linked to the counting efficiency of a radionuclide (e.g. ^{3}H or ^{14}C) in the vial. To do this, set up a series of 8 to 10 'quenched standard samples' (see Section 3.4.6) count the samples in an appropriate channel and determine the counting efficiency (E). The instrument will automatically print out the SQP(E) value; plot these figures against the counting efficiency to obtain a standard curve. This information can be stored in the counters computer (see following section).

The external standardization value of the unknown samples, can now be used to determine the counting efficiency (E). Using the above graph the real activity (d.p.m.) is then calculated by means of the relation

$$\text{d.p.m.} = \frac{\text{c.p.m.}}{E}.$$

3.4.4 Automation of quench correction

All the following stages in quench correction can be automated: counting standard samples, correlation between efficiencies and quenching parameters, recording results, counting unknown samples and calculation of the d.p.m. The correlation: efficiency–quenching parameter [for example, channels ratio or SQP(E)], is established by means of a mathematical equation of the polynomial 3rd degree type:

$$aR^3 + bR^2 + cR + d = E$$

where a, b, c, and d are numerical coefficients typical of each curve, R = quenching parameters (channels ratio, external standards, ...), and E = counting efficiency. However, instead of a polynomial, LKB and Kontron counters use the function SPLINE which may be either of the interpolated type; that is, the curve goes through all the experimental points, or smoothed (smoothing spline).

Whatever method is used, it is necessary to obtain a precise graphical representation including the experimental points and the corresponding curve. The system can only be useful if there is a good correlation between these two.

For unknown samples, efficiency is calculated from the quenching parameter and the mathematical function as previously described. This calculation must be done only if the quenching parameter is between the two furthest limits of the curve, i.e. between the first and last quenched standard. This point has to be scrupulously checked as some software does not include the necessary test.

Moreover, when unknown samples are counted, routinely check the validity of the quench correction by determining the value in d.p.m. of the 'quenched standard samples' used to form the correction curve.

3.4.5 Validity limits of a quench correction curve

A quench correction curve is only valid for a given set of circumstances that is:

(a) the particular *counter* used;

(b) the channel(s) used for the measurement of the quenched standard samples;
(c) the radionuclide contained in the standards.

If other factors vary, it may or may not entail a modification of the quench correction curve. This is a function of the make of counter. If any modification is required it is necessary that the factor(s) do not vary between the quenched standard samples and the unknown samples.

The factors referred to are:

(d) the composition of the scintillation cocktail;
(e) the nature of the quenching agent: chemical or colour;
(f) volume of scintillant contained in the vials.

In the case of quench correction by external standardization with channels ratio (generally used by older instruments), do not use plastic vials. The reason being that in the course of time the plastic becomes saturated with the scintillation product and when the external source irradiates it, some scintillations are produced in the plastic. This alters the spectrum located towards weak amplitudes, therefore a variation of the external standardization occurs with time; this, of course, has nothing to do with a quenching variation (6).

All the counters that use a spectrum analysis of the external source are less affected by the factors (d)–(f) above, or plastic vials.

3.4.6 Quenched standards

Quenched standard samples can be purchased in sets of 6 to 10 sealed vials of 20 ml or 7 ml (mini vials) containing variable quantities of quenching agent, usually carbon tetrachloride and an equal quantity of radioactivity (d.p.m.). The scintillator is usually PPO and POPOP in toluene. All of these standards are sealed under nitrogen and special care is taken to exclude moisture and organic impurities. Also available are sets of three vials, one contains ^{14}C, one ^{3}H and one 'background'. These samples are not quenched, their use is to periodically check the counters stability. All these standards are commercially available from firms selling radioactive products or counters.

For the reason given in Section 3.4.5 it is sometimes necessary to prepare a set of standards. The method is described in *Protocol 5*.

Protocol 5. Preparation of quenched standards

1. Obtain a radioactive solution calibrated by one of the following means:
   (a) buy an internal standard (d.p.m./ml) of an organic or aqueous solution;
   (b) buy internal standard capsules with for example a ^{3}H or ^{14}C compound in a solid form (e.g. from Amersham, LKB, or Lumac);
   (c) calibrate a radioactive solution by means of a quench correction curve. For this, measure the activity in a scintillation cocktail identical to the one contained in a purchased set of standard samples.

2. Put 50 μl of the radioactive solution (between 4 and 6×10^4 d.p.m. for each vial) and 5 ml of the selected scintillation cocktail into 20 vials.
3. Measure the activity with a precision better than 1% error.
4. From the 20 vials, select 10 that have the closest level of radioactivity.
5. Add quenching agent (e.g. carbon tetrachloride or chloroform) to the 10 vials in a range of concentrations appropriate to the unknown samples. Add scintillation fluid to give the same volume in all the vials.

Preserve standards in the cold (+2 to +8 °C) and from light. Commercially available standards may be preserved for about 5 years, those prepared in a laboratory for only a few months. The standards deteriorate with time because of scintillator radiolysis. The longer self-life for commercialized standards is due to the fact that they are prepared in a partially protected medium.

3.5 Dual label counting

Liquid scintillation counters allow, after proper calibration, for the simultaneous detection of two radionuclides. However, to discriminate between two radionuclides, it is necessary that the highest energies of β^- radiations emitted should be sufficiently different; that is, with a ratio equal to or higher than 3, thereby providing two counting spectra that do not entirely overlap each other (see *Table 4*). It is not possible, for example, to count ^{14}C and ^{35}S as a dual label sample.

Table 4. Examples of double-labelling

Radionuclides	E_{max} (keV) respectively
$^{3}H + ^{14}C$	18, 155
$^{3}H + ^{35}S$	18, 167
$^{3}H + ^{45}Ca$	18, 254
$^{3}H + ^{32}P$	18, 1.7×10^3
$^{14}C + ^{32}P$	155, 1.7×10^3
$^{35}S + ^{125}I$	167, 30 (x and γ)

3.5.1 Calibration of counting channels

Counters are precalibrated for the most commonly used double labels, that is ^{3}H-^{14}C, ^{3}H-^{32}P, ^{14}C-^{32}P. If the precalibration is not suitable, for example if other nuclides are involved, manual calibration is required. The principle behind the calibration of counting channels is the same whatever the radionuclides or whether there is a linear or logarithmic amplified. Described below is an example of the calibration required to measure the radioactivity in a mixture of $^{14}C + ^{3}H$.

When a counter is calibrated for single-label measurements, comparison of

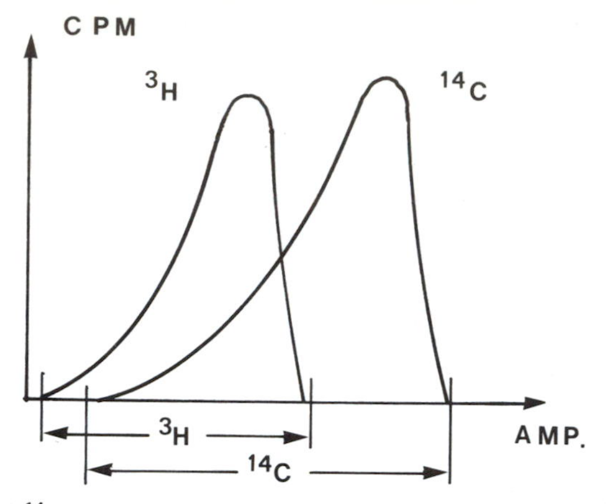

Figure 19. ^{3}H and ^{14}C spectra and corresponding channels in the case of single-label calibrations.

^{14}C and ^{3}H spectra (*Figure 19*) shows that some ^{3}H can be detected in the ^{14}C channel and vice versa. Dual label measurements are possible under these conditions but calculation involving two equations with two unknowns are needed as follows:

N_1 = d.p.m. ^{3}H × $E(^3\text{H})$ in ^{3}H channel + d.p.m. ^{14}C × $E(^{14}\text{C})$ in ^{3}H channel
N_2 = d.p.m. ^{3}H × $E(^3\text{H})$ in ^{14}C channel + d.p.m. ^{14}C × $E(^{14}\text{C})$ in ^{14}C channel

where:

N_1 = c.p.m. measured in the ^{3}H channel
N_2 = c.p.m. measured in the ^{14}C channel
$E(^3\text{H})$ = counting efficiency for ^{3}H
$E(^{14}\text{C})$ = counting efficiency for ^{14}C.

To provide data for these equations set up a ^{14}C channel that eliminates ^{3}H; the converse is not possible but set up a ^{3}H channel when the contribution of ^{14}C is reduced as far as possible (*Figure 20*).

3.5.2 Quench correction of dual label samples

If the samples to be measured show variable quenching, it will be necessary to make the corresponding corrections. When the samples are quenched, the counting spectra shift with regard to the counting window, this entails estimating counting efficiency variations by quench correction. The only quench correction method appropriate to dual labels involves the external standard: by channels ratio or spectrum analysis (Section 3.4.3). Use a set of ^{3}H and ^{14}C quenched standards to determine counting efficiency as a function of quenching parameter using the external standard (Section 3.4.3). Either plot three or four curves in the case of single or double reinjection respectively, that is:

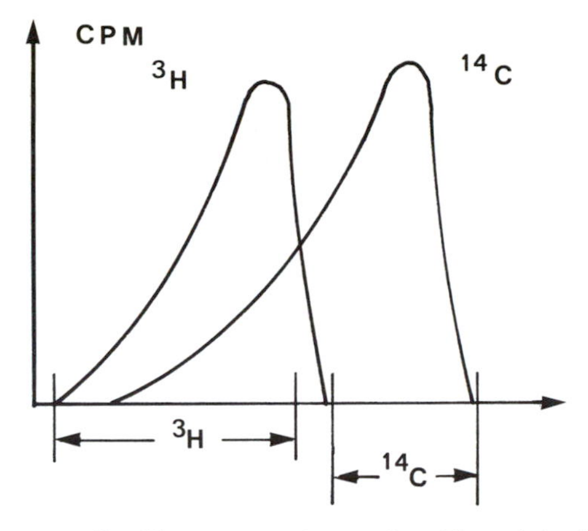

Figure 20. Calibration for ^3H–^{14}C dual label counting. The reinjection of ^3H in the ^{14}C channel has been cut out and the ^{14}C reinjection in channel ^3H has been reduced.

(a) *Single reinjection*
E(^{14}C) in ^{14}C channel, E(^{14}C) in ^3H channel and E(^3H) in ^3H) channel;

(b) *Double reinjection*
E(^{14}C) in ^{14}C channel, E(^{14}C) in ^3H channel, E(^3H) in ^3H channel, and E(^3H) in ^{14}C channel.

It is the aim of channel calibration to reduce or eliminate reinjection. This is achieved by narrowing the channel width and entails lowering the count rate. When the spectrum moves by quenching action the efficiency variation will be important. To reduce this variation, some counters include an automatic system for shifting the counting window, conditioned by the quenching rate (*Figure 21*).

The systems used by the various manufacturers (*Figure 22*) are as follows:

Beckman: AQC = Automatic Quench Compensation
Kontron: SSC = Spectral Shift Compensation
LKB: AWS = Automatic Window Setting
Packard: AEC = Automatic Efficiency Control.

An example of a d.p.m. calculation for a dual label count is shown below. The example taken involves a single reinjection calibration; that is, there is no ^3H detected in the ^{14}C channel therefore three quench correction curves are used.

(a) *Results of the sample measurements*

^3H channel	125000 c.p.m.
^{14}C channel	22000 c.p.m.
External standard	5.5

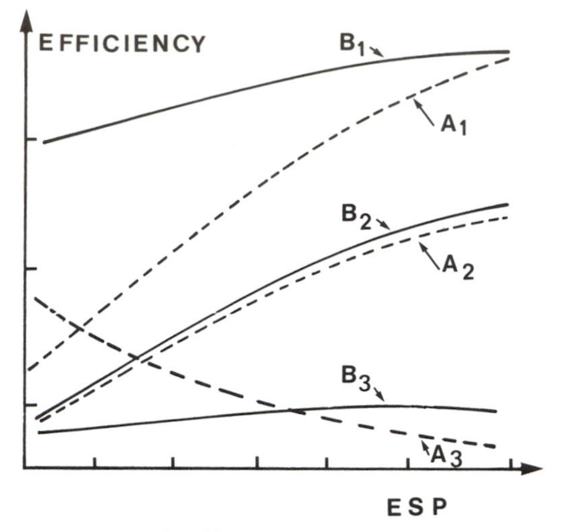

Figure 21. Correction curves for ^{3}H–^{14}C quenching obtained with fixed (A) or shifting (B) counting channels: (1) *E* ^{14}C-channel ^{14}C; (2) E ^{3}H-channel ^{3}H; (3) *E* ^{14}C-channel ^{3}H; ESP = External Standard Parameter.

(b) *Graphical determination of counting efficiencies (see Section 3.4.3)*

Efficiency of ^{14}C in ^{14}C channel = 0.52
Efficiency of ^{14}C in ^{3}H channel = 0.30
Efficiency of ^{3}H in ^{3}H channel = 0.42

(c) *Activities calculation (d.p.m.)*

d.p.m. ^{14}C = activity of ^{14}C channel in ^{14}C channel/efficiency ^{14}C
22000 c.p.m./0.52 = 42300 d.p.m. ^{14}C

Contribution of ^{14}C in ^{3}H channel:
c.p.m. ^{14}C in ^{3}H channel = d.p.m. $^{14}C \times E(^{14}C)$ in ^{3}H channel =
42300 d.p.m. × 0.30 = 12690 c.p.m. ^{14}C

Therefore c.p.m. ^{3}H in ^{3}H channel:
125000 c.p.m. − 12690 c.p.m. = 112310 c.p.m. ^{3}H
d.p.m. ^{3}H = activity of ^{3}H in ^{3}H channel/efficiency ^{3}H
112310 c.p.m./0.42 = 265404 d.p.m. ^{3}H

All the processes described above may be automated according to the same principle as that used for single-label correction. The software calculates the counting efficiencies by means of polynomes corresponding to each curve and then performs the previously described calculations.

3.6 Scintillators

As described in Section 3.1.3 a scintillation cocktail is made up of a solvent and two solutes.

3.6.1 Solvent

Solvents are aromatic compounds such as toluene, xylene, cumene, or pseudo-cumene. The scintillant properties of these compounds are very similar; however, the temperatures of their flash point and boiling point increase from toluene to pseudo-cumene. The latter compound has the least chemical and fire risks; its use is therefore advisable. 'Analytical' grade of solvent is adequate.

3.6.2 Primary solutes

PPO (2,5-diphenyloxazole) is the most used compound. It has a good solubility in the solvents and is used at 5 to 6 g/litre. It is not affected by the amines used to hydrolyse some samples. Butyl-PBD [2-(4-ter-butylphenylyl)-5-(4-biphenyl)-1,3,4-oxadiazole] is the best primary scintillator, but is quite expensive and is altered by acid and basic media. BBOT [2,5-bis-2-(ter-butylbenzoxazolyl)-thiophene] shows approximately the same characteristics as butyl-PBD.

3.6.3 Secondary solutes

Dimethyl-POPOP [1,4-bis-2-(4-methyl-5-phenyloxazolyl)-benzene] is a compound with good solubility in the solvents; it is used at a concentration of 0.1 to 0.2 g/litre. However, it may be affected by highly acid samples, in that case, replace it with POPOP although this is less soluble.

Bis-MSB [*p*-bis(O-methylstyryl)-benzene] has a quite good solubility. It may be used at high concentrations (0.5 to 1.5 g/litre) and gives the cocktail a better resistance to quenching. Other primary and secondary solutes exist but the above-mentioned are the most used. Fewer and fewer laboratories are preparing their own cocktails, replacing them with ready to use cocktails that are very often more efficient (see *Table 5*).

3.6.4 'Ecological' scintillation cocktails

Cocktails termed 'ecological' [that is non-toxic, non-hazardous, high flash point (150 °C), and biodegradable] are now available. For example, Ecoscint (Nat. Diagnostics), Safefluor (Lumac), Ultima-gold (Packard), Ready-safe (Beckman), Ecolum (ICN Radiochemical), Aquasafe (Zinsser), (see *Table 5*).

3.7 Sample preparation

Sample preparation is critical to the success of liquid scintillation counting. The goal is to disperse the sample throughout the cocktail but very few samples will dissolve directly in toluene or other scintillation fluid solvents.

3.7.1 Samples soluble in organic solvents

These compounds are dissolved in an organic aromatic solvent, if possible. Otherwise they should be dissolved in an organic solvent known not to cause quenching; this can be tested by counting a control sample.

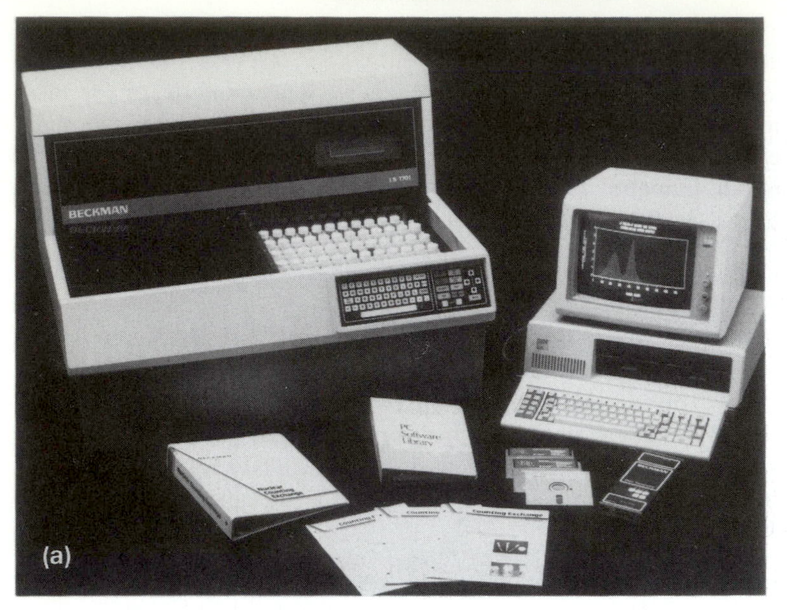

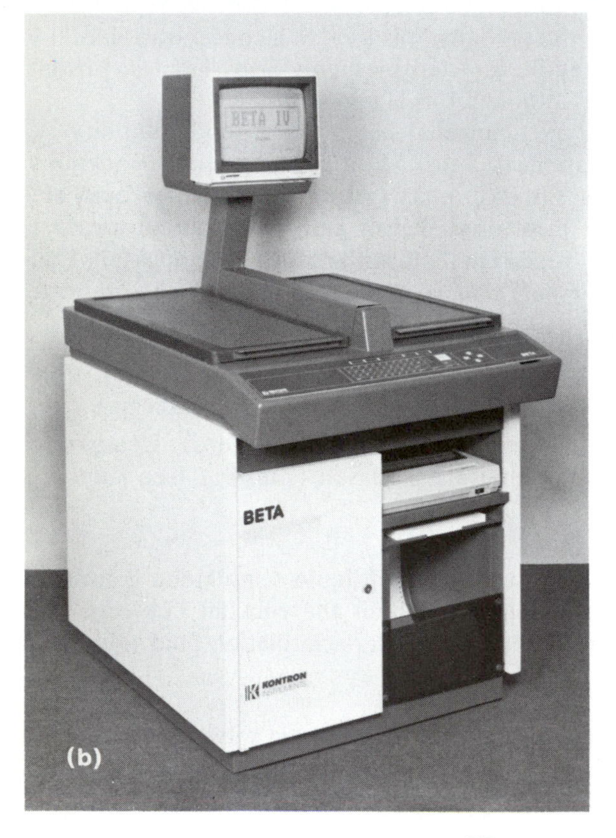

Figure 22. Liquid scintillation counters. *Liquid Scintillation counters.* (a) Beckman-LS 1701: (b) Kontron Beta IV with a linear multichannel analyzer (8192 channels); (c) LKB RackBeta 'Spectral'; (d) Packard Tri-Carb Model 2200CA Scintillation Analyzer computer assisted.

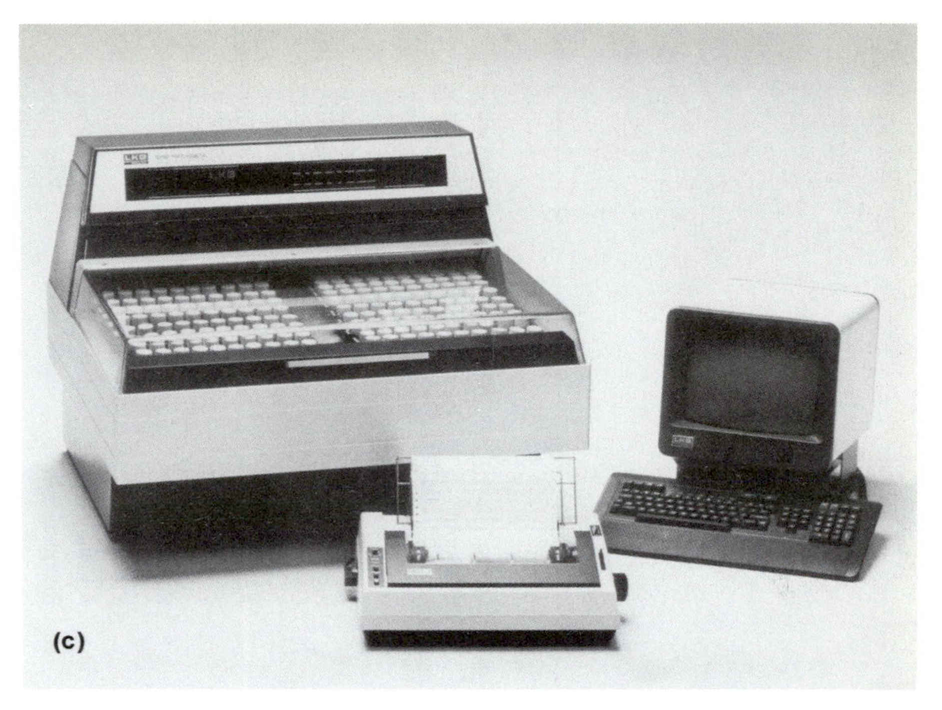
(c)

2200CA
TRI-CARB
PACKARD
(d)

Table 5. Ready-made liquid scintillation cocktails

	Amersham	**Beckman**	**Lumac**	**NEN**	**Packard**
Organic soluble samples	OCS	Ready-organic	Lipoluma	Econofluor Liquifluor Omnifluor Econofluor-2	Insta-fluor
Water <15% Water >15%	ACS II PCS II	Ready-protein Ready gel	Aqualuma + Rialuma Lumagel	Biofluor Aquasure Atomlight Formula 989	Pico-fluor 15 Instagel
0.15 M NaCl	PCS-ACS II	Ready safe	Aqualuma +	idem	Pico-fluor 40 Pico-aqua Hionic-fluor
Phosphate buffer	PCS II	Ready safe	Rialuma	Biofluor Aquasol 2 Atomlight	Pico-fluor 40
Tris–HCl buffer	PCS II	Ready safe	Aqualuma +	idem	Pico-Aqua
0.1 M HCl	PCS-ACS II	Ready gel	Rialuma	idem	Pico-Aqua Pico-fluor 40

0.1 M NaOH	PCS-ACS II	Ready gel	Aqualuma +	idem	Pico-fluor 40 Pico-aqua
0.15 M NaCl	PCS-ACS II	Ready safe	Aqualuma +	idem	Pico-fluor 40
25% sucrose	PCS-ACS II	Ready gel Ready protein	Rialuma	idem	Pico-fluor 40 Hionic-fluor
$CsCl_2$	PCS-ACS II	Ready safe	Rialuma	Biofluor Atomlight Aquasol	Pico-fluor 40 Hionic-fluor
Serum	PCS II		Rialuma Lumagel	Biofluor Atomlight Aquasol	Instagel
Urine	PCS II		Rialuma Lumagel	Biofluor Aquasure Atomlight	Instagel
Radioimmunoassay	ACS ACS		Aqualuma + Rialuma	Biofluor Atomlight Riafluor	Pico-fluor
Ecologically safe biodegradable [a]		Ready safe	Safefluor	Formula 989 Aquasure	Opti-fluor Emulsifier-safe

[a] Confirmation of suitability by local sewerage authority recommended.

3.7.2 Water, aqueous organic and inorganic samples

There are three ways to introduce these samples into scintillant fluid:

(1) *Ternary cocktails*. For 0.2 ml of sample add 2.5 ml of absolute ethanol, then 10 to 12 ml of xylene-PPO-DMPOPOP scintillant. This cocktail is the least expensive one, but the maximum content of aqueous solution is 3% (v/v).

(b) *Emulsifier-based cocktails*. Scintillation cocktails can include an emulsifier, usually Triton X-100. These cocktails will take an aqueous solution up to 50% (v/v), and are easy to prepare (simply add scintillator to Triton X-100 3:1), but ready-made cocktails for aqueous samples are more efficient. Suppliers include instructions with these cocktails that provide a phase diagram that allows the operator flexibility and freedom from toilsome calibrations.

For aqueous concentrations ranging from 10 to 20% (*E* v/v) some of these cocktails separate into two phases. It is impossible to accurately count emissions under such conditions (*Figure 23*). Beware of plastic vials as it is not easy to observe the state of the cocktail. For a given concentration, of cocktail, phase varies with temperature. This can be a problem if the counter used does not maintain the samples at a constant temperature.

Some of the 'ecological' scintillant cocktails accept aqueous solutions.

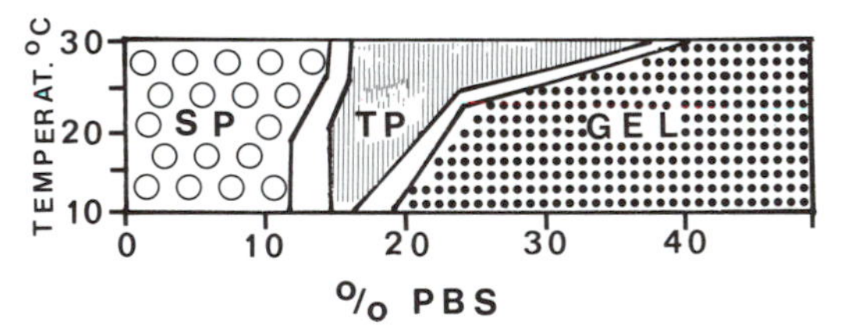

Figure 23. Phase diagram of READY-SOLV EP (Beckman). SP = single phase; TP = two phases; PBS = phosphate buffered saline.

3. Dioxane-naphthalene scintillation cocktails. In the past, these cocktails were used, but their toxicity has led to them being replaced by emulsifier-based cocktails as above.

3.7.3 Animal tissues

(a) *Tissue solubilizers*: animal tissues may be rendered soluble by means of quaternary amines such as Soluene-100 and Soluene-350 (Packard), NCS (Amersham), Protosol (NEN), BTS (Beckman), etc. Not surprisingly, all these products are highly toxic. Each product includes an instruction sheet; 1 ml is usually sufficient to solubilize about 50 mg of tissue. The hydrolysis is done at 40–50 °C within a few hours, in well-closed vials as oxidation colours the medium. The scintillation fluid is added afterwards. The hydrolysates are sometimes highly coloured; in this case, to reduce colour quenching bleach the samples before adding the scintillator as follows: add 0.5 ml of H_2O_2 at 30% (v/v); incubate for 15 min without closing the vial; add 10 ml of scintillation for aqueous solutions (that is, low in chemiluminescence; e.g. Hionic-fluor, Packard).

(b) *Combustion*: This is carried out in specific apparatus (supplied by Packard) in the presence of oxygen, where the organic matter is transformed into H_2O and CO_2. Up to 1 g of sample may be burnt in less than three minutes. The combustion products ^3H_2O or/and $^{14}CO_2$ are collected on the two separate exits of the device and immediately combined with the liquid scintillation cocktail. The following precautions must be taken:

- Use only scintillation fluids recommended by the manufacturer, otherwise there may be a risk of explosion.
- Determine the percentage of yield by burning standards samples (d.p.m.), supplied by Amersham.
- Determine the memory effect that is the carry over of radioactivity, by analysing a non-radioactive sample after the combustion of a radioactive one.
- Determine the reinjection rate, that is the activity from the '^3H' exit during the combustion of ^{14}C and vice versa.

The advantages of the combustion apparatus are that a convenient mass of samples can be analysed (500 mg), there is little quenching or chemiluminescence and it is the only method available for analysis of samples non-hydrolysable by quaternary amines. The disadvantage is the high price of the apparatus.

3.7.4 Precipitates of macromolecules and deposits on filters

Several categories of radioactive samples—for example, precipitates of macromolecules such as nucleic acids or proteins, other deposits on filters, and samples from thin layer or paper electrophoresis or chromatography—can be treated in one of two ways. Dissolve the samples off the support with a suitable solvent or tissue solubilizer and count as described above. Alternatively, introduce the samples on their filters directly into a cocktail for non-aqueous samples. The radioactive material, if it is insoluble in the scintillator (which is usually the case), will stay fixed on its filter. This is referred to as a heterogeneous counting system and two important points apply: the sample must be perfectly dehydrated before its introduction into the scintillant cocktail, and no automatic quench correction method can be used to determine the real activity (d.p.m.), therefore results may only be expressed in relative activity (c.p.m.). It is possible, however, to introduce an internal standard to the filter and recount. This will provide an approximate value for the counting efficiency but the method does not always fully account for self-absorption, particularly for ^{3}H counting.

3.7.5 Carbon dioxide

To fix $^{14}CO_2$ released by a biochemical reaction or expired by an animal, use a mineral base (2 M NaOH or 5 M KOH) or suitable organic compounds: phenylethylamine, hyamine-10X or ethanol-amine. If the quantity of $^{14}CO_2$ released is very small (that is, a few micromoles), it may be directly collected from the surroundings of the reaction by using, for example, a paper strip saturated with a material that will fix the CO_2. After the CO_2 has been collected, put the strip of paper into a scintillation vial containing a scintillant liquid such as Hionic-fluor (Packard), that provides minimum chemiluminescence.

For larger quantities (that is about a millimole) collect the CO_2 by bubbling through a saturated solution of baryte ($Ba(OH)_2$). Vacuum filter the precipitated $^{14}CO_3Ba$, wash the precipitate with water then absolute ethanol then acetone. After drying, put the $Ba^{14}CO_3$ deposits into the scintillation fluid. No counting loss through self-absorption is observed if the lower level of the counting channel is near zero and the counting efficiency is around 85%. There is no risk of chemiluminescence. Moreover, by measuring the mass of $Ba^{14}CO_3$ the specific radioactivity of the $^{14}CO_2$ can be calculated (7).

3.7.6 Agarose or polyacrylamine gels

After slicing the gel into appropriate pieces extract the radioactive product by one of the following methods:

- H_2O_2 at 30% (v/v) for 3 h at 60 °C;
- $HClO_4$ 60% (v/v) + H_2O_2 at 60 °C for 3 h;
- Protosol (NEN) at 3% for 18 h at 37 °C.

Add scintillation fluid such as Pico-fluor 40 or Hionic-fluor (Packard).

In spite of the quite corrosive properties of these reactions some of the radioactive product may not be extracted and therefore will not be detected. If necessary an extraction check can be done by combustion (see Section 3.7.3). Agarose samples may be melted in 1 ml of water and counted in an aqueous based scintillation fluid but there may be some quenching.

3.8 Choice of vials

Vials can be glass or plastic, 20 ml or 5 ml in volume. Minivials (5 ml) are useful if the radioactive sample is not heavily quenched and can be incorporated into only a few millilitres of cocktail. A saving will therefore be made on the vials price, the quantity of cocktail and the price of radioactive waste disposal.

Plastic vials are often less expensive than glass vials, especially compared with glass vials with a low background. However, because of certain counting conditions, plastic vials should be avoided if:

- the counter has no suppression of static electricity or is not refrigerated;
- quench correction is made by the calculation of channels ratio of the external standard, and if a scintillation liquid is used which may diffuse into plastic (Section 3.4.3);
- there is a risk of a two-phase separation of samples.

3.9 Causes of counting errors

3.9.1 Scintillation cocktail unsuitable for the sample

For example an aqueous sample counted in a cocktail designed for an organic sample. The consequence is an underestimation of activity.

3.9.2 Chemiluminescence

All basic samples or samples containing oxidizing agents mixed with dioxane-based scintillation fluids or emulsifier cocktails create chemical reactions that produce fluorescent photons (chemiluminescence reaction). Some of these photons are detected as if they originated from the radioactivity and the counting rate due to the chemiluminescence may go beyond 10^5 c.p.m. This rate decreases as an exponential function with respect to time ($t½$ is from 1 h to 1 day). Any sample that gives some chemiluminescence must be counted several times until a stable count-rate is obtained.

The reduction or suppression of chemiluminescence may be obtained through various means described below.

- Neutralize the basic samples with acid (HCl or CH_3COOH); this treatment may make some samples precipitate.
- Use scintillation fluids that resist chemiluminescence. For example Hionic-fluor (Packard).
- If the radionuclide is not ^3H raise the lower level of the counting channel to about 8 keV. The pulses due to chemiluminescence are of weak amplitude (0–7 keV).
- Some counters include electronic systems for the detection and correction of chemiluminescence (delayed coincidence or pulses analysis). Some systems flag the presence of chemiluminescence on the print-out and others notify it and subtract it from the reading. Even if this facility is available it is prudent to count the samples more than once.

3.9.3 Luminescence

Fluorescent photons can be emitted by the scintillator after its exposure to light (fluorescent tubes, sunlight). As with chemiluminescence, the decrease is exponential with a $t½$ of about 1 min. The pulses have the same amplitude as those of chemiluminescence. Counters that have a detection system for chemiluminescence will also detect luminescence. To avoid luminescence leave the samples in the dark for a few minutes prior to counting and repeat the reading.

3.9.4 Static electricity

Plastic vials are easily charged with static electricity. If their electrical discharge takes place in a counting chamber, counts are registered. Modern devices cancel the static electricity; this is based on air ionization by a source (α) of ^{210}Po (Beckman), electric fields (LKB, Packard), or charge collection (Kontron). As a precaution, count the samples several times.

3.9.5 Phase separation

Aqueous samples introduced into a scintillator with an emulsifier base may separate into two phases (see Section 3.7.2) and yield erroneous c.p.m. Some counters fitted with spectral analysers notify in the print-out that this has occurred. Samples must be recounted with the correct proportions of sample and scintillant.

3.10 Cerenkov counting

3.10.1 β^- interaction through the Cerenkov effect

The Cerenkov effect is an interaction between charged particles (β^-, α, etc.) and matter. This effect can be exploited for detection of β^- participles in a liquid scintillation counter without having to use scintillation fluid.

The Cerenkov effect is obtained when particles pass through a transparent substance with a speed higher than that of light in that medium. High energy β particles, induce, along their trajectory, a polarization of molecules that emit photons of 350 to 600 nm as they return to their ground state.

The minimum energy (E_{min}) in MeV of a β^- particle required to produce the Cerenkov effect is given by the relation:

$$E_{min} = 0.511\left[\frac{1}{\sqrt{1-\frac{1}{n^2}}} - 1\right]$$

where 0.511 = rest mass of a β^- particle in MeV, and n = refractive index of the medium.

For water, $n = 1.33$, and the E_{min} for Cerenkov counting of β^- is 270 keV. This energy is much higher than the β^- energy of ^3H, ^{14}C, or ^{35}S so those radionuclides cannot be detected by this means. However, ^{32}P (E_{max} = 1710 keV) has a large part of its spectrum above 270 keV and it can be detected through the Cerenkov effect.

3.10.2 Counting conditions

To measure the ^{32}P via the Cerenkov effect, the ^{3}H channel is sometimes used. However, it is better to calibrate the counter specifically (see Section 3.3).

The counting efficiency of ^{32}P in liquid scintillation fluid is nearly 100%; by Cerenkov counting in water, the efficiency varies according to whether the counting is done in glass ($E \simeq 0.5$) or in plastic vials ($E \simeq 0.6$). The difference is due to the dispersive effect of plastic on light: the photons are better distributed to both photomultipliers.

The Schifter addition (Amino-G; 7 amino-1,3-naphthalene disulphonic acid, 1 g/litre) is sometimes used but this gives little significant increase in efficiency with modern photomultipliers.

With Cerenkov counting, there is practically no chemical quenching, but there is a risk of colour quenching; in this case, bleach the sample or make a quench correction by internal standard or channels ratio.

Cerenkov counting can be obtained with samples dried on to the vials. The Cerenkov effect is then produced by the β^- interaction in the vial itself but the reproducibility of counting is not good, the coefficient of variation is 6–7%.

3.11 Detection of ^{32}P in scintillating plastic vials

It is possible to measure the activity of dry ^{32}P (for example, on filters) in plastic vials that have been soaked in scintillator. Impregnate the vials by leaving them for 48 h, filled to the top with a solution containing PPO (4 g) and DMPOPOP (0.25 g) in a litre of xylene. Then empty and dry the vials.

The counting efficiency is approximately 95% and the variation coefficient is 0.3–0.4%. The advantage of this technique is that vials may be used over and over again for quite long periods; the background reading must be checked regularly but significant savings can be made in vial usage and waste disposal (8).

3.12 Low-level counting

Low-level counting is mainly concerned with environmental monitoring, radiation protection, and any biological experiment that leads to samples that have activities close to the background of the counter (9). The lower the background, the lower will be the detectable activity. The 'critical level' is defined as the counting rate below which no activity can be quantified. This detection level (DL) is given by the relation

$$DL = 2\sqrt{2 \cdot n/t}$$

where n = background in c.p.m., and t = counting time in minutes.

The performance of a counting system, can be improved in several ways described below:

- Environment: place the counter in an area where the external irradiation level is minimal.
- Optimize the counting channels by shifting the upper and lower levels and determining the figure of merit (*FM*) after each shift:

$$FM = E^2/BK$$

  where E = counting efficiency for the radionuclide to be detected, and BK = background.
  The optimum conditions are obtained when the figure of merit value is maximum. On some counters this optimization is obtained automatically.
- Some counters include electronic of software systems that result in very low backgrounds;
- Use plastic vials.
- Avoid chemiluminescence, luminescence, and static electricity (see Section 3.9).

4. Continuous flow measurements

A radioactive sample may be a liquid or gas automatically generated with time, for example in:

- high performance liquid chromatography (HPLC);
- extracorporeal blood circulation;
- circulation of $^{14}CO_2$ expired by an animal;
- the recording of radioactive effluents.

In such cases, rather than fractionate the liquids or gas and count each fraction separately, it is better, if possible, to measure the radioactivity in a continuous flow, by making the liquid or gas run through a detector.

Detection in continuous flow can be applied to β^-, α, X, or γ emitters by using the appropriate detector and the same electronic system as that used in a scintillation counter is used: two photomultipliers, coincidence circuit, and amplitude selection. However, there is no sample changer; instead a flow-cell is placed between the photomultipliers.

The counting cell can be made of a scintillant solid material or a non-scintillant plastic tube for passage of sample and a scintillation cocktail. The different types of cells are described in *Table 6*. The cell volume varies from 50 μl to several millilitres depending on the nature of the instrument.

The main suppliers of radiochromatography and related equipment of this type are: Beckman, Berthold, Isomeas, Nuclear Enterprises, and Radiomatic. When choosing one of these counters, pay particular attention to the different types of counting cells available and their facility for adaptation. The counter is usually coupled with a microprocessor or microcomputer allowing treatment

Table 6. Continuous counting methods

Counting cells[a]	**Radionuclide and approx counting efficiency**	**Remarks**
Catheter plunged into scintillator	^{32}P: 80%	
Scintillant catheter	^{32}P >80% ^{14}C: 2–8%	Preparation required[b]
Scintillant plastic cell	^{32}P >80% ^{14}C: 2–3%	
Anthracene crystals	^{32}P: >80% ^{14}C: 18%	Retention risk of the radioactive product
Scintillant balls in plastic or glass	^{32}P: >80% ^{14}C: 20–25% ^{3}H: 0.5–1%	
CaF_2 balls	^{32}P: >85% ^{14}C: 75–85% ^{3}H: 5%	
Cocktail: effluent liquid scintillator	^{14}C: 95% ^{3}H: 50%	
Gas circulation vial filled with scintillation fluid	^{14}C: 95%	For insoluble gas in the scintillation fluid
Catheter and gamma-vial	^{125}I: 42%	

[a] The majority of counting cells are adaptable to the Radiomatic counter.
[b] Soak the catheter for 48 h occluded at both ends, in a solution of 15 g of αnpo, 1 g of DMPOPOP in 200 ml of benzene (3).

of results and the facility to simultaneously register the result of other measurements, for example UV, fluorescence, or any other analogue signal. It is also possible after calibration to carry out double-label measurements such as between ^{3}H and ^{14}C.

In gas chromatography the output may be monitored by means of a proportional counter, based on gas ionization rather than scintillation counting (e.g. Radiomatic).

5. Multidetector liquid scintillation counters

Two types of multidetector liquid scintillation counters are available, designed to count vials, or filters.

5.1 Vial counting

Such counters (e.g. Micromedic Systems†) simultaneously analyse four vials of 20 or 5 ml. The counter comprises four pairs of photomultipliers that must be calibrated in parallel. A quench correction curve is set for each detector.

5.2 Filter counting

These purpose-built counters (e.g. 1205 Betaplate LKB‡) will analyse glass-fibre filters or nylon transfer membranes, for example Southern blots. The filters are introduced into a plastic bag with 4 to 10 ml of scintillation fluid. This is then placed into a rigid cassette which is positioned in the counter comprising 6 detectors (2×6 photomultipliers). The cassette is automatically moved, and counters accept a number of cassettes.

The counting efficiencies are 26 to 54% for ^{3}H and 81 to 96% for ^{14}C and ^{32}P depending on the filter used.

Acknowledgements

The author is particularly grateful to Severine Izabelle and Robert Slater for their help in preparing the English version. Thanks are due also to Drs Françoise Simonnet and José Combe for their critical reading of this manuscript.

† Micromedic Systems Inc. = Tauris™.
‡ LKB-Wallac Pharmacia '1205 Betaplate'.

Further reading

Briere, M., Jouve, B., and Paulin, R. (1970). *Mésures en Radioactivité*. Ed. Herman, Paris.

Knol, G. F. (1979). *Radiation, Detection and Measurement*. John Wiley, New York.

Tait, W. H. (1980). *Radiation Detection*. Butterworths, London.

Liquid scintillation counters

Dyer, A. (1980). *Liquid Scintillation Counting Practice*. Heyden, London, New York, Rheine.

Horrocks, D. L. (1974). *Application of Liquid Scintillation Counting*. Academic Press, New York and London.

Peng, C. T. (1977). *Sample Preparation in Liquid Scintillation Counting*. Amersham, England.

Simonnet, G. and Oria, M. (1980). *Les mésures de radioactivité à l'aide des compteurs à scintillateur liquide*. Eyrolles, Paris.

References

1. Beynon, R. I. (1984). In *Microcomputers in Biology* (eds C. R. Ireland and S. P. Long). IRL Press, Oxford.
2. Freedman, A. J. and Anderson, E. C. (1952). *Nucleonics* **10,**(8) 57.
3. Eldridge, J. S. and Crowther, P. (1964). *Nucleonics* **22**(6), 56–59.
4. Mantel, J. (1972). *Int. J. App. Radiat. Isotopes* **23,** 407–13.
5. Freedman, A. J. and Anderson, E. C. (1952). *Nucleonics* **10,**(8), 57.
6. Horrocks, D. L. (1975). *Int. J. Appl. Radiat. Isotopes* **26,** 243–56.
7. Simonnet, F. and Bocquet, C. (1983). *Health Physics* **44**(2), 160–4
8. Simonnet, F., Combe, J., and Simonnet, G. (1987). *Appl. Radiat. Isotopes* **38**(4), 311–12.
9. *Methods in Low-level Counting and Spectrometry* (1981). IAEA, Vienna.
10. Currie, L. A. (1968). *Analytical Chemistry* **40**(3), 586–93.

4

Radioisotope detection using X-ray film

RON A. LASKEY

1. Introduction

During the last two decades gel electrophoresis has steadily replaced other methods for analytical fractionation of macromolecules. This chapter describes and discusses methods of detecting radioisotopes within flat samples such as acrylamide or agarose gels, nitrocellulose filters, thin layer chromatography plates and paper chromatograms. For all of these samples X-ray film provides a convenient non-destructive means of isotope detection which combines high sensitivity with high resolution.

For some applications such as detection of ^{14}C or ^{35}S on the surface of thin layer chromatograms direct autoradiography on X-ray film is ideally matched to the radioactive emissions. However, for the great majority of combinations of isotope and sample, sensitivity can be greatly improved by converting emitted β particles or γ rays to light, using either X-ray intensifying screens for high energy emitters such as ^{32}P, or organic scintillators for weak β emitters such as ^{3}H, ^{14}C, or ^{35}S. Although sensitivity can be increased in this way, conversion of ionizing radiation to light has several compensating disadvantages. First, it decreases resolution by causing secondary scattering. Second, it requires exposure at low temperature (−70 °C) because the response of film to light is fundamentally different from its response to ionizing radiation for reasons which are explained in Section 7. Third, this different response of film to light means that unlike direct autoradiography, methods which convert to light are not quantitative unless the film is pre-exposed to an instantaneous flash of light to bypass a reversible stage of image formation (see Sections 6.3 and 7).

This chapter starts with a guide to selection of the most appropriate method for a range of isotopes and applications. It then describes procedures for individual methods and evaluates their merits and drawbacks. These sections are followed by several general considerations such as choice of materials and exposure conditions. Finally the principles which underly these detection methods are summarized in the belief that this information can help the user to avoid pitfalls.

2. Choosing a class of detection method: autoradiography versus fluorography or intensifying screens

Table 1 summarizes the suitability of various detection methods for different combinations of radioisotope and sample type. *Table 2* summarizes their sensitivities.

Table 1. Choosing the appropriate method

Isotope	Sample	
^{3}H (0.0186 MeV)	Acrylamide gels Agarose gels Paper chromatograms Thin layer chromatograms Nitrocellulose filters	Fluorography (Section 5)
^{14}C or ^{35}S (0.156 or 0.167 MeV)	Acrylamide gels Agarose	Fluorography for maximum sensitivity (Section 5) or direct autoradiography for extreme resolution (Section 3)
	Thin layer chromatograms Paper chromatograms Nitrocellulose filters	Direct autoradiography (Section 3). [There may be slight enhancement by fluorography (Section 5)]
^{32}P (1.71 MeV) or any γ-emitting isotope	All types of flat sample	Intensifying screen for maximum sensitivity (Section 4) or direct autoradiography for maximum resolution (Section 3)

Fluorography or intensifying screens require 'screen-type' X-ray film exposed at −70 °C (Section 6) and only pre-flashed film gives quantitative images and maximum sensitivity (Sections 6 and 7). In contrast, direct autoradiography does not require 'screen-type' film, exposure at −70 °C or pre-flashed film for quantitative accuracy or sensitivity (Sections 2 and 7).

2.1 Direct autoradiography gives high resolution but limited sensitivity

For direct autoradiography an X-ray film is held as close as possible to the sample in a light-tight container and exposed at any convenient temperature (see below for details). This simple method gives optimum resolution with moderate sensitivity for isotopes with emission energies equal to or greater than ^{14}C. It produces quantitative images in which the absorbance of the film image is directly proportional to the amount of radioactivity for all film absorbances up to 1.5.

2.2 Fluorography increases sensitivity for weak β emitters

Direct autoradiography is ideally suited to unquenched emissions from ^{14}C or ^{35}S. This situation is achieved when these isotopes are on the surface of a thin layer chromatography plate. However, when these or weaker β emitters such

Table 2. Sensitivities of film detection methods for commonly used radioisotopes

Isotope	Method	Detection limit d.p.m./cm^2 for 24 h	Relative performance compared to direct autoradiography
^3H (0.0186 MeV)	Direct autoradiography (Section 3)	$>8 \times 10^6$	1
	Fluorography using PPO (Section 5)	8000	>1000
^{14}C or ^{35}S (0.156 or 0.167 MeV respectively)	Direct autoradiography (Section 3)	6000	1
	Fluorography using PPO	400	15
^{32}P	Direct autoradiography (Section 3)	525	1
(1.71 MeV)	Intensifying screen (Section 4)	50	10.5
^{125}I (0.035 MeV γ rays, plus X-rays and electrons)	Direct autoradiography (Section 3)	1600	1
	Intensifying screen (Section 4)	100	16

Data from refs. 1, 3, and 4.

as ^3H are located within the lattices of acrylamide or agarose gels or similar samples, then the β particles are absorbed within the sample ('quenched') and fail to reach the film.

The solution to this problem is impregnation of the sample with an organic scintillator. This procedure is called fluorography. Beta particles from the sample excite the scintillator producing ultraviolet light which is not absorbed by the sample, but which can escape to produce a *photographic* image on the film. Several fluorographic reagents and procedures have been described. They differ in sensitivity, image quality, preparation time, cost, and convenience. Their relative merits are described in Section 5. Because a photographic image is produced, rather than an autoradiographic image, films must be exposed at $-70\,°C$. They must also be pre-exposed if quantitative images are required, or if extreme sensitivity is required in long exposures (Sections 6.3 and 7).

2.3 Intensifying screens increase sensitivity for high energy β particles or γ-rays (1)

Gamma rays or high energy β particles such as those from ^{32}P pass through and beyond an X-ray film, so that most of their emission energy is not recorded by the film but wasted. This problem can be overcome by placing the film between the sample and a high density fluorescent 'intensifying screen'. Emissions which pass through and beyond the film excite the screen causing it to emit UV light which superimposes a photographic image over the autoradiographic image. As with fluorography, the production of a *photographic* image from the very low intensities of light involved requires exposure at −70 °C. Furthermore, contrary to widespread misunderstandings images obtained using intensifying screens are not quantitative or fully sensitive to small amounts of radioactivity unless the reversible stage of image formation is by-passed by pre-exposing the film to a brief flash of light (see Section 7).

Although intensifying screens greatly increase the sensitivity of detection for ^{32}P- or γ-emitting isotopes (*Table 1*) they result in decreased resolution. This is because both the primary emissions and the secondary emissions of light diverge from the source. For many purposes the slight loss of resolution is unimportant. However, for uses such as DNA sequencing the loss of resolution becomes a serious disadvantage and direct autoradiography should be considered as an alternative.

3. Direct autoradiography

Direct autoradiography without scintillators or intensifying screens is the method of choice when spatial resolution is more important than absolute sensitivity. In addition there are also rare examples of circumstances when an X-ray film can capture most of the emissions from one face of the sample. For example, ^{35}S or ^{14}C emissions from a thin layer chromatography plate are essentially unquenched and escape from the surface of the plate. Therefore direct autoradiography is roughly as efficient as fluorography for these samples (2). However, this would not be true of weaker β-emitters such as ^{3}H or for more highly quenched samples such as ^{14}C or ^{35}S in polyacrylamide gels. For each of these cases fluorography is substantially more efficient (2–4).

Any type of X-ray film can be used for direct autoradiography, but maximum efficiency is obtained when a 'direct' type of film (as opposed to a 'screen-type' film) is used. 'Direct' film types such as Kodak Direct Exposure Film or Amersham Hyperfilm β max have a high silver halide content to maximize their absorption efficiency. In the case of ^{3}H, self-absorption within the sample is a very severe problem, but provided that this is not excessive, then direct autoradiography of ^{3}H can be performed using a special purpose film such as Hyperfilm-^{3}H (Amersham International) which lacks the 'anti-scratch' plastic coating used on most films.

Essentially the same procedure for direct autoradiography is applicable to all kinds of flat, dry sample, though the choice of film for optimal sensitivity varies as described above and in Section 6. For wet samples such as acrylamide gels, ^{32}P can be detected efficiently without drying, but resolution is increased by drying, because the thickness of the sample determines the image resolution. However, in the case of weaker emissions such as ^{14}C, ^{35}S, or ^{3}H, drying greatly increases sensitivity as well as resolution, by decreasing self absorption within the sample. Procedures for drying gels are discussed in Section 6.4.

3.1 Procedure for direct autoradiography

The sample to be exposed should be dry, though in the case of ^{32}P wet samples can also be exposed at the expense of resolution. In the case of ^{32}P the sample may be covered by thin plastic film such as Saran or Clingfilm, but this should not be used for direct autoradiography of weaker β emitters such as ^{14}C, ^{35}S, or ^{3}H as they are effectively absorbed by it. To enable the exposed film to be oriented correctly, mark opposite corners of the sample with different symbols using 'radioactive ink'. This is usually ink to which waste radioisotope has been added. For this purpose isotopes with long half-lives are most convenient (e.g. ^{14}C) but others are satisfactory. After exposure orient and align the film correctly by placing the exposed symbols on the film over the ink marks on the sample. Remove a sheet of X-ray film (selected as described above) from the box and place it over the sample in a dark-room with minimum intensity dark red safe lights. Then firmly clamp it to the sample to achieve the closest possible contact. The most convenient way to achieve this is by using a medical X-ray cassette obtainable from most medical X-ray film manufacturers. A cheaper but less convenient alternative is to clamp the film and sample between glass plates using spring clips and to expose in a light-tight envelope.

Films can be exposed at ambient temperature for direct autoradiography. There is no gain in sensitivity by exposing at lower temperatures. Nor is there any advantage in pre-exposing the film. Exposure at −70 °C can have advantages when wet samples such as acrylamide gels are exposed since it can prevent diffusion of solute molecules. When ^{32}P or γ emitting isotopes such as ^{125}I or ^{131}I are used, care should be taken to keep samples and exposures away from film stocks to avoid fogging. Similarly stacking exposures together can lead to 'phantom' images from adjacent samples. One way of minimizing this problem is to use X-ray intensifying screens as barriers either between cassettes or inside cassettes at ambient temperature. If exposed at ambient temperature, the screen will not contribute to the film image and so will not decrease the resolution.

After exposure remove and process the films in a dark room following the manufacturer's instructions. Most types of X-ray film can be processed either by hand or using an automatic film processor. However, 'direct' films such as Amersham Hyperfilm β-max (see above) require unusually long processing cycles. In addition films which lack protective anti-scratch layers such as Hyperfilm-^{3}H are not suitable for automatic processing but should only be processed by hand.

4. Use of X-ray intensifying screens to increase sensitivity for ^{32}P and γ-ray emitters

High energy β particles such as those emitted by ^{32}P, or γ rays such as those emitted by ^{125}I or ^{131}I, are too penetrative to be absorbed efficiently by X-ray film. Consequently, most of their energy is wasted during direct autoradiography as it passes through and beyond the film.

This problem can be overcome by converting any emissions which pass beyond the film into light. The light produced passes back through the film superimposing a photographic image over the autoradiographic image. This conversion is achieved routinely in medical radiography by placing a high density fluorescent screen behind the film. Calcium tungstate is used widely though other fluors are discussed below. Radiation which passes beyond the film is absorbed more efficiently by the dense screen, producing light which generates the secondary photographic image.

Although this procedure is used routinely in medical radiography at least one additional step is required for its use in the longer exposures required for radioisotope detection (1). This arises because the initial step of photographic image formation by light is reversible as explained in Section 7. However, this problem can be partly overcome by exposing the film at −70 °C (1,5) and completely overcome by combining low temperature exposure with pre-exposure of the film to a brief flash of light to bypass the reversible stage of image formation. This latter step is essential if image quantitation is required (1, 6).

Figure 1 shows that the advantage of using X-ray intensifying screens is increased sensitivity, but that there is a disadvantage of decreased resolution.

To obtain the increased sensitivity which intensifying screens offer it is necessary to use a 'screen type' X-ray film (see Section 6.1). The spectral sensitivity of these films is matched to the emission spectra of intensifying screens. Many manufacturers make suitable films such as Kodak, Fuji, Du Pont, or Amersham. In general the 'fastest' (most sensitive) films are suitable, but these also give the most grainy images and they have the shortest shelf-lives. Most 'screen-type' films are sensitive to UV or blue light. However some are sensitized to green emissions from rare-earth intensifying screens. These are not recommended for use with other types of screen and there are problems with rare-earth screens as described below.

There is a wide choice of intensifying screens available (see Section 6.1), but some of the most sensitive screens are not suitable for long exposures at −70 °C especially when pre-exposed film is used, because they contain low levels of endogenous radioisotopes which cause the film to blacken in the absence of a sample. This problem was observed for screens consisting of europium activated barium fluorochloride or terbium activated rare earth oxysulphides. It should be noted that this problem is not encountered when these products are used in medical radiography since much shorter exposures

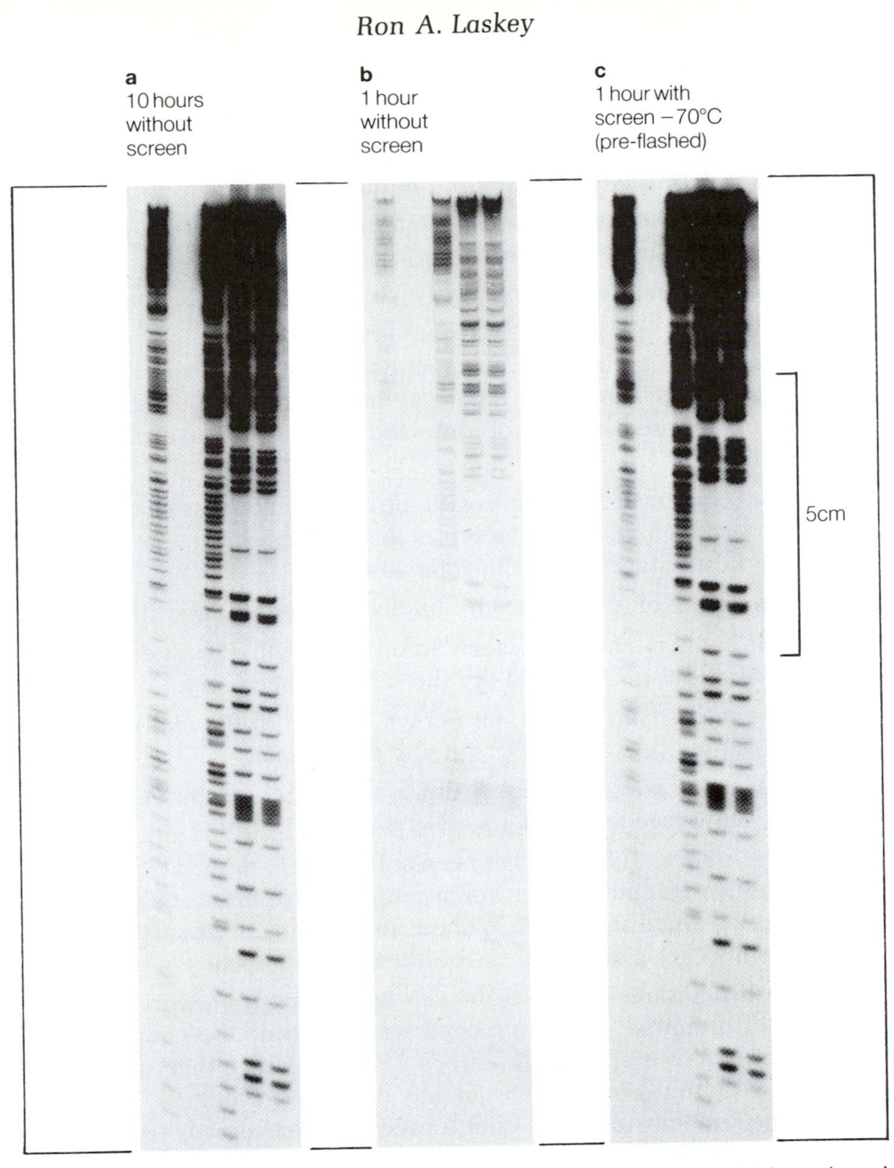

Figure 1. Effect of a single calcium tungstate intensifying screen on sensitivity and resolution of ^{32}P. Note that the screen increases sensitivity by approximately tenfold (compare b with c) but decreases resolution (compare c with a). All exposures were recorded on Fuji RX film. The gel containing ^{32}P- labelled DNA fragments was kindly provided by Dr S. E. Kearsey.

are used at room temperature. The problem arises because sensitivity to the light produced from trace amounts of radioactivity is far greater at −70 °C. This point is the key to the success of the intensifying screen method and is explained further in Section 8.

Protocol 1. Procedure for indirect autoradiography using intensifying screen (1)

1. Work in a dark room with dark red illumination.
2. If the sample is wet cover it with Clingfilm or Saran. Gels can be dried as described in Section 6.5.
3. Select only 'screen-type' X-ray film such as Kodak XAR-5, Amersham Hyperfilm-MP, Fuji RX or Du Pont Cronex 4 (see Section 6.1).
4. For quantitative accuracy or maximum sensitivity for small amounts of radioactivity, pre-expose the film to an instantaneous flash of light. See Section 6.3 for pre-flashing procedure and Section 7 for an explanation of its effect.
5. Mark two corners of the sample with different symbols using radioactive ink (see Section 3.1). For wet samples this can be achieved by writing on adhesive labels stuck to the Clingfilm cover.
6. Place one sheet of X-ray film (pre-flashed if required) over the sample.
7. Select an X-ray intensifying screen according to the criteria described in Sections 4 and 6.1. Clean and dry the screen if necessary.
8. Place the intensifying screen, face down, on top of the film so that the film is enclosed between the screen and the sample.
9. Clamp sample, film, and screen tightly together, ideally in a medical X-ray cassette (see Section 3.1 for alternatives).
10. Expose at $-70\,°C$. If the film has been correctly pre-flashed exposure can also be performed at ambient room temperature, but at only 50% the efficiency obtained at $-70\,°C$. Without pre-flashing, exposure at $-70\,°C$ is essential to achieve the benefit of intensifying screens.
11. For translucent samples such as dry gels on filter paper, sensitivity can be increased further by placing a second screen outside the sample so that the order is screen 1, gel, film, screen 2 (5). Although this can increase sensitivity up to twofold, it substantially decreases resolution. This point should be remembered when using a cassette which already contains two screens for a high resolution sample.
12. After exposure, develop the film according to the manufacturer's instructions.

5. Fluorography of weak β emitters

The efficiency of detection of radioisotopes which emit weak β particles (e.g. ^{3}H, ^{14}C, or ^{35}S) is limited by the extent of absorption of the β particle in the sample. The extent of this problem depends both on the isotope's emission energy and on the extent to which the radioactive molecules are buried within the sample. For example, the very weak emissions from ^3H are essentially unable to escape from within the matrix of an acrylamide gel, and are therefore most unsuitable for direct autoradiography, whereas the stronger emissions from ^{14}C or ^{35}S can be detected efficiently by direct autoradiography of thin-layer plates (2), but not of acrylamide gels (3), because most of the emission energy is absorbed within the gel and therefore unable to reach the film.

Fluorography is designed to overcome this problem, by impregnating the sample with an organic scintillator. This is excited by β particles from the isotopically labelled sample so that it emits UV light which can escape more freely from the sample to expose the film. Because the energy of an emitted β particle is divided into smaller quanta of light, it is necessary to expose film at −70 °C to obtain a fluorographic image (7, 8). In addition, a quantitative response of the film and maximum sensitivity for small amounts of radioactivity are only obtained when the reversible stage of image formation is bypassed by pre-exposing the film to a brief flash of light (4). The physical principles underlying this effect are discussed in Section 7, together with a cautionary illustration of how easily this effect is underestimated.

The conversion of β particles to light has additional consequences. First, it makes it necessary to use a screen-type X-ray film such as Kodak XAR 5, Fuji RX, or Amersham Hyperfilm-MP in order to achieve maximum sensitivity. Second, it decreases resolution because in addition to the dispersed spread of β particles, the secondary light emissions are also dispersed.

Most of the applications of fluorography concern acrylamide gels and several procedures exist to achieve this. However, fluorographic procedures have also been devised for several other types of sample such as agarose gels, thin-layer plates, paper, or nitrocellulose sheets. To simplify the choice of method the description of fluorographic procedures is divided. Section 5.1 considers general points which apply to a wide range of procedures. Section 5.2 considers the various procedures for fluorography of acrylamide gels and Section 5.3 considers fluorography of other types of sample.

5.1 General technical procedures for fluorography of acrylamide gels by all methods

Protocol 2.

1. Unstained gels can be fixed by soaking in 7–10% acetic acid, or 7% acetic acid, 10% methanol, 83% H_2O for 15–30 min. This minimizes diffusion of bands during fluorographic impregnation. Diffusion of protein bands during impregnation is usually negligible, but smaller molecules such as oligonucleotides generated in DNA sequencing reactions could diffuse significantly.

2. Gels which have been stained in Coomassie Blue can be processed directly for fluorography as the blue stain does not absorb fluorographic emissions too severely. However, most fluorographic procedures remove stain, so stained gels should be photographed before fluorography.

   Gels which have been stained in ethidium bromide must have the stain removed before fluorography, because ethidium bromide absorbs fluorographic emissions efficiently. This is easily achieved by soaking the gel in ethanol or methanol for 30 min before proceeding. Similarly, silver staining quenches fluorographic emissions, so silver should be removed, such as by immersion in photographic fixer before proceeding.
3. Acrylamide gels which have been dried must be rehydrated by soaking in water for approximately 1 h before proceeding with fluorographic impregnation.
4. After fluorographic impregnation by one of the methods described in Section 5.2, gels should ideally be dried before exposure as described in Section 6.4. Although this is not essential for fluorography, drying the gel increases both sensitivity and resolution. In addition, it prevents diffusion of labelled bands.
5. After drying, the orientation of the gel should be marked using radioactive ink (ink to which waste radioisotopes has been added), marking two corners with different symbols. This will allow the relative orientations of the film and the gel to be determined by super-position.
6. Fluorographic exposures require 'screen-type' X-ray film such as Kodak XAR-5, Amersham Hyperfilm-MP, Fuji RX, etc. (see Section 6.1). These are sensitized to the wavelengths of light emitted by intensifying screens and, coincidentally, fluorographic scintillators.
7. For fluorographic exposure films should be held in close contact with the dried gel. This is achieved most conveniently by enclosing them in an X-ray film cassette. Note that it is not necessary to include an intensifying screen, but nor is it necessary to remove it if one is present.
8. Fluorographic exposures must be performed at −70 °C or below to overcome the reversible stage of latent image formation in the film. An exception is that film which has been hypersensitized by pre-exposure to a flash of light (see next paragraph and Sections 6.3 and 7) can be exposed at room temperature with approximately 50% the efficiency which would be obtained at −70 °C.
9. To achieve maximum sensitivity and quantitative accuracy of fluorography film should be hypersensitized by pre-exposure to an instantaneous flash of light as described in Section 6.3 and explained in Section 7.
10. After exposure film should be processed in a darkroom with a safelight according to the manufacturer's instructions.

5.2 Procedures for fluorography of acrylamide gels

The various methods which are available for fluorography of acrylamide gels differ in cost, convenience, time, efficiency, and image quality. They are summarized in *Table 3*. One method uses sodium salicylate (9), two use the organic scintillator PPO (2,5-diphenyloxazole), and at least one of the commercial methods uses a derivative of PPO which is more soluble in water. The commercial products are the most convenient to use, but they are also relatively expensive. In spite of commercial claims to the contrary, the author receives repeated comments from users that the original method (3) which uses a solution of PPO in dimethyl sulphoxide (DMSO) yields the best combination of sensitivity and resolution. The two disadvantages of this method are the long procedure time and the potential hazard posed by the skin-penetrating properties of DMSO. An alternative method has been described which uses glacial acetic acid in place of DMSO as the solvent for PPO (10). This offers some saving of time over DMSO, but in other respects the relative merits of these methods are essentially similar. The final method which is described below uses an aqueous solution of sodium salicylate (9).

Table 3. Methods for fluorography of acrylamide gels

Method	Advantages	Disadvantages
1. Commercial (e.g. Amplify[a] Enlightening[b] or Enhance[b]	Convenience and speed	Relatively expensive (see note concerning effect of gel drying on sensitivity in Section 5.2.1)
2. PPO in DMSO(3)	Sensitivity. Cost. Image quality	Tedious procedure Potential hazard
3. PPO in acetic acid(10)	Sensitivity. Cost. Image quality	Potential hazard
4. Sodium salicylate(9)	Sensitivity. Cost. Speed	Grainy image Potential hazard

[a]Amplify is a registered trade mark of Amersham International.
[b]Enlightening and Enhance are registered trade marks of New England Nuclear.

When comparing claimed sensitivity and impregnation times of these various methods, it is important to note that time can be saved at the expense of sensitivity and that some claims have been based on comparisons with suboptimal versions of alternative methods.

5.2.1 Procedures for impregnating acrylamide gels with commercial products such as Amplify, Enlightening, or Enhance

Precise details for impregnating gels with these materials are provided by the manufacturers together with the products.

The fastest of these impregnation protocols are those for Amplify (Amersham International) and Enlightening (New England Nuclear). For these the gel is simply soaked in the product for 15–30 min and then dried under vacuum. Excess product can be reused for further gels. Although these procedures are extremely convenient, their sensitivity can be affected by the gel drying procedure, as it is possible to suck some of the scintillant back out of the gel under vacuum. Methods which precipitate the scintillant before drying, such as the longer procedure for Enhance (New England Nuclear) or the PPO methods in *Protocols 3* and *4*, avoid this problem.

Protocol 3. Procedure for impregnating acrylamide gels with PPO using dimethyl sulphoxide (3)

1. Before fluorographic impregnation, unstained gels can be fixed. Dried gels can be rehydrated as described in Section 5.1.
2. Soak the gel in approximately 20 volumes of dimethyl sulphoxide for 30 min. Use a fume-hood and wear rubber gloves to prevent skin contact with dimethyl sulphoxide for this and later stages. Swirl the gel gently in a sealed container such as a plastic food storage box.
3. Repeat step 2 with fresh dimethyl sulphoxide to remove all water from the gel.
4. Keep both stocks of dimethyl sulphoxide separately for reuse in the same sequence with other gels.
5. Soak the gel for 3 h in 4 volumes of a 22% w/v solution of PPO (2,5 diphenyl oxazole) in dimethyl sulphoxide. Use a sealed container to prevent water absorption and agitate the gel gently. Three hours is the optimum time, but this can be decreased significantly with slight loss of resolution (see ref. 3 for details).
6. Soak the gel in water for 1 h to remove dimethyl sulphoxide and to precipitate PPO in the gel. This step decreases drying time and ensures high resolution by forming very small crystals of PPO in the gel. In addition it prevents removal of the scintillator from the gel by suction during drying, a problem which affects efficiency of some fluorographic methods. Gels will shrink during immersion in dimethyl sulphoxide, but re-swell during the water step.
7. Dry the gel and expose it to pre-flashed screen-type X-ray film as described in Section 5.1.
8. To recover excess PPO from solutions in DMSO, pour the solution into 3 volumes of 10% ethanol in water. Filter after 10 min, wash with water and air-dry (or dry under vacuum) for reuse. Use of 10% ethanol can be replaced by water, but the PPO crystal size will be smaller, so that filtration, rinsing, and drying will take longer.
9. The optimum concentration of PPO varies with the acrylamide concentration of the gel. The procedure described is optimal for gels which contain

more than 10% acrylamide. 10% PPO should be used for gels which contain less than 5% acrylamide and 15% PPO should be used for gels which contain between 5% and 10% acrylamide.

Protocol 4. Procedure for impregnating acrylamide gels with PPO using acetic acid (10)

1. Soak the gel in undiluted acetic acid for 5 min.
2. Soak the gel in 20% PPO (w/v) in acetic acid for 1.5 h. Gentle agitation in a sealed container is recommended.
3. Soak gel in water for 30 min.
4. Dry and expose the gel to pre-flashed screen-type X-ray film at −70 °C as described in Section 5.1.
5. Recover excess PPO for reuse by addition of five volumes of water. Collect the PPO crystals by filtration, rinse and air-dry.

Protocol 5. Procedure for impregnating acrylamide gels with sodium salicylate (9)

1. Before impregnation gels may be stained or fixed as described in Section 5.1, but gels fixed in acid should be soaked in water for 30 min to prevent precipitation of salicylic acid.
2. Soak the gel in 10 vols of 1 M sodium salicylate, pH 5–7 for 30 min.
3. Dry and expose the gel to pre-flashed screen-type X-ray film at −70 °C as described in Section 5.1.

Excess salicylate solution can be reused, but prolonged storage causes brown discoloration presumably due to oxidation. To avoid this salicylate can be recovered as salicylic acid by precipitation with equimolar HCl followed by washing with water and drying overnight on a Buchner funnel with suction.

Salicylate is toxic and like DMSO it is readily absorbed through the skin. Therefore wear gloves when handling gels in salicylate.

5.3 Fluorography of other samples excluding acrylamide gels

Fluorography is suitable for many other types of sample including agarose gels, thin layer plates, nitrocellulose filters, or paper. Although impregnation procedures differ from those used for acrylamide gels, the principles of film choice and exposure conditions are the same as those described in Section 5.1.

5.3.1 Fluorography of agarose gels

Although the principles of fluorography are similar for agarose and acrylamide gels, there are four important additional considerations for agarose. First, agarose gels are more fragile and they break easily in response to mechanical agitation. Second, they are frequently stained with ethidium bromide which must be removed or it will quench fluorographic emissions. The ethanol (or methanol) procedures described here remove ethidium bromide from nucleic acids in the gel. Third, agarose gels melt when heated; therefore extra care is necessary when drying them. Fourth, scintillant is easily sucked out of the gel during drying. This makes it desirable to precipitate the scintillant in the gel before drying.

Procedures for fluorography of agarose gels

Agarose gels can be impregnated with scintillator using Amplify (Amersham International), but fluorographic efficiency varies with the level of vacuum used to dry the gel as this may suck the scintillant out of the gel, and not just its solvent. Alternatively, the PPO in acetic acid method (10) described in Section 5.2.3 can be used. However, in the author's experience, PPO in ethanol provides an efficient method (6) which also removes ethidium bromide which is bound to nucleic acids in the gel.

The recommended procedure (6) is as follows:

Protocol 6. Procedure for fluorography of agarose gels

1. Soak the gel in approximately 10 volumes of 100% ethanol or methanol (95% ethanol is unsatisfactory) in a sealed container to prevent water absorption. Agitate very gently, taking care not to break the gel.
2. Transfer the gel to fresh ethanol for a further 30 min to remove water completely.
3. Retain these two ethanol stocks in sealed bottles for reuse in the same sequence with future gels.
4. Transfer the gel to 4 volumes of a 3% w/v solution of PPO (2,5,diphenyl oxazole) in absolute ethanol (or methanol) for approximately 3 h. If the gel appears white at this stage due to precipitation of PPO, return it to ethanol to remove residual water, before continuing in 3% PPO in ethanol.
5. Transfer the gel to water to precipitate PPO in the gel. This prevents suction of PPO from the gel during drying and ensures that very small crystals of PPO are formed. Large crystals reflect fluorographic emissions distorting the image.
6. Dry the gel under vacuum essentially as described in Section 6.5 except that the gel must not be overheated or it will melt before drying. This will be revealed as a very blurred image when compared to the ethidium bromide stain.

7. Expose the dried film to pre-flashed screen-type X-ray film at −70 °C as described in Section 5.1.

Any of the immersion steps can be extended overnight as nucleic acids will not diffuse during this procedure.

5.3.2 Fluorography of paper chromatograms (11)

Protocol 7. Procedure for fluorography of paper chromatograms (11)

1. Soak the paper in a 7% w/v solution of PPO in ether (or alternative solvent such as ethanol or acetone).
2. Dry the paper in air and expose to pre-flashed screen-type X-ray film as described in Section 5.1.
3. Uniformity of impregnation should be checked by observing under ultraviolet illumination to ensure a uniform level of fluorescence.

5.3.3 Fluorography of thin layer chromatograms (2)

As explained earlier, fluorography of ^{14}C and ^{35}S on thin layer chromatograms is not much more efficient than direct autoradiography, though there is a large gain in efficiency by fluorography of ^{3}H (2). For the most widely used method simply pour a 7% w/v solution of PPO in ether rapidly over the dried chromatogram (2). Alternatively, stand the dried plate in a 7% solution of PPO in acetone in a chromatography tank and allow the solution to ascend (4). To detect molecules which are soluble in ether or acetone, but not in water; stand the dried plate in Amplify (Amersham International) and allow it to ascend the plate.

A further increase in sensitivity has been reported for the following more complex procedure (12).

Protocol 8. Procedure for fluorography of thin layer chromatograms (2)

1. Melt 2-methylnaphthalene in a water bath (m.p. 34–36 °C) and add 0.4% w/v PPO.
2. Pour this mixture into a warm Pyrex dish on a hot-plate in a fume cupboard and immerse the dried thin-layer plate until soaked (usually less than 1 min).
3. Drain the plate and allow the methyl naphthalene to solidify before exposure.

When the plates treated by any of the above procedures are dry expose them to pre-flashed screen-type X-ray film at −70 °C as described in Section 5.1.

5.3.4 Fluorography of nitrocellulose filters (13)

Protein or nucleic acid blots on nitrocellulose filters can be revealed conveniently by fluorography. Although the method was described for ^{3}H, ^{14}C, or ^{35}S (13), it is not clear from the literature how much it increases sensitivity for ^{14}C and ^{35}S. By analogy with thin layer plates the gain for these two isotopes compared to direct autoradiography may be much less for nitrocellulose than for gels.

Protocol 9. Procedure for fluorography of nitrocellulose (13)

1. It is essential that the filter to be impregnated must be completely dry.
2. Soak the dried filter in 20% w/v PPO in toluene or ether.
3. Air-dry and check impregnation is complete by uniform fluorescence under ultraviolet illumination. If fluorescence is patchy re-dry the sample thoroughly and repeat the impregnation procedure.
4. Expose to pre-flashed screen-type X-ray film at −70 °C as described in Section 5.1.

---

6. General technical considerations for radioisotope detection by X-ray film

The sections which follow briefly survey general points which may apply to many of the individual methods described above. For this reason they also repeat several points made earlier in the text.

6.1 Choice of films and screens

The techniques described in this chapter can all be performed with medical X-ray film, but it is important to realize that this is available in two types 'direct' and 'screen-type', neither of which is suitable for all of the methods described here.

Although direct autoradiography without scintillators or intensifying screens can be performed on most types of film, it is most efficient when 'direct' film types are used such as Kodak Direct Exposure Film or Amersham Hyperfilm-β max. These have high contents of silver halide which increase absorption efficiency and hence sensitivity. Furthermore direct autoradiography of ^{3}H requires film which lacks the anti-scratch plastic coating, e.g. Hyperfilm-^{3}H from Amersham.

In contrast, only screen-type film should be used to record from X-ray intensifying screens or fluorography as its absorbance spectrum is matched to the emission spectrum of intensifying screens and fortuitously to the emission spectrum of scintillators like PPO. It should be noted that some films are specifically designed to be sensitive to the green light emitted from lanthanum oxysulphide intensifying screens. We have found this type of film to be less satisfactory for radioisotope detection (1).

Within these constraints a wide choice of films is available, though it should be noted that the most highly sensitive films give the grainiest images and therefore relatively lower resolution. Kodak XAR 5 and Amersham Hyperfilm-MP both yield good results and are widely used with intensifying screens or fluorography. Fuji RX and Cronex 4 (Du Pont) are also suitable.

A wide range of X-ray intensifying screens is also available from manufacturers such as Du Pont, Fuji, Kodak, or CAWO. In general screens made of calcium tungstate (e.g. Du Pont, Cronex Lightning Plus, Fuji Mach II, or CAWO) give the best combination of high resolution and low background for radioisotope detection. Screens containing either europium-activated barium fluorochloride or terbium-activated rare-earth oxysulphides (lanthanum, gadolinium, or yttrium) may offer greater sensitivity, but with decreased resolution and greatly increased background. Although these problems are not revealed during medical radiography, for which the screens are designed, they become acute in long exposures when the reversible stage of image formation is manipulated by exposing at −70 °C or by pre-exposure of the film. The screens which have been most satisfactory for radioisotope detection in the author's experience are Cronex Lightning Plus (Du Pont) for sensitivity, or CAWO (CAWO Photochemische Fabrik, Schrobenhausen FRG) for image quality.

6.2 Conditions for exposure and processing of film

All films should be handled in dark-rooms with dark red safelights. Avoid exposing film to luminous clocks or dials. The shelf-life of sensitive films can be prolonged by storage at 4 °C in sealed bags. It is important to store film stocks well away from radioisotopes.

For maximum resolution and sensitivity it is important that films are clamped in the closest possible contact with the sample and with the screen when a screen is used. The most convenient means of achieving this is a medical X-ray cassette, but glass plates can be clamped together in a light-tight envelope as an alternative. Avoid using adhesive tape because its removal generates light which produces artefactual images.

There is no advantage in exposing direct autoradiographs at −70 °C. Room temperature is just as efficient. However, fluorographs and images from intensifying screens require exposure at −70 °C because the response of film to light is a multi-hit process as explained in Section 7. The exact temperature optimum has not been reported, but it appears to be between −40 °C and −90 °C and probably close to −70 °C or −80 °C. If a low temperature freezer is not available, then intensifying screens or fluorography can be performed at room temperature, provided that the film has been correctly pre-exposed as described in Section 6.3.

When cassettes are stacked in a freezer, films may acquire 'phantom' images from adjacent cassettes. This is a problem for ^{32}P and a severe problem for γ emitters such as ^{125}I or ^{131}I. However, this problem can be minimized by ensuring that there is an intensifying screen between consecutive films in the freezer as screens absorb radiation much more efficiently than film.

Before developing films which have been exposed at $-70\,°C$, they should be allowed to warm. This prevents condensation from forming on films and causing them to stick to the rollers of processing machines.

Development procedures should follow the film manufacturer's instructions. Small bench-top film processors are widely available and extremely convenient if there are many samples to process.

6.3 Pre-exposure of film to hypersensitize it for use with intensifying screens or fluorography

Once ionizing radiation has been converted to light by a fluorographic scintillator or an intensifying screen, the kinetics of the film response become fundamentally different from the response to β particles or γ rays (see Section 7). In summary a back reaction occurs which cancels the latent image produced by earlier photons. Exposing films at $-70\,°C$ slows this back reaction. Pre-exposing the film to a flash of light bypasses the back reaction (4). Hence it increases sensitivity for *small* amounts of radioactivity, though not for large amounts, and in particular it allows quantitation of images because all photons contribute equally to the image of pre-flashed film. Pre-exposure has no effect when large amounts of radioactivity are used in short exposures or when direct autoradiography is performed without intensifying screens or scintillators (1,4).

For pre-exposure to hypersensitize a film it is essential that the flash should be short, in the order of 1 msec. This can be achieved by attenuating the output from a photographic flash gun. Longer flashes only increase the background fog level of film without hypersensitizing it.

Provided that the flash is of the order of 1 msec, then the increase in background fog level can be used as a convenient index to monitor hypersensitivity. The background absorbance should be increased to between 0.15 and 0.2 (A_{540}) above the absorbance of untreated (but developed) film.

The intensity of the flash from a photographic flash gun can be attenuated by wavelength filtration. Thus orange filters (Wratten numbers 21 or 22) taped to the units decrease the output to approximately the correct level. Further adjustments can be made by adding neutral density filters, or by varying the aperture in an opaque mask or by varying the distance of the flash unit from the film. Trial exposures can be made on a single film by changing the position on the film of a clear window in an opaque mask. The fog levels achieved can be measured using a densitometer or by placing pieces of the film in a spectrophotometer.

It is important to note that the film will only yield a linear response to the amount of radioactivity when the fog level has been raised between 0.1 and 0.2 (A_{540}) above that of untreated film. Whereas unflashed film under-represents small amounts of radioactivity, film which has been pre-flashed to densities above 0.2 over-represents small amounts of radioactivity (*Figure 2*).

6.4 Procedures for drying acrylamide or agarose gels

It is not essential to dry gels before exposing them to X-ray film for any of the methods described in this chapter. Nevertheless, drying the gel increases both sensitivity and resolution. In addition, it prevents diffusion of labelled bands. Commercial gel dryers are available from several manufacturers (such as Bio-Rad). They are designed to heat the gel on a porous polythene support pad, under vacuum. The gel should be placed on a sheet of filter paper such as Whatman 3MM before placing it on the gel dryer. It should then be covered with cling film before drying.

Agarose gels will melt if they are heated excessively before most of their water has been removed; therefore heating should be delayed when drying agarose gels.

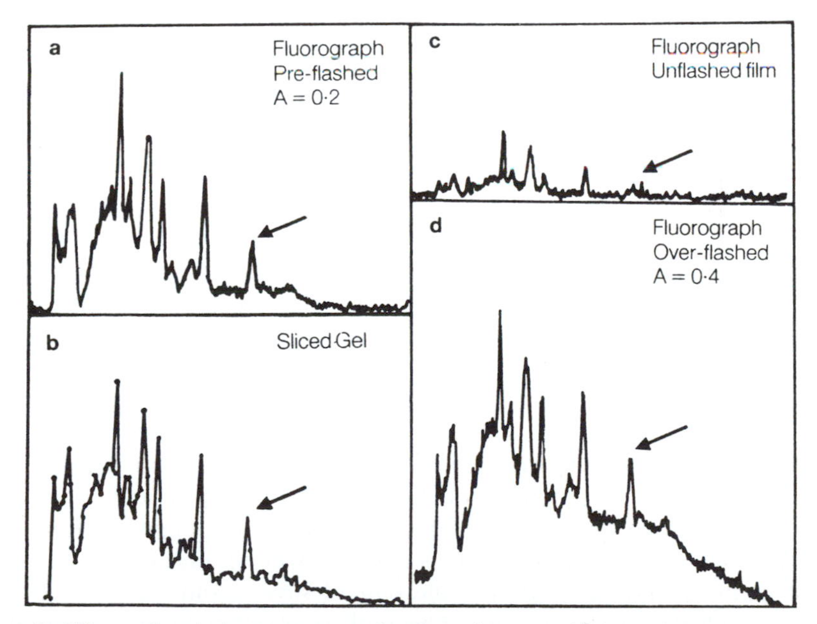

Figure 2. Effects of varied pre-exposure levels on images of ^3H distribution obtained by fluorography. Note that the arrowed peak is lost when unflashed film is used and exaggerated when overflashed film is used. (Data from ref. 3.)

If acrylamide gels crack during drying it may help to lower the bis-acrylamide concentration for future gels towards the ratio:

$$\% \text{ bisacrylamide} = \frac{1.3}{\% \text{ acrylamide}}.$$

Gels made by this formula do not crack when dried under continuous vacuum (14), but 20% acrylamide is required to give good protein resolution with this formula.

7. The underlying principles of radioisotope detection by X-ray film

Photographic emulsions are composed of silver halide crystals (grains of the film) each of which behaves independently. To produce a developable image each silver halide crystal requires several photons of visible light (~ 5 in average emulsions), each of which produces an atom of metallic silver. These then catalyse the reduction of the entire silver halide crystal by the developer.

A single hit by a β particle or γ-ray can produce hundreds of silver atoms rendering the grain fully developable. Hence direct autoradiography is a linear 'single hit' process in which all emissions are recorded equally until the film is saturated.

However, once the ionizing radiation is converted to multiple photons of light by an intensifying screen or fluorographic scintillant, the response of film is fundamentally different. Each photon produces only a single atom of silver. Although two or more silver atoms in a silver halide crystal are stable, a single silver atom is unstable and it reverts to a silver ion with a half-life of about 1 sec at room temperature. We have suggested previously (1, 4, 15) that this is the reason why exposure at −70 °C is necessary for the low light intensities produced by fluorography and intensifying screens. Lowering the temperature slows the thermal reversion of the single silver atom, increasing the time available to capture a second photon and thus produce a stable pair of silver atoms.

The probability of a second photon being captured by a grain before the first silver atom has reverted is greater for large amounts of radioactivity, and hence higher photon flux, than for small amounts. Hence small amounts of radioactivity are under-represented for both fluorography and intensifying screens, even when exposed at −70 °C (*Figure 2*). Pre-exposing film to an instantaneous flash of light overcomes this problem because it provides many of the grains of the film with a stable pair of silver atoms. Thereafter each photon which arrives has an equal chance of contributing to the growth of the latent image. Consequently, correctly pre-flashed film responds linearly to the amount of radioactivity from intensifying screens and fluorographs. For the same reason pre-exposure largely (but not completely) bypasses the need to expose at −70 °C.

There is confusion in the literature over the need to pre-flash in order to obtain maximum sensitivity from intensifying screens. This has arisen because the effect of pre-exposure is negligible when tested using large amounts of radioactivity in short exposures. Only *small* amounts of radioactivity are under-represented when radiation is converted to light and there is no effect for large amounts. The practical consequence of this confusion is that serious errors arise when faint bands are compared with dark bands in long exposures of 1 week or more using intensifying screens with untreated film. Under these conditions errors of eightfold have been observed, using ^{32}P and intensifying screens for peak comparisons on unflashed film. Hence the purpose of this final section is to stress the importance of understanding the underlying principles to ensure success with the practical approach.

Acknowledgements

I am grateful to Amersham International for permission to reproduce illustrations from Booklet 23 Radioisotope Detection by fluorography and X-Ray Intensifying Screens (Laskey 1984, see ref. 15).

References

1. Laskey, R. A. and Mills, A. D. (1977). *FEBS Lett.* **82,** 314.
2. Randerath, K. (1970). *Anal. Biochem.* **34,** 188.
3. Bonner, W. M. and Laskey, R. A. (1974). *Eur. J. Biochem.* **46,** 83.
4. Laskey, R. A. and Mills, A. D. (1975). *Eur. J. Biochem.* **56,** 335.
5. Swanstrom, R. and Shank, P. R. (1978). *Anal. Biochem.* **86,** 184.
6. Laskey, R. A. (1980). *Methods Enzymol.* **65,** 363.
7. Luthi, U. and Waser, P. G. (1965). *Nature, Lond.* **205,** 1190.
8. Koren, J. F., Melo, T. B., and Prydz, S. (1970). *J. Chromatog.* **46,** 129.
9. Chamberlain, J. P. (1979). *Anal. Biochem.* **98,** 132.
10. Skinner, M. K. and Griswold, M. D. (1983). *Biochem. J.* **209,** 281.
11. Shine, J., Dalgarno, L., and Hunt, J. A. (1974). *Anal. Biochem.* **59,** 360.
12. Bonner, W. M. and Stedman, J. D. (1978). *Anal. Biochem.* **89,** 247.
13. Southern, E. M. (1975). *J. Mol. Biol.* **98,** 503.
14. Blattler, D. P., Garner, F., Van Slyke, K., and Bradley, A. (1972). *J. Chromatog.* **64,** 147.
15. Laskey, R. A. (1984). Radioisotope detection by fluorography and intensifying screens. *Review Booklet 23.* Amersham International.

In vivo labelling

A. Radiolabelling in animals

ROBIN GRIFFITHS

1. Introduction

Following their development, radioisotope tracer techniques have become widely used in many applications; particularly in the tracing of dynamic processes such as flow, transport or chemical reactions (1–4). With the recent development of external detection capabilities such as positron emission computed tomography (5), tracer techniques became an integral part of the new methods for measuring many physiological or biochemical parameters in man (6, 7).

The tracer technique requires the introduction of an appropriate (in chemical and/or physical characteristics, depending on the application) tracer, in small enough quantity that the process to be measured is not perturbed. Usually the dynamic process to be traced is assumed to be at steady state. Inference about the system or process is then made based on tracer kinetic measurements.

2. The administration of radiolabelled compounds *in vivo*

Whether an introduced radiolabelled compound is designed to enter the metabolic system of an animal or to interfere with it, there are a number of factors which mediate the success of the exercise. Essentially, these are: how effectively the compound is presented to the largest tissue, how efficiently the compound is incorporated into the biochemical matrix, and how the matrix is affected by its presence, and also what other factors are present which indirectly affect these processes.

2.1 Routes of administration of labelled precursors

A labelled molecule, selected as a tracer, may be administered in various ways, depending on the type of animal that is used in the experiment and depending on the physical state of the tracer molecule, whether gas, liquid, or solid.

2.1.1 Culture or nutrient medium

Any living organism may ingest a tracer compound, which in gas form may also be a component of the atmospheric environment. Bacteria, single cells, or small multicellular organisms can be labelled by exposure to a nutrient medium containing the tracer. An example of the technique as used in drug metabolism studies using hepatocytes is shown in *Figure 1*. In general, these techniques involve the use of only small quantities of radiolabelled compounds (typically $\sim 10\,\mu$Ci, 370 kBq) as exposure of the organism is continuous and sampling is direct.

Labelling of parasites can also be achieved directly, as in the case of cestodes living in the bile duct, which are incubated with radiolabelled material by injecting tracer into the ligated bile duct (8).

2.1.2 Oral or intragastric administration

In more highly developed animals, ingestion or oral administration by gavage of a tracer may be desirable for particular studies on the absorption mechanism in the gastrointestinal (GI) tract. If the tracer is to be distributed throughout the animal, this route of administration is unreliable. Besides metabolic alterations of the labelled molecule in the gut, the gut wall, or in the liver at the 'first pass' from the hepatic portal system, the absorption efficiency in the gastrointestinal tract is difficult to predict accurately. The method is not recommended except in exceptional circumstances where the absorption of the tracer is guaranteed; for example, tritiated water after oral administration is completely absorbed from the GI tract within about 40 min. (9). This is administered directly by gavage into the stomach of the animal on test.

2.1.3 Subcutaneous or intramuscular injection

In order to become efficiently distributed in an animal the labelled compound must quantitatively enter the blood circulation. This is achieved most easily by injection of the tracer solution into a suitable vein, artery, or a cardiac ventricle. Subcutaneous or intramuscular injections are sometimes used as substitutes for these routes on the assumption that they also represent a 100% bioavailability of the tracer. In these cases make up the tracer in a small volume ($\sim$1 ml) of an isotonic matrix (usually isotonic saline) and inject directly into a well-perfused area. In most animals the skin at the back of the neck or the hind thigh muscle is convenient. Remember that the presentation of tracer to the systemic circulation in such formulations will be first-order (at best), and may be considerably delayed. Because of the delay in absorption and the potential localized metabolism of the tracer molecule, intravenous injection is preferred.

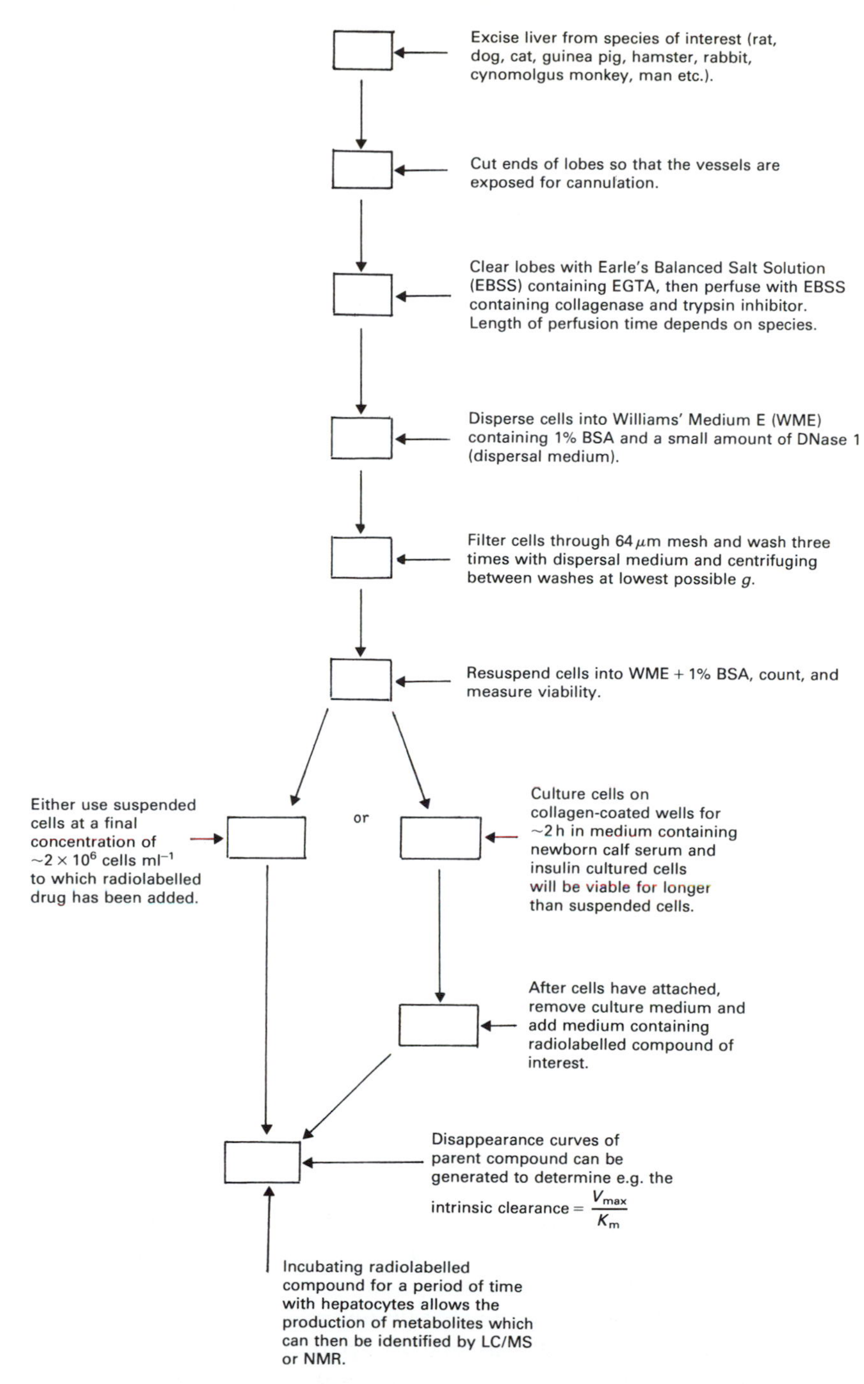

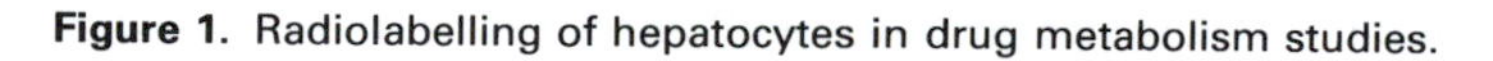
Figure 1. Radiolabelling of hepatocytes in drug metabolism studies.

2.1.4 Intravenous injection

The importance of prompt distribution of a tracer in the vascular system is seen in the case of injected ^{3}H-thymidine which is removed from the circulating blood within minutes to be incorporated by proliferating cells or to be catabolized (10). When mice were killed 15 sec after intravenous injection of ^{3}H-thymidine, labelled cells could already be demonstrated in the bone marrow (11).

A similar rapid clearance from the circulation after intravenous injection was observed for labelled amino acids, only 3% of which remained in the blood 10 min after injection (12).

2.1.5 Intraperitoneal injection

Intraperitoneal injection requires the diffusion of the labelled precursor through the peritoneum into the circulation. This route of administration has been found particularly useful in small animals, and it may at times be as adequate as intravascular injection. However, an intraperitoneal solution bathes all the organs of the abdomen, and is therefore as likely to enter the liver, the spleen, or even the gut as the systemic circulation. It cannot therefore be assumed that such administration will result in rapid or complete availability.

2.1.6 Injection into the cerebrospinal fluid

Administration of labelled molecules into the cerebrospinal fluid has also been used. Altman and Chorover (14) have shown that this mode of tracer administration is rather poor at supplying the brain with labelled nucleosides. Labelling was observed close to the wall of the cerebral ventricle only, in a region not more than about 3.6 mm deep. We have also found in our laboratory that intraventricular ^{14}C-imipramine (which actually concentrates within the brain to produce a brain/blood ratio of about 5 at a steady state) enters the brain by leaving the cerebrospinal system with the bulk flow of the CSF via the arachnoid villi, and approaching the brain with the systemic circulation. It would seem that the most effective way of presenting radiolabelled compounds to the brain is with the circulating blood, so long as the blood–brain barrier may be passed (15).

2.2 Distribution of labelled compounds *in vivo*

The distribution of a labelled compound in an animal is seldom truly uniform and depends, in addition to the mode of administration, on the circulatory system transporting the compound equally well to all parts of the body, and on the efficiency of diffusion within the tissue of interest. In their early studies with the autoradiographic technique, Lacassagne and Lattes (16) noted the lack of uniform distribution of injected polonium in tumour tissue. There was more tracer observed in the vascularized part of the tumour than in the centre, where there were also signs of necrosis.

Single cells within a given tissue become labelled according to their position relative to the vascular bed. Some cells are close to capillary and hence have a first chance to incorporate a labelled molecule that was transported via the circulating blood. Other cells are more remote and can incorporate only those tracers which were not utilized by the first cells. However, when the size of intracellular pools is adequate, labelled precursors may redistribute from cell to cell and thus balance unequal blood supply and provide a more uniform distribution. A non-uniform tracer distribution is particularly obvious in organs with dense cellularity and relatively poor vascularization.

The diffusion of a tracer in a dense cellular mass is also influenced by the metabolic activity and specific uptake mechanisms of the cells. For example, in the brain, radiolabelled adenine diffuses to a lesser extent than thymidine, because of the greater demand of the cells for adenine. Where even distribution of radiolabel is critical, it is often necessary to adopt a mode of administration which is continuous; for example, constant rate intravenous infusion, such that a steady state condition may be established. Where the perfusion time required to reach steady state is long, specialised techniques need to be adopted to deal with conscious ambulatory animals.

A method allowing the continuous introduction of tracer to any vessel in the conscious rat is shown in *Figure 2*. (A list of equipment and suppliers is given in *Table 1*.) In the case illustrated the dose is directed into the lower vena cava, and simultaneous blood samples can be taken from the external jugular vein. These procedures can be easily carried out with the minimum of stress to the animal.

Protocol 1. Cannulation of the femoral vein

1. Carry out the surgery under halothane anaesthesia. Although aseptic technique is not strictly necessary it is advisable to sterilize the cannulae in advance and to minimize the risk of infection by soaking all the surgical instruments in an antiseptic. The cannula used for insertion into the femoral vein is made of polythene and should be bevelled slightly at the tip. It is jointed at the distal end to large bore polythene tubing to which a syringe needle is attached and the cannula filled with sterile saline (0.9% w/v solution).
2. Make an incision in the rat's inner thigh approximately 3 cm long in order to expose the femoral vein.
3. Ligate the vein and make a small hole in it so that the cannula can be pushed in a distance of approximately two inches when the tip should be positioned in the vena cava.
4. Tie the cannula in and check its patency by drawing blood back into it with a syringe.
5. Externalize the cannula at the distal end at the nape of the neck and close the wound with sutures.

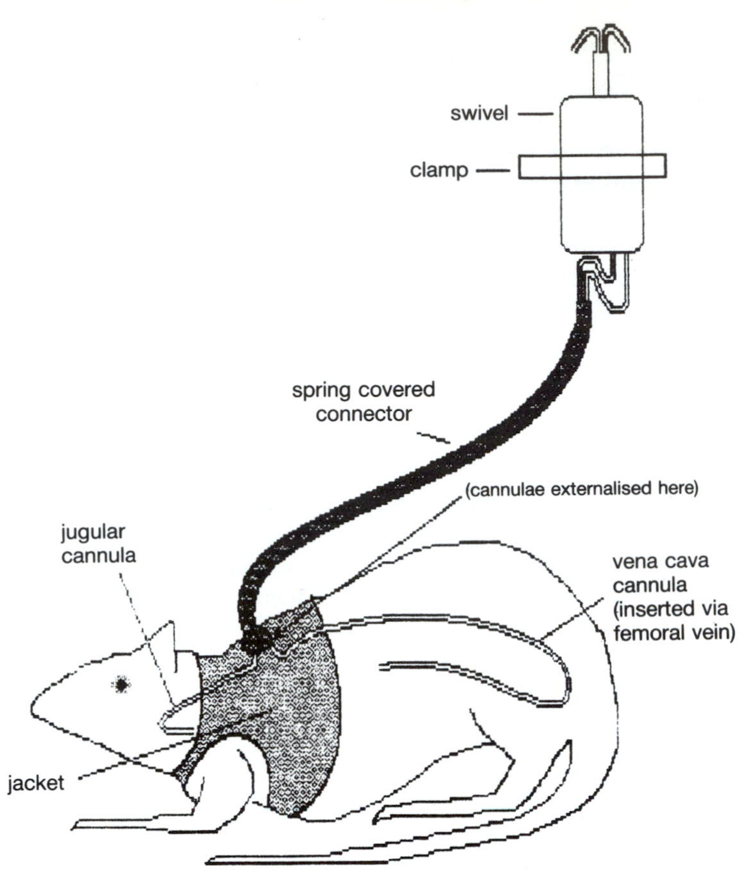

Figure 2. Tether allowing the continuous i.v. infusion of tracer to the conscious ambulatory rat.

The cannula used for insertion into the jugular vein is constructed in two parts.

Protocol 2. Infusion of tracer to the conscious ambulatory rat

1. Take a length of silastic tubing approximately 2.5 cm long which is positioned in the vein.
2. Joint to polythene tubing at the point where the cannula emerges from the vein so that the cannula can be securely tied.
3. Externalize the distal end of the tubing at the nape of the neck.
4. Pass the externalized cannulae up a metal spring covered connector in order to protect them from the rat. The flexible connector is held by a rat jacket which is made from elastic material.

5. Attach the connector to a swivel which is anchored outside the rat's cage, so that the animal is free to move throughout the cage, effectively unrestrained.
6. Immediately following surgery administer antibiotic (e.g. Penbritin, 10 mg/0.1 ml, i.m.) to combat any infection following the surgery.

Table 1. Equipment specification and suppliers

Equipment	Size	Supplier
Rat and dog jackets	Designed for individual animals	Alice King Chatham Medical Arts, 5043 Onaknoll Avenue, Los Angeles, CA 90043, USA
Teachers and swivels		Alice King Chatham Medical Arts, 5043 Onaknoll Avenue, Los Angeles, CA 90043, USA
Polythene cannula	i.d. 0.28 mm o.d. 0.61 mm	Portex Ltd, Hythe, Kent, UK
Silastic cannula	i.d. 0.020 inch o.d. 0.037 inch	Dow Corning Corp., Medical Products, Midland, Michigan, USA
Vascular access ports	Skin parallel Model LC1	Norfolk Medical, 7307 N Ridgeway, Skokie, Illinois 60076, USA
Ambulatory infusion pump	Model MS 26	Graseby Medical Ltd, Colonial Way, Watford, Herts, UK

When not in use fill the cannula with sterile 50% dextrose and heparin solution (100 units/ml) and seal. The cannula should be flushed every 2–3 days with the sterile solution.

In the dog, a similarly protracted infusion of tracer can be given using a silicone rubber cannula inserted permanently into the jugular vein at the neck, the other end being attached to a skin parallel access port located subcutaneously between the shoulder blades (see *Figure 3*). This consists of a

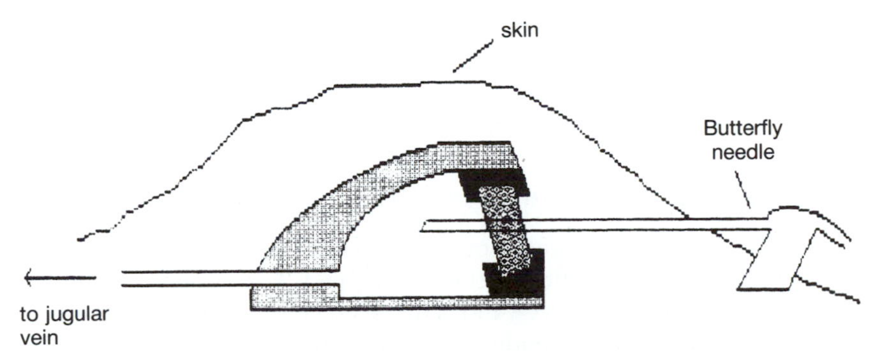

Figure 3. Skin-parallel vascular access port, for use in dogs.

rigid reservoir body fitted in a silicone rubber base and an inlet septum in the centre of the reservoir body. Fixed to the reservoir body is a self-sealing rubber septum designed to accept multiple punctures while maintaining its leak-tight integrity. Embedded in the flexibile silicone rubber base is a mesh which can be used as a suture anchor to secure the port to the subcutaneous tissue.

Implant the vascular access port under halothane and nitrous oxide anaesthesia, under aseptic conditions, with the rubber septum facing towards the tail. Insert the tubing leading from the access port into the external jugular vein so that the tip lies close to the heart. The tubing can be positioned by injection of a radio-opaque solution (e.g. omnipaque). Once installed, the vascular access port must be filled with 50% dextrose and heparin solution (500 units/ml) while not in use. The dog should be allowed to recover for a minimum of 2 weeks prior to use.

During use, care should be taken not to introduce infection into the port. The port should be flushed weekly.

Carry out constant rate infusions of tracer by inserting a butterfly needle through the rubber septum into the reservoir of the port and connecting the butterfly needle to an ambulatory infusion pump installed in the pocket of a jacket worn over the forequarters of the dog. Care should be taken to ensure that the jacket covers the site of the vascular access port so that the butterfly is protected.

A tracer enters the cell through the cellular membrane, which discriminates against large or charged molecules. Large molecules may be taken up by the cell by pinocytosis, a process by which the large molecule or small particle is engulfed by cytoplasmic protrusions covered by the membrane. There may be specific sites on the cell wall for which the compound must have some affinity before pinocytosis will take place.

Small, uncharged molecules such as nucleosides or amino acids pass the cell membrane readily by diffusion or facilitated transport, but charged nucleotides (phosphorylated nucleosides) do not pass freely. Hence, the concentration of nucleosides and free amino acids in the extracellular space tends to equilibrate through the cell membrane with the soluble fraction in the cell. As mentioned above, the entrance of thymidine into the cell and its subsequent phosphorylation and incorporation into DNA may occur within seconds.

3. Precursors and species differences

The suitability of a labelled precursor as a tracer in living cells depends on the enzymes with which the particular cell is equipped. For example, thymine is one of the four bases in the DNA molecule in all cells. However, not all cells accept its deoxyriboside, thymidine, as a precursor from the extracellular environment. It is known that various rodents such as the woodchuck and the ground squirrel do not utilize injected ^{3}H-thymidine at all, whereas other rodents such as the red squirrel utilize ^{3}H-thymidine for DNA synthesis in the cells in the ileum but not in the tongue (17).

4. The acid-soluble 'pool'

A useful concept in considering the uptake of labelled precursors into macromolecules is that of the 'soluble pool'; that is, the acid-soluble small molecular fraction in a cell. The various small molecular components are present in this fraction in varying amounts, which may be constant relative to each other. The constituents of the pool are continuously interchanged, catabolized or anabolized to serve as building blocks for larger molecules, and the majority of the energy needed for these reactions is also delivered by the 'soluble pool'. Some constituents of the pool are renewed rapidly; others turn over more slowly.

Synthesis and breakdown of molecules, interchanges, and energy-providing reactions are all balanced by the interplay of enzymes and their control mechanisms. Whilst, in principle, the types of enzymatic reactions are the same or very similar in the various cell types, the quantity of enzymes and their substrates varies greatly among cells. In general, the homeostasis of the soluble pool is under tight control by a variety of feed-back mechanisms expressed by substrate and product inhibition and induction of enzyme activity. Any change in the environment of a cell quickly alters the steady state of reactants in the soluble pool. For this reason cells which are taken out of their physiological environment are liable to react differently to a labelled compound than those *in vivo*. Similarly, if the concentration of labelled compound is sufficient to perturb the homeostatic balance of concentrations or the pool size of endogenous substances, then the compound is not a true tracer and the conclusions drawn from such experiments must be circumspect.

4.1 The size of the acid-soluble 'pool'

The usefulness of a particular precursor in labelling a specified macromolecule can be greatly affected by the size of the soluble pool for the precursor. The size of the pool is a function of the rates of production and utlization of the precursor (38). Where the pool size is small, it can generally be assumed that the incorporation of a radiolabelled precursor will be rapid. Thymidine, for example, can be traced more easily than other nucleosides, since in most higher organisms it is incorporated only into DNA and the pool of thymidine is very small in most cells [about 2 μg per gram of spleen, (18)]. Other nucleosides have a less specific precursor relationship, in addition their easy interconversion and multiple functions within the cell cause the renewal time within the soluble fraction to be much slower, so that the pool size is larger than that for thymidine (19).

The size of the thymidine pool varies with different cell systems, as was demonstrated for cells of the intestine, spleen, and thymus in the rat (18). The pools in spleen cells and intestinal cells could be saturated more easily with thymidine than the pool in thymus cells. In addition, the size of the pool of small molecular nucleotides may be expected to vary with the stage of the cell cycle between two mitoses.

Injury and disease greatly affect metabolic equilibrium. Thus, examining the effects of radiation on DNA synthesis one must take into account changes in pool size, and tracer data must describe not only the specific activity of the DNA but also that of the acid soluble pool. Indeed, nucleoside pool sizes have been shown to increase dramatically as a result of irradiation injury and infection (20).

5. Tracer kinetic measurement techniques

In analysing data from any tracer kinetic study it is necessary to assume a model of the system under test. The analysis is only as good as the validity of the assumptions. It is often the case that the amount of tracer in blood or urine is the only measurement available, particularly in studies in man. The data analysis for this kind of measurement has to make the largest assumptions; for example, to treat the blood as a single compartment communicating with other body organs which are also regarded as single compartments (21). Regarding an organ or a group of organs as a uniform pool is a gross approximation which may not be valid. More sophisticated experiments using external detection techniques (for example, scintillation probes) have measured tracers in organs and parts of organs (22, 23, 24). However, the non-uniformity of the detection efficiency in depth, and the overlap of many tissues, mean that the measurements are often difficult to quantify. Quantitative autoradiography—or indeed total organ dissection—makes it possible to measure the concentrations of isotopes in tissues of animals labelled *in vivo*. These studies are necessarily of the 'one animal per point' type, and therefore involve statistical considerations in describing the dynamics of tracer behaviour. Recently, with the development of emission tomography, and particularly positron emission computed tomography (positron CT), the amount of radioactivity in local regions of the body can be quantitatively and non-invasively measured. This allows the description of tracer dynamics within an organ or even within a tissue within an organ. (See, for example, ref. 37.)

5.1 Principles of radioisotopic methods for assay of biochemical processes *in vivo*

The rate of a chemical reaction *in vivo* can be measured by determining the rate of disappearance of one or more of the reactants, or the formation of one or more of the products. To derive the rate of the total reaction from measurement of the radiolabelled (tracer) species, it is necessary to know the integrated specific activity (i.e. the ratio of labelled to total molecules) of the precursor pool. Occasionally, the labelled species exhibits a kinetic difference from the natural compound (isotope effect) and an appropriate correction has to be made for this. Direct measurement of the specific activity of the precursor pool is generally impossible, and so this must be done indirectly from measurements in the blood supplied to the tissue. Measure the specific activity in the arterial blood and calculate the precursor specific activity from it by correcting for the lag in the equilibration of the precursor pool in the tissue with that in the plasma (it is necessary to know the kinetics of the equilibration process between the precursor pools in the tissue and plasma).

Autoradiography and emission tomography, which measure only the total concentration of the radioactive molecules, cannot distinguish among the various chemical species which may be labelled, neither the precursor nor any of the metabolites. Strategies must therefore be developed that ensure that the radioactivity is contained exclusively in the precursor and/or in one or more of the products specific to be assayed.

5.2 Principles of radioisotopic methods for measuring peripheral circulation

In studying blood flows by tracer methods, the system is assumed to be at steady state, at least over the period necessary to make the measurements. The system is also usually assumed to be time-invariant (two identical tracer applications, separated in time, give the same response), and linear (double the dose gives double the response). These two properties of the system are necessary for using either compartmental analyses or the so-called 'model independent' methods which use the convolution principle (see below).

To trace blood flow, the tracer must mix with the blood and must enter the system in exactly the same way as the substance to be traced; that is, tracer concentration at any time is the same everywhere in a cross-section of the inlet. No fraction of the tracer must be permanently retained by the system.

6. Basic concepts in dynamic tracer data analysis

6.1 Mean transit time (or mean time, or turnover time)

This is the mean time required for a particle to traverse the distance between two locations in a system or to be transported between two phases (28). Many dynamic tracer studies consist essentially of the determination of a mean transit time ($\bar{t}$). The basic equation is as follows:

$$\bar{t} = \frac{V}{F} \qquad (1)$$

where F is the flow rate of the tracer through the system and V is the volume of distribution of the tracer between the points of injection and sampling at the input and output respectively. Provided the amount of tracer injected is known, the volume **V** can be determined by the appropriate dilution formula, and if t is also known the flow rate can be derived.

In many cases it is useful to note that if two or more systems are in series, the mean transit times are additive; that is,

$$\bar{t}_{(+2)} + \bar{t}_1 + \bar{t}_2. \tag{2}$$

6.2 Extraction, clearance, and capillary diffusion capacity

The extraction of the tracer, E, is defined as the fraction of the inflowing amount, that is taken (removed) from the blood in a single passage through the system (i.e. uptake divided by delivery)

$$E = \frac{F(C_{\text{in}} - C_{\text{out}})}{F \cdot C_{\text{in}}} \tag{3}$$

where F is flow rate as before, and C is the concentration of tracer.

$$E = \frac{C_{\text{in}} - C_{\text{out}}}{C_{\text{in}}}. \tag{4}$$

The proportion of tracer unextracted by the system is termed the transmitted fraction T, i.e. $T = 1 - E$.

Clearance, Cl, is defined as the flow of reference fluid carrying the amount of tracer taken up per unit time, i.e.

$$Cl = \frac{F(C_{\text{in}} - C_{\text{out}})}{C_{\text{ref}}} \tag{5}$$

where C_{ref} is the concentration in the reference. If C_{in} is used as the reference fluid, then eqn (4) and (5) can be combined to

$$Cl = E \cdot F. \tag{6}$$

Note that Cl only has a physical meaning in special circumstances, for example, when the substance is completely extracted from the reference fluid ($E = 1$). An example is glomerular inulin clearance. Since no exchange of inulin occurs across the tubular membrane the urinary excretion of inulin equals the amount ultrafiltered from arterial plasma water. Thus, using plasma water as reference fluid inulin clearance equals the glomerular water filtration rate, GFR.

In the case of a substance cleared from the blood by diffusion across a capillary membrane offering the main barrier to exchange where the outside concentration remains essentially zero, the concentration of trace decreases mono-exponentially along the capillary as shown by Crone (29) and Renkin (30)

$$C_{out} = C_{in}.e^{-PS/F} \tag{7}$$

where PS is the permeability–surface area product defined as the membrane's tracer flux per unit concentration gradient.

Solving eqn (7) for PS gives

$$PS = \frac{F(C_{in} - C_{out})}{(C_{in} - C_{out})/(\ln C_{in} - \ln C_{out})}. \tag{8}$$

The denominator defines the so-called logarithmic average concentration, C^*.

Using C^* as concentration in the reference fluid:

$$PS = \frac{F(C_{in} - C_{out})}{C^*} = Cl, \tag{9}$$

so that the clearance calculated using C^* equals the capillary diffusion capacity, PS, which may be defined as the virtual volume of fluid totally cleared at the capillary membrane.

It should be noted that there are two tacit assumptions in using these parameters to monitor uptake of tracer; namely, uptake must be linear (first-order) and there must be no back diffusion.

6.3 Non-compartmental analysis ('model independent' analysis)

In this type of analysis, the system is regarded as a 'black box', with a single outflow and a single inflow. Tracer enters at the input in zero time as a spike or delta function, (i.e. an ideally brief bolus) having mixed with the tracee. A frequency distribution of transit times $h(t)$ is observed at the output. This is the fractional rate at which the tracer leaves the system at time t, measured as the concentration at time t divided by the total area under the concentration–time curve, and having the dimensions of reciprocal time. If measurements are continued until all the tracer is recovered at the output, the mean transit time is given by

$$\bar{t} = \int_0^\infty t \cdot h(t) \cdot dt. \tag{10}$$

If Q is the quantity of tracer injected and $C(t)$ the concentration at the output, the flow rate F is given by

$$F = \frac{Q}{\int_0^\infty C(t) \cdot dt}. \tag{11}$$

This is the Stewart–Hamilton formula (31). If the cumulative fraction of the tracer recovered at the output is measured as a function of time, one obtains the function $H(t)$, which is the integral from $t=0$ to $t=t$ of the function $h(t)$. The fraction of the tracer still remaining in the system at time t is therefore equal to $1 = H(t)$. If measurements are continued until all the tracer is washed out of the system, the mean transit time is given by

$$\bar{t} = \int_0^\infty (1 - H(t)) \cdot \mathrm{d}t; \tag{12}$$

that is, it is the area under the fractional retention curve.

The equations above assume a spike or bolus injection input. In fact, any continuous function can be used as input. In these cases the convolution principle is used to consider the input as series of ideally brief boluses. The function observed at the output $0(t)$ is then the convolution integral of the frequency distribution of the system under study $h\ (t)$ and the input function $i(t)$;

$$0(t) = i(t) * h(t) \tag{13}$$

where * denotes the operation of convolution.

The principle of convolution is basic to non-compartmental tracer analysis (see refs. 2, 33 for further reading).

Writing eqn (13) as a full convolution integral:

$$0(t) = \int_0^t h(t - \tau) \cdot i(\tau) \cdot \mathrm{d}\tau. \tag{14}$$

If any two of the functions $0(t)$, $i(t)$, $h(t)$ can be measured, the third can be derived by mathematical convolution or deconvolution. This can be done on the observed experimental data without having to assume any particular mathematical form for any of the functions.

The particular feature of this type of analysis is that it involves the minimum assumptions possible in analysing data from tracer experiments. The accuracy of the measurement depends almost exclusively on the number and accuracy of the experimental data. However, eqns (10), (11), and (12) require that the integral from zero to infinite time is measurable, that is the complete passage of the tracer through the system must be recorded. This may not always be possible because of the difficulties of extrapolating from the last datum point measured to infinity.

6.4 Compartmental analysis

By definition, a compartment is a system or part of a system in which the ratio of tracer to tracee is constant at any given time. Mixing is instantaneous within a compartment, and each compartment is kinetically distinct (described by numerically different parameters) from all others in the system. The basic equation for a single compartment is

$$\frac{\mathrm{d}Q}{\mathrm{d}t} = kQ(t) \tag{15}$$

where Q is the quantity of tracer, as before; that is, the rate of outflow of tracer at any time t is proportional to the amount of tracer remaining.

Integrating eqn (15),

$$C(t) = C(0)e^{-kt} \quad (16)$$

where k is a first-order rate constant, which is inversely related to the half period and the mean transit time

$$k = \frac{0.693}{T_{1/2}} = \frac{1}{\bar{t}} = \frac{F}{\bar{V}}. \quad (17)$$

Knowing the amount of tracer injected, the volume of distribution can be calculated from the value of $C(0)$ at zero time.

7. Practical applications exemplifying the principles of tracer kinetic measurement

7.1 Blood-flow measurements using the indicator dilution technique

There are many permutations possible on this theme, and there is insufficient space here to give more than an indication of some of the basics involved in the simplest kind of study. However, there are several papers and monographs available which cover these procedures in detail (13, 29, 38).

7.1.1 The constant infusion method (Stewart principle)

If blood flows in a vein at the constant rate F ml/min and tracer is infused upstream at the constant influx rate of j_{in} mg/min [concentration C_{st} mg/ml, inflow rate (slow) F_{st} ml/min], then

$$j_{in} = F_{st} \cdot C_{st} \text{ mg/min.} \quad (18)$$

Downstream, blood samples are taken until the tracer concentration has reached a constant maximum, steady state, value $C_{out}(\infty)$. The steady-state outflux, $j_{out}(\infty)$ equals $(F + F_{st})C_{out}(\infty)$. As outflux equals influx

$$F_{st} \cdot C_{st} = (F + F_{st})C_{out}(\infty) \text{ ml/min} \quad (19)$$

Assuming F_{st} is very small relative to the system's flow rate F, then $(F + F_{st}) = F$, then

$$F = \frac{F_{st} \cdot C_{st}}{C_{out}(\infty)} \text{ ml/min.} \quad (20)$$

In other words, for sufficiently small tracer-solution infusion rates F_{st}, the flow is as many times greater than the tracer solution infusion flow, as the *dilution factor* $C_{st}/C_{out}(\infty)$. This classical tracer dilution method was first applied by Stewart in 1897 (32) for blood-flow measurement, using NaCl as indicator and conductivity as a measure of concentration.

If the tracer concentrations have not reached steady state it is almost always necessary to correct for recirculation; that is, radiolabel which has made a complete circuit and adds to the flux of indicator passing the outlet. The blood concentration will therefore be continuously rising. A method of correcting for recirculation exists if bilaterally symmetrical organs such as kidneys or forearms are studied. Then sampling also from the outlet of the non-injected side gives the concentration to be subtracted from that of the injected side. This difference's constant maximal value as measured after the initial transient should be used in eqn (20).

Other means of correcting for recirculation exist, notably by using the convolution theorem and knowing the recirculation at the inflow (usually the arterial) site. However, the effect of recirculation can be minimized by (a) infusing the tracer as close as possible upstream to the mixing site and sampling close downstream of the site; (b) using an indicator that recirculates minimally or not at all (heat/cold is the only indicator for which recirculation can be ignored; inert gases show little recirculation, as they leave the circulation at the tissues and at the lungs).

7.1.2 Bolus injection method

If blood-flow is F and an amount, Q_0, of a tracer is injected into the system giving rise to a tracer concentration in the blood at time t of $C_{\mathrm{out}}(t)$, then assuming that the tracer does not change F and the volume injected is insignificant compared to the flow rate, the rate of tracer leaving the system is

$$\frac{\mathrm{d}Q}{\mathrm{d}t} = F \cdot C_{\mathrm{out}}(t). \tag{21}$$

When all tracer has left the system and provided no recirculation occurs, then the total amount entering, Q_0, equals the sum of all amounts leaving

$$Q_0 = F\int_0^{\infty} C_{\mathrm{out}}(t)\mathrm{d}t, \tag{22}$$

or

$$F = \frac{Q_0}{\int_0^{\infty} C_{\mathrm{out}}(t)\mathrm{d}t} = \frac{\int_0^{\infty} j_{\mathrm{in}}(t)\mathrm{d}t}{\int_0^{\infty} C_{\mathrm{out}}(t)\mathrm{d}t} \tag{23}$$

where the sign (∞) denotes any time after T, the longest transit time, i.e. the time at which the tracer has left the system completely. The area in the denominator of eqn (23) can be obtained by a single cumulative sampling from time zero to time T, provided that the entire amount of injected tracer has passed the sampling site before time T. If so, then $\int_0^{\infty} C_{\mathrm{out}}(t)\mathrm{d}t = \int_0^{T} C_{\mathrm{out}}(t)\,\mathrm{d}t = C_{\mathrm{out}} \cdot T$, where C_{out} is the average concentration in the cumulative sample.

It can be seen in eqn (23) that $F \cdot T$, the cumulative volume flowing out from the start of sampling to time T, is as many times greater than the injected standard volume, V_{st}, as the measured dilution factor C_{st}/C_{out}.

The indicator dilution methods described above have been widely used. In particular, the bolus injection method has been used for determination of cardiac output injection at the right side of the heart and following the concentration of the tracer as a function of time in a systemic artery. The main difficulty with this method is correcting for recirculation. Often mono-exponential extrapolation to time infinity is used, which underestimates the area under the outflow curve and hence cardiac output correspondingly is overestimated. Correction for recirculation may be improved by sampling of the recirculating blood from the inlet as this makes it possible to calculate the first part of the combination of recirculation by convolution (33).

7.2 Blood-flow measurements and the bolus fractionation method, using microspheres

If an amount, Q_0, of an indicator is injected as a bolus and assuming complete mixing between the indicator and blood at the inlet, then the fraction of the dose $^{m}\text{tissue}/F$ of the labelled blood flow F_0 that goes to the tissue in question is

$$\frac{^{F}\text{tissue}}{F_0} = \frac{^{m}\text{tissue}}{Q_0}. \tag{24}$$

The indicator must be completely retained by the tissue. The best degree of tissue retention is obtained by using radioactively labelled inert particles of uniformly small size: microspheres. These are injected into the vena cava or the left auricle to label the entire cardiac output [(F_0 in eqn (24)] uniformly.

Rearrangement of eqn (24) gives

$$\frac{^{m}\text{tissue}}{^{F}\text{tissue}} = \frac{^{m}\text{blood sample}}{^{F}\text{blood sample}} = \frac{Q_0}{F_0}. \tag{25}$$

Thus,

$$^{F}\text{tissue} = {^{m}\text{tissue}}.\ \frac{^{F}\text{blood sample}}{^{m}\text{blood sample}} \tag{26}$$

^{m}tissue is the amount of tracer retained in the tissue and ^{m}blood sample is the amount in the blood sample collected at the rate of ^{F}blood sample ml/min.

Recently, a tracer (isopropyl-iodo-amphetamine, labelled with ^{125}I with high initial extraction and prolonged tissue retention, has been described by Winchell (34). This can be used in the same way as microspheres, and can be injected intravenously. The prolonged retention allows time for external detection, for example by emission computed tomography.

Highly diffusible tracers with a short physical half-time compared to the shortest transit times in the system studied (e.g. krypton-81 with $t½ = 13$ sec) constitutes another related approach. With such tracers the steady-state distribution of the indicator in the organ studied during constant infusion, equals (effectively) that after a bolus injection of microspheres. This approach has been used for determination of blood flow distribution in lungs and brain (35).

7.3 Measurement of energy metabolism in the CNS—the deoxyglucose method (36)

This method has been used extensively in animals with autoradiography in order to measure the rates of glucose utilization simultaneously in all areas of the CNS in animals. It has been adapted to man by the use of [^{18}F] fluorodeoxyglucose and positron emission tomography. The method is based on the measurement of product accumulation and exemplifies many of the principles involved in the use of tracer techniques to monitor metabolic processes *in vivo*.

Glucose and 2-deoxyglucose are competitive substrates for both blood–brain transport and hexokinase-catalysed phosphorylation. Unlike glucose-6-phosphate, however, deoxyglucose-6-phosphate is not further metabolized and remains trapped in the cerebral tissues for at least 45 min. The quantity of [^{14}C]DG-6-P accumulated in the brain at any time is equal to the rate of [^{14}C]DG phosphorylation by hexokinase in that tissue in that time. This integral can be related to the amount of glucose phosphorylated, knowing the time courses of the relative concentrations of [^{14}C]DG and glucose in the precursor pools and the Michaelis–Menten kinetic constants for hexokinase with respect to both substrates. With cerebral glucose consumption at a steady state, the amount of glucose phosphorylated during the time stipulated equals the steady state flux of glucose through the hexokinase-catalysed step, times the duration of the interval. The net rate of flux through this step equals the rate of glucose utilization.

The assumptions of the method are:

- steady state glucose concentrations;
- homogeneous tissue compartment in direct kinetic exchange with the plasma;
- concentrations of free [^{14}C]DG essentially equal to zero.

It is necessary to obtain first-order rate constants for carrier mediated transport of [^{14}C]DG from plasma to tissue, and from tissue to plasma and also for phosphorylation by hexokinase, from a separate group of animals. The method essentially isolates the first step in the pathway of glucose metabolism. It measures therefore *in vivo* the net rate of glucose phosphorylation, and in a steady state, the net rate of the glycolytic pathway itself.

B. Radiolabelling of plants

STEPHEN A. BOFFEY

8. Introduction

The radiolabelling of plant material is carried out for so many reasons that it would be futile to attempt an exhaustive list of applications. The wide variety of these applications and of the plant tissue used also makes it impossible to present a few well-defined protocols that can be used without modification in any new situation. However, this section starts with an outline of the most common applications of *in vivo* labelling of plants; it then proceeds to a discussion of some of the points that should be considered when designing any such labelling experiment, and finally presents a few protocols which can be used as a starting point by the reader.

By supplying a radioactively labelled precursor, biochemical pathways may be investigated by monitoring the appearance of labelled products, which may be subsequently identified. The best known example of this application to plant processes is its use by Calvin to work out the cycle of reactions involved in CO_2 fixation which now bears his name (this was, admittedly, first carried out using algae). More frequently the technique is used to monitor the activity of a single enzyme in a plant by supplying labelled substrate and measuring the rate of appearance of label in the immediate product.

The locations within the plant of various processes can be determined by feeding appropriate labelled precursors. The products of the process being studied are isolated from different regions of the plant, and their degrees of labelling are compared. For example, by feeding labelled thymidine and isolating nuclear and organelle DNA from various parts of a plant, the regions in which nuclear or organelle DNA replication are most active can be determined.

For a systemic insecticide, fungicide, or herbicide to be effective it must enter a plant and be carried to its site of action. Such uptake and movement can readily be followed by use of radioactively labelled chemicals. The effects of additives, such as surfactants, or chemical modifications on uptake and transport can be assessed, and the products of breakdown can be detected. Such monitoring is essential to the agrochemical industry.

Our knowledge of factors influencing the transport of newly synthesized carbohydrate from its source, the leaves, to its sink, for example a storage organ, has been enhanced by the use of $^{14}CO_2$ applied to selected leaves of a plant, followed by monitoring of the levels of ^{14}C in other parts of the plant at regular intervals (39).

There is considerable interest in the processes that regulate the expression of plant genes. A particular gene may be expressed only in, for example, the leaf or developing seed, or it may be expressed at only one stage of development. A convenient way to obtain a 'snapshot' of the proteins being synthesized in a particular tissue at a particular time is to supply one or more labelled amino-acid to the plant for a short time. Analysis by gel electrophoresis and autoradiography of the proteins extracted from the plant tissue will reveal a pattern of bands corresponding to proteins which were being synthesized during the labelling period; pre-existing proteins will not be detected.

When combined with autoradiography of tissue sections, labelling of plants can be very informative. This method has, for example, been used to determine if all the chloroplasts in a cell replicate their DNA or if this activity is restricted to a subpopulation of chloroplasts (40). Pulse-labelling, with or without a non-radioactive 'chase', followed by autoradiography of sections from tissue harvested at intervals after the pulse has revealed the involvement of endoplasmic reticulum and Golgi apparatus in such processes as cell wall synthesis and storage protein packaging (41). Sometimes the plant is used as a convenient 'factory' for producing a labelled compound, particularly when careful choice of labelled precursor can ensure that the desired product is the only one labelled. In such cases, efficiency of uptake and high specific radioactivity of the precursor are of paramount importance, as is a knowledge of the most active site of synthesis within the plant.

9. Choice of method

Ideally, the method chosen for supplying label to plant cells should ensure that all cells have equal access to the label and that they continue to behave as they would in an intact, untreated plant. Unfortunately, these ideals are not attainable. Whilst cells isolated from plants have equal access to label in their medium, they do not, in many respects, behave as they would in the intact plant; on the other hand, the intact plant may present barriers to the passage of label into it, and will usually generate a very uneven distribution of label within itself. Depending on the application, these two extremes will be more or less suitable, and often a compromise must be used. Some of the methods available for labelling plant cells *in vivo* are summarized below.

9.1 Intact plants

Whole plants may be labelled through their roots, using hydroponic culture, through their leaves and stems by surface application of the label (particularly relevant to agrochemicals), or through leaves by use of $^{14}CO_2$. Compounds taken up by the roots will remain there unless carried to the remainder of the plant in the xylem as a result of transpiration, when they will tend to accumulate in the mature leaves, which are most actively respiring. Label applied to leaves can move to other parts only if it passes into the phloem, in which it will be carried to the most active sinks, such as developing fruits.

The final distribution of label will therefore be very uneven, and will depend on the method of application and the stage of development of the plant. This is not a problem when the experiment is carried out in order to investigate how a compound becomes distributed within the plant. However, if the aim is to compare the rates of a process in different regions of the plant by supplying a labelled substrate, the widely varying concentrations of the substrate within the plant will seriously distort results.

9.2 Seeds and seedlings

Seeds can readily be sterilized using relatively high concentrations of sodium hypochlorite (1–6% available chlorine for 10 min), and, after thorough washing in sterile distilled water, can be allowed to germinate under sterile conditions. Radiolabel can be supplied at any stage during germination. Alternatively, the seeds can be kept longer under sterile conditions to produce sterile seedlings for long, steady-state labelling experiments.

9.3 Plant organs

If the aim of an experiment is to examine a process in the leaf, then it may be sensible to label only leaves. This can be achieved by cutting a leaf from the whole plant (under water, to avoid breaking the transpiration stream) and then placing its cut end in a small volume of radioactive solution. In order to ensure rapid uptake of label, conditions should be chosen to favour rapid transpiration; namely, a flow of air over the leaf, adequate light and temperature.

This method overcomes the potential problem of a barrier to uptake in the roots, leaving only the membranes between xylem vessels and leaf cells to be crossed. It also reduces the quantities of radioactive material needed and produced as waste. However, the plant is subjected to the trauma of cutting, and the resulting wound responses may modify the behaviour of the plant cells. For this reason it is common to leave the cut leaf in non-radioactive solution for a few hours to allow a return to 'normal' metabolism prior to addition of label.

The distribution of label within the leaf is not completely uniform, being highest in the vicinity of veins. In the leaves of monocotyledonous plants there will be a tendency for label to accumulate near their tips, where the more mature, actively respiring cells are located. For many purposes such an uneven distribution of label *within one tissue type* does not matter.

A similar approach can be used in the labelling of roots, though there will obviously be no transpiration current to aid uptake of label. Roots are usually immersed in the minimum possible volume of radioactive solution, and diffusion or active pumping across the cell membranes is relied on for uptake of label. The same caution about the shock of cutting applies to roots as to leaves.

9.4 Leaf discs

A logical progression from using cut leaves is to use discs of leaf tissue, cut from the leaf lamina using a cork borer. These discs are then either floated on a small volume of radioactive solution or are placed on radioactive medium solidified with agar. Uptake of label is mainly by diffusion, and so there is a steep gradient of substrate concentration from the circumference of each disc to its centre. Veins and stomatal pores also provide relatively easy pathways for the passage of label. During an incubation lasting several hours, cells farther from the cut edge than about 1 mm may receive very little label, though there will be slightly elevated levels in the vicinity of the larger veins and stomatal openings.

Hence it may seem that, for the highest possible levels of incorporation, one should use very small leaf discs. However, this is only true up to a point, since cells near the edge will tend to be the most damaged, and may therefore be dead or abnormal in behaviour. Such cells make up a large proportion of very small discs. In practice, discs about 5 mm in diameter appear to be most satisfactory. Discs have the advantage that several can be cut from adjacent areas of one leaf, thus reducing the effects of biological variability, and, if sterile, they can be incubated with label for several days, allowing high levels of labelling to be achieved.

10. General considerations

10.1 Membrane transport

It is important to remember that, regardless of the labelling method used (with the exception of $^{14}CO_2$ labelling), failure to obtain incorporation of label may occur because of the inability of the labelled compound to cross a cell or organelle membrane. For example, labelling of DNA requires the use of nucleoside precursors rather than nucleotides, since the latter are not readily transported across cell membranes. It is therefore well worth assaying a sample of tissue for total radioactivity (both free and incorporated label) at the end of the labelling period to ensure that adequate label has entered it.

10.2 Sterility

Wherever possible, plant material should be sterile. In most cases it is not convenient to grow the plant under sterile conditions, but a reasonable degree of sterility can be achieved by a brief (~5 min) wash in hypochlorite solution (1% available chlorine, with a drop of detergent to aid wetting) followed by thorough rinsing in freshly distilled water. Total sterility is rarely necessary, except during prolonged incubations. It may be possible to include antibiotics in the radioactive medium, but this should only be done if the antibiotics are known not to affect the plant process being studied.

10.3 Choice of isotope

The isotope used will depend on its availability in the compound to be supplied as precursor, and on the method of analysis of the plant material. For highest final specific radioactivity, choose ^{32}P or ^{3}H; for best resolution in autoradiography choose ^{14}C or ^{35}S.

The quantity of label supplied, and the period of labelling must generally be determined by trial and error. Clearly, a long incubation with very high concentrations of label will give the highest specific radioactivity in products. However, long incubations can cause problems: the products of reaction may be unstable (an unstable, major product may eventually be less intensely labelled than a stable minor one), bacterial activity may become more significant, the tissue may change its properties. Considerations of finance and safety normally place an upper limit on the concentration of label used, though it would obviously be pointless to have concentrations greater than needed to saturate the system. It should be remembered that washing off free label becomes harder the greater the initial concentration of label.

11. Specific examples

It must be stressed that the following examples can be no more than a guide to the types of method available. The choice of isotope, labelled compound, specific radioactivity of label, concentration of label, and period of labelling must all be determined empirically. It is essential to keep variation in the plant material to a minimum, since even small changes in temperature, light, humidity or age of plant can have a marked effect on the efficiency of labelling.

11.1 Labelling of intact plants using $^{14}CO_2$

Various procedures are given in *Protocols 3* and *4*.

Protocol 3. Whole plants

1. Place the plant in a fume cupboard. If the whole plant is to be exposed to radioactive CO_2, place a small, shallow container (e.g. a 10 ml beaker) on the soil surface and then cover the whole plant with a large polythene bag, ensuring that the bag is sealed with tape around the plant pot. Then use a syringe to inject about 0.74×10^6 Bq (20 μCi) $Na_2{}^{14}CO_3$ through the bag into the shallow container.
2. Start the labelling by injecting about 0.1 ml 1 M HCl into the container, thus releasing $^{14}CO_2$ into the space within the polythene bag. Seal holes caused by injection using adhesive tape.

3. Bright tungsten lighting in the fume cupboard will ensure that a large proportion of the label is taken up by the more mature, actively photosynthetic leaves within about 30 min.

Protocol 4. Individual leaves

1. To label individual leaves, use a similar procedure, but support the shallow container just above the leaf to be labelled. This is most easily achieved by glueing the container to a piece of stiff wire, which can be anchored in the soil and bent to place the container in the desired position.
2. Enclose both leaf and container within a polythene bag, sealing the mouth of the bag securely but not tightly around the petiole (which should be wrapped in cotton wool for protection) using adhesive tape.

Note that, with both these methods, it is essential to open the polythene bag *in the fume cupboard* at the end of the labelling period, to ensure the safe removal of free $^{14}CO_2$.

11.2 Germinating seeds and seedlings

Seeds and seedlings can be germinated as in *Protocol 5*.

Protocol 5. Labelling seeds

1. Soak seeds in sodium hypochlorite solution (usually 6% available chlorine can be survived by seeds) including a drop of detergent for 5–10 min at room temperature.
2. Wash the seeds well with sterile distilled water.
3. Place the washed seeds on moist, sterile paper in a closed box, and place at the appropriate temperature for germination.
4. Once germinated the seeds can be labelled by transfer to sterile filter paper soaked in radioactive solution, or, if the seeds are large enough, the paper can be dispensed with and the seeds allowed to lie in a small volume of solution in a sterile Petri dish with their radicles dipping into the solution.

11.3 Root apices

This method has been described for spinach (42) but should be applicable to a wide range of plant types.

Protocol 6. Labelling roots

1. Germinate sterilized seeds in vermiculite at 16 °C for 5 days in darkness.
2. Select seedlings with roots between 1.5 and 2 cm in length, wash thoroughly with sterile distilled water and place these in an aqueous solution of label (e.g. [^{3}H] thymidine, 0.2×10^6 Bq/ml) for between 1 h and 24 h.

11.4 Cut pea shoots

Although described for peas, this method is well-suited to many dicotyledonous plants (43).

Protocol 7. Labelling shoots

1. Harvest pea seedlings and then immediately recut their stems under sterile distilled water to ensure a continuous column of water in the xylem vessels.
2. Place each shoot in a glass vial containing 200 μl sterile distilled water, with each vial held firmly in a test-tube rack or similar support.
3. Place the vials in a current of air and ensure that the shoots are well-illuminated, without risking overheating.
4. Allow some time (e.g. 3 h) for the shoots to recover from being cut, then add label. Using 2×10^6 Bq of [^{3}H] L-leucine (2×10^{12} Bq/mmol) per vial for a period of three hours, incorporations of the order of 3×10^5 c.p.m./mg protein can be achieved.

11.5 Cut wheat seedling leaves

The following method can be used with seedlings of most monocotyledonous plants (44).

Protocol 8. Leaves of monocotyledonous plants

1. Harvest leaves by cutting just above soil level, then rinse well in sterile distilled water and cut the leaves again under water to ensure a continuous column of water in the xylem vessels.
2. Take a wire test-tube rack and place a 0.5 ml microcentrifuge tube in each tube position, using Plasticene or a pre-drilled sheet of expanded polystyrene to hold the tubes firmly in place.
3. Dispense 100 μl of sterile distilled water into each tube, and then place the cut end of one leaf in each tube, arranging the leaves so that the upper part of the wire rack gives them support.

4. Place the rack in a draught of air (e.g. near the front of a half-open fume-cupboard) and provide illumination by a 100 W tungsten bulb, 40 cm above the leaves.
5. After a period of about 30 min to allow recovery from cutting, add label to each microcentrifuge tube. Easily detectable levels of radioactive nucleic acid or protein should be obtained using between 37 and 370 kBq of the appropriate precursor per tube over a period of 2–3 h, depending on the specific activity of the label.

11.6 Leaf discs, using aqueous label

This method can be used with virtually any dicotyledonous plant and the larger leaves of some monocotyledonous plants.

Protocol 9. Leaf discs

1. Where possible, use leaves from plants grown under sterile conditions, otherwise rinse the leaves very thoroughly with sterile distilled water. For prolonged incubations it will be necessary to surface-sterilize the leaf using a solution of sodium hypochlorite (1% available chlorine) containing a trace of detergent to aid wetting. Sterilize for no more than 2 min, then rinse thoroughly with sterile distilled water. Leaves, especially when cut, are easily damaged by excessive exposure to hypochlorite.
2. Cut discs from the leaf lamina using a sharpened, sterile cork-borer, cutting on a glass sheet sterilized with alcohol.
3. Using a laminar flow hood, add label (e.g. 2×10^6 Bq [^{3}H]thymidine, 0.8–1.2 $\times 10^{12}$ Bq/mmol, in 1 ml) to 20 ml sterile nutrient agar medium in a 9-cm Petri dish and allow to soak in.
4. Place the leaf discs on the agar with their lower, stomata-containing surfaces facing down.
5. Provided sterile conditions are maintained, incubation at 23 °C under fluorescent light (6 mW cm^{-2}) can be carried out for at least five days, giving sufficient incorporation for autoradiography or scintillation counting (45).

11.7 Leaf discs, using $^{14}CO_2$

Protocol 10. Leaf discs and $^{14}CO_2$

1. Place leaf discs with their lower surfaces facing upwards on filter paper, saturated with sterile distilled water, in the bottom of a plastic Petri dish.

In the centre of the dish place a shallow receptacle (e.g. a small watch-glass). The Petri dish lid should have a small hole bored through its centre.

2. Place the lid on the Petri dish, sealing it with tape round its circumference, and put the dish into a fume cupboard under a light (e.g. fluorescent light, $6\,mW\,cm^{-2}$).

3. Use a syringe to place about $0.8 \times 10^6\,Bq\,Na_2{}^{14}CO_2$ into the shallow receptacle.

4. Start the labelling by injecting about 0.1 ml 1 M HCl into the receptacle, thus releasing $^{14}CO_2$; immediately seal the hole in the lid using adhesive tape. Significant quantities of $^{14}CO_2$ should have been fixed after only 10 min.

5. At the end of the labelling period remove the Petri dish lid *in the fume cupboard* to ensure that any free $^{14}CO_2$ is safely vented.

References

1. Sheppard, C. W. (1962). *Basic Principles of the Tracer Method*. John Wiley, New York.
2. Rescigno, A. and Segre, G. (1966). *Drug and Tracer Kinetics*. Blaisdell, Waltham, Mass.
3. Jacquez, J. A. (1972). *Compartmental Analysis in Biology and Medicine*. Elsevier/North Holland.
4. Lassen, N. A. and Pearl, W. (1979). *Tracer Kinetic Methods in Medical Physiology*. Raven Press, New York.
5. Phelps, M. E., Hoffman, E. J., and Mullani, N. A. (1975). *J. Nucl. Med.* **16,** 210.
6. Phelps, M. E. (1981). *Semin. Nucl. Med.* **11,** 32.
7. Phelps, M. E., Massiotta, J. C., and Huang, S. C. (1982). *J. Cereb. Blood Flow Metabol.* **2,** 113.
8. Dvorak, J. A. and Jones, A. W. (1963). *Exptl. Parasitol.* **14,** 316.
9. Pinson, E. and Langham, W. H. (1957), *J. Appl. Physiol.* **10,** 108.
10. Rubini, J. R., Cronkite, E. P., Bond, V. P., and Fliedner, T. M. (1960), *J. Clin. Invest.* **39,** 909.
11. Feinendegen, L. E. and Bond, V. P. (1962). *Cell Res.* **27,** 474.
12. Borsook, H., Deasy, C. L., Haagen-Smith, A. J. Keighley, G., and Lowy, P. H. (1950). *J. Biol. Chem.* **187,** 839.
13. Lassen, N. A. and Henriksen, O. (1987). In *Tracer Kinetics and Physiologic Modelling* (eds. R.M. Lambrecht and A. Rescigno), pp. 236–47. Springer-Verlag, Berlin.
14. Altman, J. and Chorover, S. L. (1963). *J. Physiol.* (*London*) **169,** 770.
15. Bradbury, M. (1979). *The Concept of a Blood Brain Barrier*. Wiley-Interscience, Chichester and New York.
16. Lacassagne, A. and Lattes, J.-S. (1924). *Compt. Rend.* **90,** 352.
17. Adelstein, S. J., Lyman, C.P., and O'Brien, R.C. (1964). *Comp. Biochem. Physiol.* **12,** 223.

18. Potter, R. L. and Nygaard, O. F. (1963). *J. Biol. Chem.* **238,** 2150.
19. Gentry, G. A., Morse, P. A., Jr., and Potter, V. R. (1965). *Cancer Res.* **25,** 517.
20. Newton, A., Dendy, P. P., Smith C. L., and Wildy, P. (1962). *Nature, Lond* **194,** 886.
21. Welch, T. J. C., Potchen, E. J., and Welch, M. J. (1972). *Fundamentals of the Tracer Method.* Saunders, Philadelphia.
22. Obrist, W. D., Thompson, H. K., Wang, H. S., and Wilkinson, W. E. (1975). *Stroke* **6,** 245.
23. Ter-Pogossian, M. M., Eichling, J., David, D., Welch, M., and Metzger, J. (1969). *Radiology* **93,** 31.
24. Raichle, M. E., Larson, K. B., Phelps, M. E., Grubb, R. L. Jr., Welch, M. J., and Ter-Pogossian, M. M. (1975). *Am. J. Physiol.* **228,** 1936.
25. Frackowiak R. S. J., Lenzi G. L., Jones. T., and Heather J. D. (1980). *J. Computer-Assisted Tomography* **4,** 727.
26. Sokoloff, L., Reivich, M., Kennedy, C., Des Rosiers, M. H., Patlak, C. S., Pettigrew, K. D., Sakurada O., and Shinohara, M. (1977). *J. Neurochem* **28,** 897.
27. Sokoloff, L. (1981). *J. Cereb. Blood Flow Metab.* **1,** 7.
28. Brownell, G. L., Berman, M., and Robertson, J. S. (1968). *Int. J. Appl. Radiat. Isotopes* **19,** 249.
29. Crone, C. (1963). *Acta Physiol. Scand.* **58,** 292.
30. Renkin, E. M. (1959), *Am J Physiol.* **197,** 1205.
31. Kinsman, J. M., Moore, J. and Hamilton, W. R. (1928). *Am. J. Physiol.* **89,** 322.
32. Stewart, G. N. (1897). *J. Physiol.* **22,** 159.
33. Bassingthwaighte, J. B. (1967). *Proc. Staff. Meet. Mayo Clin.* **42,** 137.
34. Winchell, H. S., Horst, W. D., and Braun L. *et al.* (1980). *J. Nucl. Med.* **21,** 947.
35. Bianchi, P. A., (1962). *Biochem. Biophys. Acta* **55,** 547.
36. Sokoloff L. and Smith, C. B. (1983). In *Tracer Kinetics and Physiologic Modelling* (eds. R. M. Lambrecht and A. Rescigno), pp. 202–34. Springer-Verlag, Berlin.
37. Huang Sung-Cheng, Carson R. E., and Phelps, M. E. (1983). In *Tracer Kinetics and Physiologic Modelling* (eds. R. M. Lambrecht and A. Rescigno), pp. 301–3. Springer-Verlag, Berlin.
38. Feinendegen, L. E. (1967). *Tritium Labelled Molecules in Biology and Medicine.* Academic Press, New York and London.
39. Jennings, S. M., (ed.) (1977). *Symp. Soc. Exptl. Biol.* **31.**
40. Rose, R. J., Cran, D. G., and Possingham, J. V. (1974) *Nature, Lond.* **251,** 641–2.
41. Northcote, D. H. (1971). *Symp. Soc. Exptl. Biol.* **25,** 51–69.
42. Possingham, J., Chaly, N., Robertson, M., and Cain, P. (1983) *Biol. Cell* **47,** 205–12.
43. Bouthyette, P. and Jagendorf, A. T. (1978). *Plant & Cell Physiol.* **19,** 1169–74.
44. Boffey, S. A., Ellis, J. R., Sellden, G. and Leech, R. M. (1979) *Plant Physiol.* **64,** 502–5.
45. Possingham, J. V. and Smith, J. W. (1972) *J. Exptl. Bot.* **23,** 1050–9.

6

In vitro labelling

A. Nucleic acids

MARTIN W. CUNNINGHAM, DENNIS W. HARRIS, and
CHRISTOPHER R. MUNDY

1. Introduction

Radioactive tracers are uniquely suitable for labelling of biological molecules, because their behaviour in enzyme-catalysed reactions mimics that of their non-radioactive isotopes. Progress in nucleic acid research has depended on the availability of suitable radioactive labels, and on methods for incorporating them into nucleic acids.

The list of techniques involving the use of labelled nucleic acids to analyse cell functions grows almost daily. In some cases, a particular combination of label and labelling method is required. The purpose of this chapter is to describe the various labels and labelling methods and to clarify their roles and characteristics.

1.1 Applications of nucleic acid labelling

Nucleic acids are labelled for the following reasons:

- *Monitoring of reactions*. A great deal of time and trouble can be saved by judicious monitoring of enzyme-catalysed reactions.
- *Structural analysis of nucleic acids*. For example, DNA and RNA sequencing and restriction mapping of DNA.
- *Production of hybridization probes*. Complementary nucleic acid sequences can be detected in mixed populations of molecules using labelled DNA or RNA.

The methods used for generation of hybridization probes exemplify the enzymic reactions employed in all of the above applications, and we will focus on these methods throughout the chapter. The somewhat different requirements of sequencing and mapping will also be discussed.

Table 1. Applications for labelled hybridization probes

Southern blots	Detection of gel-fractionated DNA molecules following transfer to a membrane
Northern blots	Detection of gel-fractionated RNA molecules following transfer to a membrane.
Dot blots	Detection of unfractionated DNA or RNA molecules immobilized on a membrane.
Colony/plaque blots	Detection of DNA released from lysed bacteria or phage and immobilized on a membrane.
S1/RNase mapping	Positional mapping of termini of target molecules.
In situ hybridization	Detection of DNA or RNA molecules in cytological preparations.

The applications of hybridization probes are listed in *Table 1*. In some cases, it is possible to use any one from a variety of labels and labelling methods. In other applications, however, only specific labels or labelling methods give the desired result. The applications differ with respect to the following criteria.

- *Required sensitivity*. The amount of target nucleic acid present per unit area, for example in the detection of small quantities of DNA on a Southern blot.
- *Required resolution*. The required distance between the point at which label has been deposited and the point at which it is detected. The closer this distance, the greater is the resolution.
- *Required label distribution*. Label may be distributed uniformly throughout the probe or at its ends.

The level of sensitivity that can be achieved is limited by the number of detectable nuclear distintegrations per unit time. This is determined by the specific activity of the probe and by the nature and energy of the emitted particles or photons (see *Table 2*). A very high specific activity ^{32}P-labelled probe, for example, provides maximum sensitivity because the high-energy beta-particles emitted by ^{32}P can be detected efficiently using autoradiography with intensifying screens (see Section 2.2). Conversely, however, resolution varies inversely with the energy of emission, because higher energy particles or photons produce a signal at greater distances from the site of label deposition. The optimum balance between sensitivity and resolution for a number of applications is discussed in Section 2.

Some applications require end-labelled nucleic acid to eliminate irrelevant signal. For example, labelled DNA is subjected to base-specific chemical cleavage during the Maxam and Gilbert sequencing reactions (1). Only those molecules which have a common terminus are useful for generation of the sequencing ladder, and the terminus is identified by end-labelling. Uniformly labelled DNA would not give an intelligible result.

Table 2. Characteristics of radionuclides used in nucleic acid labelling

Radio-nuclide	$t_{1/2}$	Type/max energy of emission (MeV)	Specific activity range of nucleotides	Labelling methods	Typical spec. act. of probe (d.p.m./μg)	Detection limit[a] (d.p.m./cm^2)	Equivalent amount of probe (pg/cm^2)
^{32}P	14.3 d	β1.71	400–6000	nick-translation	5×10^8	50[b]	0.1
			Ci/mmol	random primer	5×10^9		0.01
			14.8–222	phage RNA pol	1.3×10^9		0.04
			TBq/mmol	end labelling	5×10^6		10
^{35}S	87.4 d	β0.167	400–1500	nick-translation	1×10^8	400[c]	4
			Ci/mmol	random primer	7×10^8		0.57
			14.8–55.5	phage RNA pol	1.3×10^9		0.3
			TBq/mmol				
^{125}I	60 d	γ/0.035	1000–2000	nick-translation	1×10^8	100[b]	1
		β/0.035	Ci/mmol	random primer	1.5×10^9		0.067
			37–74	direct iodination	2×10^8		0.5
			TBq/mmol				
^{3}H	12.35 y	β/0.018	25–100	nick-translation	5×10^7	8000[c]	160
			Ci/mmol	random primer	1.5×10^8		53.3
			925–3700	phage RNA pol	5×10^7		160
			GBq/mmol				

[a] Data from ref. 2. Represents an increase in absorbance of film of 0.2 A_{540} units after 24 h exposure, using 1.5 mm gel containing the radioactive isotopes as the source.
[b] Detection using an intensifying screen.
[c] Detection by fluorography using Amplify (Amersham).

2. Choice of radiolabel

Any experiment involving a radiolabelled nucleic acid will place constraints on the labelling method and radionuclide used, and will require a particular combination of label distribution, specific activity, energy of emission, and spatial resolution. Nucleic acid labels can be used to detect target sequences in tissues or cells, in gels or on membranes, or in solution. In each case, a different combination of the factors mentioned above will apply and the balance of these, to a large extent, will be determined by the detection method employed.

In molecular biology applications the most widely used detection methods are autoradiography and fluorography (2). The details of these techniques are described in Chapter 4 although here we present particular details relevant to the detection of the radioisotopes ^{32}P and ^{35}S. These two radionuclides are widely used in the techniques of filter hybridization and nucleic-acid sequencing. Other isotopes used in nucleic acid labelling are tritium and ^{125}I. These, however, are more commonly used in the technique of *in situ* hybridization described in Chapter 7.

In both filter hybridization and nucleic acid sequencing, a certain degree of resolution is important, since information about the relative positions of particular nucleic acid fragments is required. In addition, since the particular sequence of interest may be present in small quantities, there is usually a requirement for high sensitivity. So there are two major factors which must be considered in relation to the radioisotope: resolution and sensitivity. Clearly, an isotope which is capable of immediately exposing an entire sheet of film when present as a point source is of little value in either application. On the other hand, a radiolabel which is of such low energy or is present in the sample at such low distribution that the film takes days or weeks to expose is also of little use. However, each application will require a different balance between these parameters. For instance, ultimate sensitivity may be compromised for the sake of resolution in dideoxy DNA sequencing (3), whilst the converse may be true when the detection of a single-copy gene on a Southern blot is the aim. Broadly speaking, the choice of label is a trade-off between sensitivity and resolution, in combination with other factors such as probe stability, safety, ease of use, and cost.

2.1 Filter hybridization

The most common applications of filter hybridization are (a) the detection of single-copy genes or low abundance messenger RNAs in Southern (4) and Northern (5) blotting respectively, (b) the detection of particular sequences in cDNA or genomic libraries on colony or plaque lifts, and (c) the detection/quantitation of particular sequences in dot or slot blotting (see *Figure 1*). These all require a degree of sensitivity which usually assumes higher priority

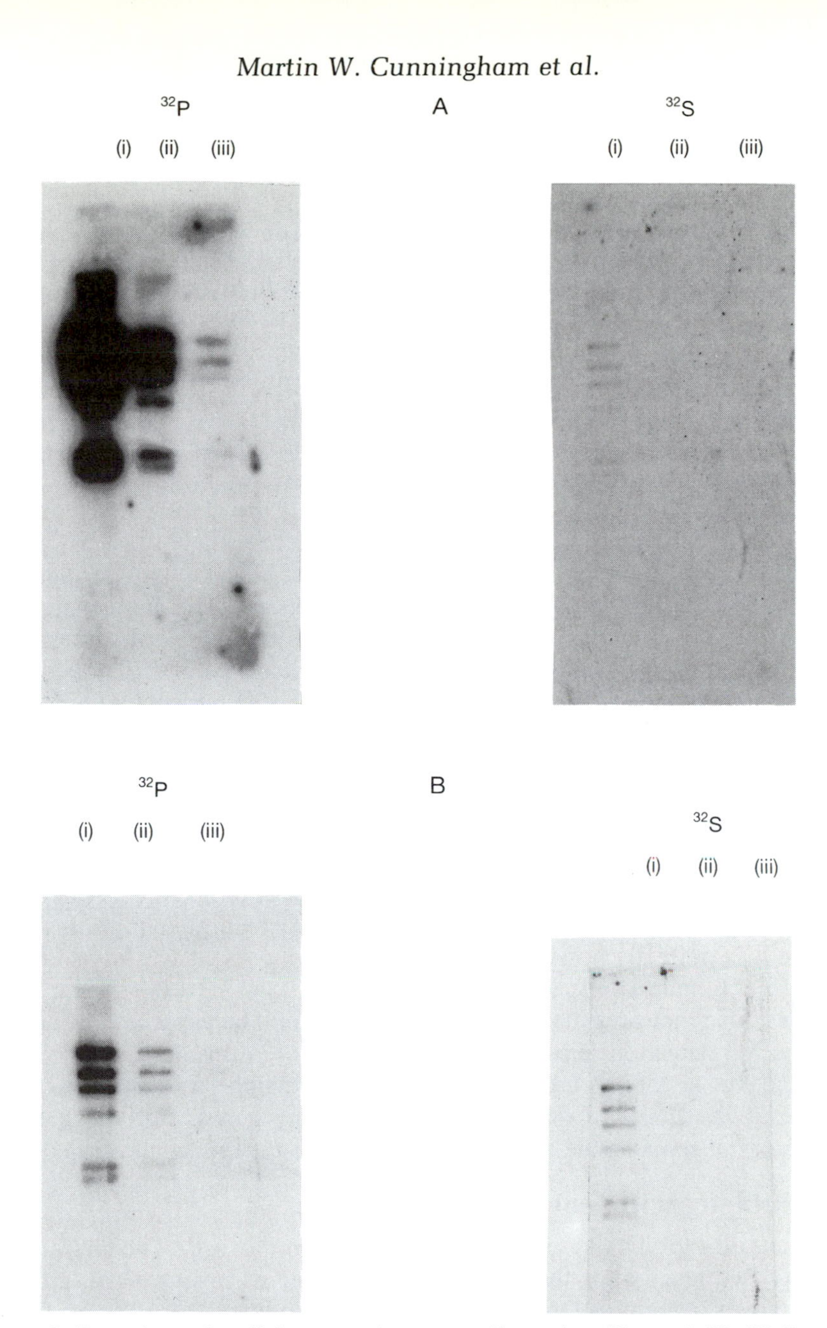

Figure 1. Detection of radioisotopes by autoradiography. Blots of *Hin*dIII-digested lambda DNA containing (i) 1 ng, (ii) 100 pg, and (iii) 10 pg DNA were hybridized using nick-translated probes labelled with ^{35}S- or ^{32}P-dCTP to a specific activity of 2×10^8 d.p.m./μg in each case. All hybridizations were for 1 h at a probe concentration of 100 ng/ml using Rapid Hybridization buffer (Amersham). The same filters were exposed in 'A' to Hyperfilm MP with intensifying screens for 48 h at −70 °C and in 'B' to Hyperfilm-β max for 48 h at room temperature.

than the resolution achieved. The generally available amounts of target sequences range from 20–100 pg in colony and plaque hybridization down to 1–5 pg in the detection of single-copy mammalian genes by Southern blot hybridization.

The important features which determine sensitivity in the detection of the available radionuclides by autoradiography are given in *Table 2*. It should be noted here that the detection limits given relate to the radionuclides themselves under ideal conditions, and take no account of factors which would affect a typical hybridization experiment, such as probe reannealing or target availability. For instance, it has been estimated (N. Dyson, pers. comm.) that in some protocols as little as 20% of an immobilized nucleic acid may ultimately be available for hybridization. Despite these reservations, however, it is still possible to draw general conclusions about the ability of a particular radioisotope to fulfil a given sensitivity requirement.

Consider, for example, the case of detection of a single-copy sequence in human genomic DNA, following Southern blotting. The human genome comprises approximately 3×10^9 deoxyribonucleotide base-pairs. Assuming an average molecular weight of 660 daltons per base-pair, it may be calculated using Avogadro's number ($\sim 6 \times 10^{23}$) that a single complete genome weighs approximately 3.3×10^{-12} g. Similarly, a single copy of a 1 kb sequence weighs approximately 1×10^{-18} g. The number of copies of the genome in 10 μg of DNA is 3×10^6. It follows, therefore, that a single-copy 'gene', 1 kb in length, must account for approximately 3 pg of a 10 μg sample. If, following Southern blotting, this sequence is present in a band of, say 0.5×0.1 cm, the figures in *Table 2* indicate that probes uniformly labelled with phosphorus-32 would easily meet the imposed detection requirements. In practice it is perfectly possible to detect single-copy sequences in as little as 0.5 μg total human DNA.

Table 2 also indicates that both ^{125}I and ^{35}S should be capable of achieving the required detection sensitivities. In the case of ^{35}S this has certainly been achieved, but in routine use it is much more unreliable, as it is necessary to operate at the limits of detection for single-copy genes. For the present, therefore, ^{32}P remains the radionuclide of choice for this application.

2.2 Nucleic acid sequencing

Two methods for DNA sequencing are in common use. The first, called 'dideoxy' sequencing, or the Sanger method after its inventor (6), employs the Klenow fragment of DNA polymerase I to elongate DNA chains from a defined primary site on a single-stranded (bacteriophage M13) template. Nucleotide analogues which are incapable of acting as a substrate for elongation by virtue of the absence of an hydroxyl group at the 3′ position are included and are incorporated randomly into the growing chains, producing a set of prematurely terminated fragments. Four such 'dideoxy' nucleoside triphosphates are used in separate reactions, one for each base. One of the four 'normal' 2'-deoxynucleotides labelled with ^{35}S or ^{32}P is included in each reaction, so that all of the fragments are uniformly labelled. Reaction products are visualized by autoradiography following polyacrylamide gel electrophoresis.

In this method, the researcher is faced with a choice of radioisotope. 2'-deoxynucleoside triphosphates are available labelled with ^{32}P. Alternatively, phosphorothioate analogues labelled with ^{35}S may be used. These analogues are incorporated at high efficiency by Klenow polymerase (see Section 3.2). The properties of the two isotopes are compared in *Table 2*. The advantage of ^{35}S over ^{32}P is that it offers improved resolution so that more sequence data can be read from each autoradiogram. This must be offset against the fact that its lower energy of emission requires that sequencing gels be dried prior to autoradiography. In addition, average exposure times are likely to be increased relative to ^{32}P.

The alternative DNA sequencing method, that of Maxam and Gilbert (1), relies on base-specific chemical cleavage. This method requires fragments that are end-labelled at a single, defined end. Because of this, the achievable density of label is very low and thus the highest specific activity possible is required. By and large, this means that ^{32}P is the radionuclide of choice. Although ^{35}S has been used, significantly longer exposure times are necessary.

3. Labelling methods

All of the methods to be described in this section are, from a practical point of view, straightforward enzyme catalysed reactions. However, it is still useful to mention some general points that may help towards obtaining good, reproducible labelling. The reactions should be set up on ice, except when the buffer contains spermidine; this compound may cause precipitation of nucleic acid. Reagents should be added in the order described in the protocols. Label is added penultimately, to reduce handling of radioactive solution, and is followed by addition of enzyme. Successful results necessitate thorough mixing of all reagents before addition of enzyme. This can be achieved by a brief (2 sec) spin in a microcentrifuge. Brief vortexing before the spin can also aid mixing, but is not usually necessary. Enzyme is then added and mixed by gently pipetting two or three times. This technique ensures effective mixing of undenatured enzyme that is frequently added in a very small volume and in a viscous glycerol-containing solution.

Poor labelling can occur for a variety of reasons. One of the most common causes is impure DNA (or RNA); it is therefore often useful to compare the efficiency of labelling to that obtained with a pure, control DNA. Commercially available plasmids or phage DNA are useful in this case. If impure substrate is the cause of poor labelling, a further phenol extraction or dialysis followed by ethanol precipitation can be of benefit. Frequently, it is possible to improve labelling efficiency, even with impure DNA, by the use of more enzyme or by extending the reaction time. It is also necessary to ensure that efficient labelling is theoretically possible by determining both the amount of label present and the maximum amount of label that can be incorporated. This point is explored further in this section but if, for example, in an end-labelling reaction, there is ten times more label present than there are ends available for labelling, then a maximum incorporation of only 10% is possible.

When establishing a labelling method or using a new preparation of substrate nucleic acid, it is valuable to determine the efficiency of labelling. A variety of methods are available including precipitation by trichloroacetic acid (TCA), filtration through DE81 paper or nitrocellulose, and thin layer chromatography. *Protocol 1* gives a procedure for the most commonly used of these techniques, TCA precipitation, and an alternative protocol using DE81 paper. If required, unincorporated label can be removed as described in *Protocol 2*. In our experience, this is not necessary for most filter hybridization applications if incorporation is 60% or above, but we would advise that this is verified under the user's conditions before adoption.

Protocol 1. Determination of percentage incorporation of radiolabel

TCA precipitation

Materials

Cellulose acetate, cellulose nitrate or glass-fibre filters (2.4 cm diameter)
8–10% trichloroacetic acid (TCA). **Care**: corrosive acid
Carrier DNA or RNA (calf thymus, or herring sperm DNA, or yeast total RNA at 1 mg/ml)

Method

1. For each sample take four scintillation vials.
2. Into two, place cellulose acetate, cellulose nitrate, or glass fibre filters. Spot 2–10μl (10^3–10^5 c.p.m.) of sample on to each. This will give the total number of counts in the sample. For most labelling reactions an initial dilution of 1 μl reaction mixture into ~100 μl H_2O or 0.2 M EDTA will be necessary to give counts in the required range.
3. Pipette an equal volume of the sample into two 5-ml tubes on ice. Half-fill the tubes with 8–10% TCA. Using a Pasteur pipette, add four drops of DNA or RNA, and fill the tubes with TCA solution. The carrier should produce a white cloudy precipitate on contact with the TCA solution. Leave the tubes for 30–60 min on ice.
4. Place two filters in position on a filtration apparatus (e.g. Millipore). Connect the apparatus to a vacuum pump, apply the vacuum and wet the filters with ~2 ml of 8–10% TCA.

5. Pour the TCA precipitates into the well. Replace the tube on ice and refill with TCA to rinse. Pour the liquid into the well once the initial solution has been drawn through.
6. Rinse the walls of the well with at least 2 ml of TCA. When all the liquid has been drawn through, dismantle the apparatus and transfer the filters to the remaining scintillation vials. This will give the amount of precipitable radioactivity in the sample.
7. Dry all four filters (e.g. under an infra-red lamp for approximately 30 min). Avoid overheating the filters. Allow the vials to cool and add 10 ml of scintillation fluid. If ^{32}P has been used, it is possible to count directly without scintillant (Cerenkov counting).
8. Cap the vials and determine radioactivity by liquid scintillation counting.
9. Calculate the percentage incorporation as follows:

$$\text{Percentage incorporation} = \frac{\text{Precipitated counts}}{\text{Total counts}} \times 100$$

$$\text{Specific activity} = \frac{\text{Total incorporated counts}}{\text{Mass of hybridizable nucleic acid}}.$$

DE81 chromatography

Materials

Whatman DE81 paper (2.4 cm diameter)
0.5 M Na_2HPO_4

Method

1. Spot aliquots containing 10^3–10^5 c.p.m. (see method A) on to four 2.4 cm discs of Whatman DE81 paper. Designate two filters A and two B.
2. Wash the B filters six times, 5 min per wash, in 0.5 M Na_2HPO_4. Then wash twice in H_20 (1 min per wash) and in 95% ethanol (1 min per wash).
3. Dry A and B filters using an infra-red lamp and count by liquid scintillation. The percentage incorporation is (B/A × 100)%.

For ^{32}P-labelled probes it is possible to quantify the radioactivity by Cerenkov counting without scintillation fluid. In this case the same filter can be counted initially before washing for total counts, and after washing for incorporated counts (see Chapter 3, Sections 3.10 and 3.11).

Protocol 2. Removal of unincorporated radionucleotides

Two methods are described for the removal of unincorporated nucleotides

Ethanol precipitation

Materials

0.2 M EDTA, pH 8.0
5 M NaC1, or 3 M sodium acetate pH7
Carrier DNA (calf thymus, or herring sperm DNA at 1 mg/ml)
Ice-cold ethanol
TE buffer (10 mM Tris–HCl, pH 8.0, 1 mM EDTA)

Method

The method given is for a labelling reaction of 50 μl; for other volumes the amounts of reagents added should be scaled up or down accordingly.

1. To a 50 μl labelling reaction, add the following:
   20 μl 0.2 M EDTA, pH 8.0
   20 μl 5 M NaCl, or 3 M Na acetate, pH 7
   20 μl carrier DNA[a]
   400 μl ice-cold ethanol[b]
2. Leave to precipitate at −80 °C for 30 min or at −20 °C overnight.
3. Spin in a microcentrifuge for 15 min. Transfer the radioactive supernatant to a waste bottle for disposal.
4. Wash the pellet in ice-cold absolute ethanol, centrifuge, pour off or aspirate the supernatant and dry the resulting pellet under vacuum.
5. Redissolve the labelled DNA pellet in an appropriate volume of TE buffer for use as a probe.
6. If desired a small aliquot of the solution can be removed for scintillation counting to calculate the specific activity of the probe.[c] When using ^{32}P, it is frequently sufficient to monitor remaining activity using a Geiger counter.

Sephadex spin columns

Materials

1-ml disposable syringe
Sephadex G-50 (Pharmacia) equilibrated in TE buffer
TE buffer (10 mM Tris, pH 8.0, 1 mM EDTA)

Method

1. Plug the bottom of a 1-ml disposable syringe with a small amount of sterile glass wool, preferably siliconized to prevent adsorption of nucleic acid.

2. In the syringe, prepare a column of Sephadex G-50 equilibrated in TE, pH 8.0.
3. Insert the syringe into a centrifuge tube and spin at 1600 *g* for 4 min.
4. Continue to add Sephadex suspension until the packed volume is 0.9 ml.
5. Add 100 μl TE buffer and recentrifuge at exactly the same speed and for exactly the same time.
6. Repeat step 5. The volume collected should now be ~100 μl. If the volume is greater than 150 μl, repeat step 5.
7. Apply the sample to the column in a total volume of 100 μl.
8. Recentrifuge at exactly the same speed and for exactly the same time. The unincorporated nucleotides remain in the syringe, while the labelled probe is eluted in ~100 μl. To minimize high-level solid radioactive waste, the syringe can either be stored until the isotope has decayed and disposed of as low-level waste, or it can be thoroughly washed and the label disposed of as liquid waste.
9. If desired a small aliquot of the eluate can be removed for scintillation counting to calculate the specific activity of the probe. When using ^{32}P, it is frequently sufficient to monitor remaining activity using a Geiger counter.

[a] Carrier DNA is optional, but improves recovery of probe at low concentration (e.g. during random primer labelling). Alternative carriers are total RNA (e.g. from yeast) or glycogen. As little as 1 μg may be used effectively.

[b] For larger volumes it is adequate to use 3 volumes of ice-cold ethanol, to allow precipitation to be carried out in a single microcentrifuge tube.

[c] This assumes complete removal of unincorporated nucleotides. Although the above procedure does remove most unincorporated nucleotides, for a more accurate value the aliquot should be precipitated with TCA before counting (see *Protocol 1*). For more complete removal of unincorporated label (e.g. when the expected level of labelling is low), an additional precipitation step can be included or, alternatively, precipitation can be carried out by the addition of 0.5 volumes of 7.5 M ammonium acetate instead of NaCl or Na acetate, followed by 2.5–3 volumes of ethanol. Cool and spin as above.

If a labelling method is performed routinely and unincorporated nucleotides are usually removed, then it is often sufficient simply to monitor the incorporated label in the probe either quantitatively by scintillation counting or, in the case of ^{32}P, semi-quantitatively with a Geiger counter. *Protocol 2* describes two methods for effectively removing unincorporated nucleotides, ethanol precipitation and Sephadex G-50 chromatography. Enzyme can be removed by a prior phenol extraction but, for most filter hybridization applications, this is not necessary. Virtually all the enzymes used for labelling require a divalent metal cation as cofactor, so the chelating agent EDTA is generally added to terminate the reaction effectively.

Estimation of probe specific activity after precipitation or G-50 chromatography assumes complete removal of concentrated nucleotides and no loss of probe. For nick-translation and most end-labelling procedures, there is little or no net synthesis of DNA during the reaction so probe specific activity (d.p.m./μg) is simply the total incorporated d.p.m. divided by the initial mass of DNA labelled. If there is net synthesis of probe during the labelling reaction, its specific activity will also depend on the specific activity of the labelling nucleotide. If the initial template does not participate in the subsequent hybridization, then probe specific activity depends solely on the specific activity of the labelling nucleotide. Total incorporated counts simply indicate the amount of probe synthesized. This is the situation in RNA polymerase-catalysed reactions if, before hybridization, the DNA template is removed by DNase treatment or the single-stranded RNA probe is not denatured. The most complex case is random primer labelling where net synthesis occurs but newly synthesized probe is also diluted by template. In this case total incorporated counts should be used to calculate the mass of DNA synthesized. Probe specific activity is then equal to total incorporated dpm divided by the final weight of nucleic acid (probe plus original template).

We would certainly advise that the efficiency of a labelling reaction is always monitored. Although simple, the reactions are not foolproof.

3.1 Nick-translation

Although predating the other uniform labelling methods described on the following pages, nick-translation remains the most common means of labelling DNA for hybridization. The technique can be used with a variety of radioactive labels to generate probes suitable for most hybridization applications. For example, using ^{32}P-labelled deoxynucleoside triphosphates (dNTPs), it is relatively simple to generate probes of a specific activity ($>1 \times 10^8$d.p.m./μg) high enough to detect single-copy genes on Southern blots of mammalian DNA (Section 2.1).

The nick-translation reaction (7, 8) involves the simultaneous action of two enzymes, pancreatic deoxyribonuclease I (DNase I) (9) and *Escherichia coli* DNA polymerase I (DNA pol I) (10). DNase I introduces nicks at random points in both strands of a DNA duplex. This produces a free 3′-hydroxyl and a free 5′-phosphate group at each nick. The 5′–3′ exonuclease activity of DNA polymerase I then progressively removes nucleotides working from the free 5′ ends within the duplex. Simultaneously, the 5′–3′ polymerase activity of DNA pol I successively adds nucleotides to the free 3′-hydroxyl ends, using the complementary DNA as a template. Thus, the initial nick is actually 'translated' along the DNA molecule in a 5′–3′ direction.

Because DNase I introduces nicks randomly, the net effect is the production of a uniformly labelled population of molecules. The process is illustrated diagrammatically in *Figure 2*. By the inclusion of one or more radiolabelled

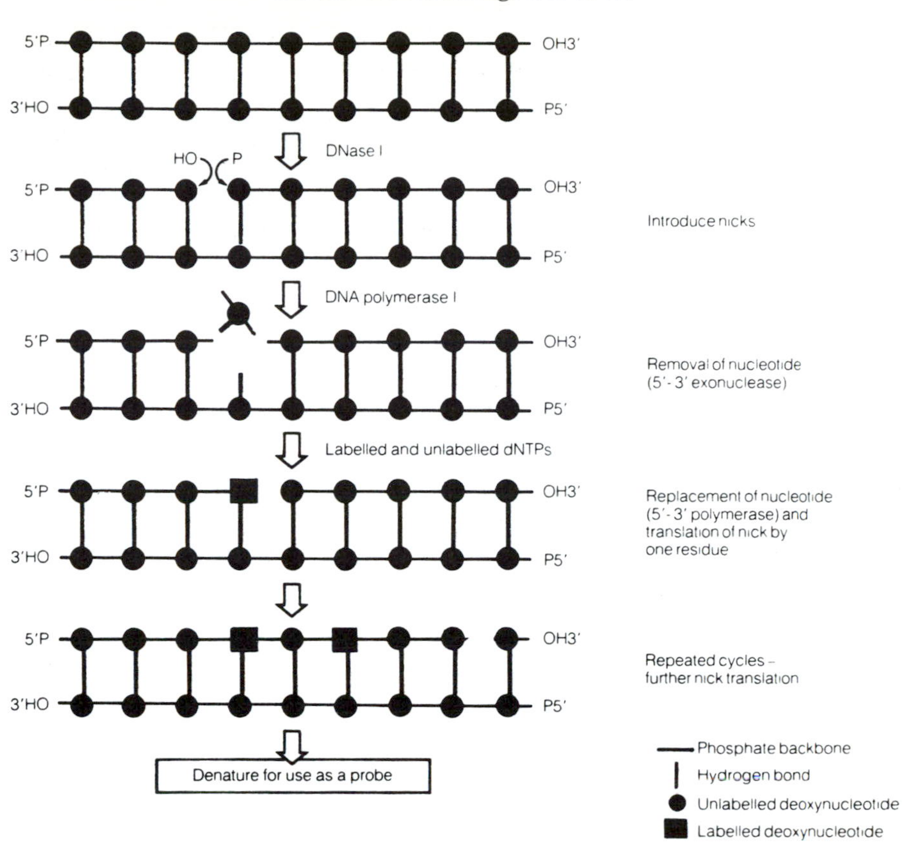

Figure 2. The nick-translation reaction.

nucleoside triphosphates in the reaction, the DNA can be labelled to high specific activity.

It is possible to manipulate most of the parameters of the nick-translation reaction in order to alter probe size, specific activity, or yield or to control the efficiency of incorporation of labelled precursor. For this reason there is a wide variety of protocols in existence. The procedure in *Protocol 3* is appropriate for many situations, and those parameters that can be altered are discussed on the following pages. The example given in the protocol uses ^{32}P-labelled dNTPs but ^{3}H- and ^{35}S-nucleotides are also incorporated efficiently and the use of different radionucleotides is discussed below. Efficiencies of incorporation of over 60% should be achievable. If labelled nucleotide of a specific activity other than 3000 Ci/mmol is to be used, refer to Section 3.1.5.

Protocol 3. Labelling DNA by nick-translation

Materials

10 × nick-translation buffer (600 mM Tris–HCl, pH 7.8, 100 mM $MgCl_2$, 100 mM 2-mercaptoethanol)
TE buffer, pH 8.0 (10 mM Tris–HCl, pH 8.0, 1 mM EDTA)

300 μM dATP	in TE buffer, pH 8.0
300 μM dCTP	in TE buffer, pH 8.0
300 μM dGTP	in TE buffer, pH 8.0
300 μM dTTP	in TE buffer, pH 8.0

Enzyme mix (0.04 units/ml DNase I and 500 units/ml DNA polymerase I).
[α-^{32}P] dNTP at 3000 Ci/mmol, 110TBq/mmol.

Method

1. Prepare an appropriate nucleotide mix from the 300 μM stocks above by mixing 3.3 μl of three non-radioactive dNTPs, excluding that to be used as label. Alternatively, if the same labelled nucleotide is frequently used, it may be simpler to prepare a 5 × nucleotide mix containing each dNTP at 100 μM except that to be used as label. For example, if [α-^{32}P] dCTP is to be used, the 5 × nucleotide mix contains 100 μMd ATP, 100 μM dGTP and 100 μM dTTP.
2. Add the following to 10 μl of the nucleotide mix on ice in the order given

• DNA to be labelled	50–500 ng
• Water	to 30 μl
• 10 × buffer	5 μl
• [α-^{32}P] dNTP	10 μl (100 μCi, 3.7 MBq)
• Enzyme mix	5 μl
• Total volume	50 μl

   Mix gently by slowly pipetting up and down once or twice. Spin briefly (2 sec) in a microcentrifuge to collect contents at the bottom of the tube.
3. Place the tube in a constant temperature bath at 15 °C for approximately 2 h.
4. Terminate the reaction and remove most unincorporated labelled nucleotides by ethanol precipitation or chromatography on Sephadex (see *Protocol 2*)
5. Denature the probe by heating to 95–100 °C for 5 min before hybridization.

3.1.1 Nature and quantity of substrate

The procedure given in *Protocol 3* is appropriate for between 50 and 500 ng double-stranded DNA. To label more DNA, it is advisable to increase the total reaction volume proportionately. Most published protocols use relatively large amounts of DNA (0.5–1 μg) which can give a high input of contaminating material if the DNA substrate is impure, leading to reduced efficiency of labelling. Since there is some evidence that DNase I is sensitive to inhibitors in agarose (11), this can be a particular problem if it is necesary to prepare a probe free of vector sequences by gel purification. This problem can be offset to a large extent by using less DNA in the reaction (<100 ng) thus reducing the level of added contaminants. We have efficiently labelled 25–50 ng DNA in low melting point agarose using ^{32}P-labelled nucleotides.

As implied by its mechanism, the nick-translation reaction is not appropriate for single-stranded DNA, but it is equally efficient with both linear and circular double-stranded molecules. In general, the length of the DNA has little effect on the efficiency of labelling although, for relatively short molecules (<100 bp), it may be necessary to increase DNase I concentration in order to introduce enough nicks for efficient labelling. However, this will reduce final single-stranded probe size (see following section).

3.1.2 DNase I concentration

The concentration of DNase I determines the frequency of nicks and hence the final single-strand length of the labelled products. In common with most standard methods, the procedure given in *Protocol 3* results in single-strand probe lengths of 200–500 nucleotides. Lower concentrations of enzyme can be used to produce longer probes and higher concentrations for shorter probes. The latter can be of particular use in *in situ* hybridization to facilitate penetration of the target material. In our hands, when using high probe concentration for rapid filter hybridization (see Section 4.1.2), optimum sensitivity and minimum background are obtained with a relatively narrow range of single-strand probe lengths (150–500 bases).

It should be noted that, if higher concentrations of DNase I are used, reaction times can be shorter because 3′-ends become more rapidly available for DNA polymerase I. A short pre-treatment with a high concentration of DNase I forms the basis of a recently described method for rapid nick-translation that leads to highly efficient incorporation of labelled nucleotides (12).

A further consideration is that, with probes labelled to high specific activity with ^{32}P, strand breakage due both to bombardment with high energy β-particles and to radioactive decay of ^{32}P in the phosphate backbone occurs rapidly, with concomitant reduction in average probe length.

3.1.3 Reaction temperature

Nick-translation reactions are normally carried out at low temperature

(~15 °C). This is because higher temperatures can lead to rapid degradation of the probe due to increased activity of both DNase I and the 5′-3′ exonuclease activity of DNA pol I. Investigations of the reaction mechanism at higher temperatures (7) have demonstrated that nick-translation is to some extent replaced by strand displacement. In this case an elongating strand displaces residues from the duplex instead of excising them by the 5′–3′ exonuclease activity. Under these conditions it is possible for template switching to occur, so that the polymerase begins to copy a displaced strand. This produces regions of intrastrand complementarity and may lead to non-uniform labelling.

3.1.4 Time-course

Maximum incorporation can be obtained at incubation times varying between 30 min and several hours, depending on the precise protocol used. *Figure 3* shows a typical time-course using the procedure given in *Protocol 3* with 500 ng substrate DNA. Apparent reaction rate, as measured by the percentage incorporation of labelled precursor, increases with addition of more enzyme (DNase and/or DNA pol I), or when concentration of labelled nucleotides is reduced, for example by using label of higher specific activity (see Section 3.1.5).

Nick-translation reactions should not be left for longer than is necessary

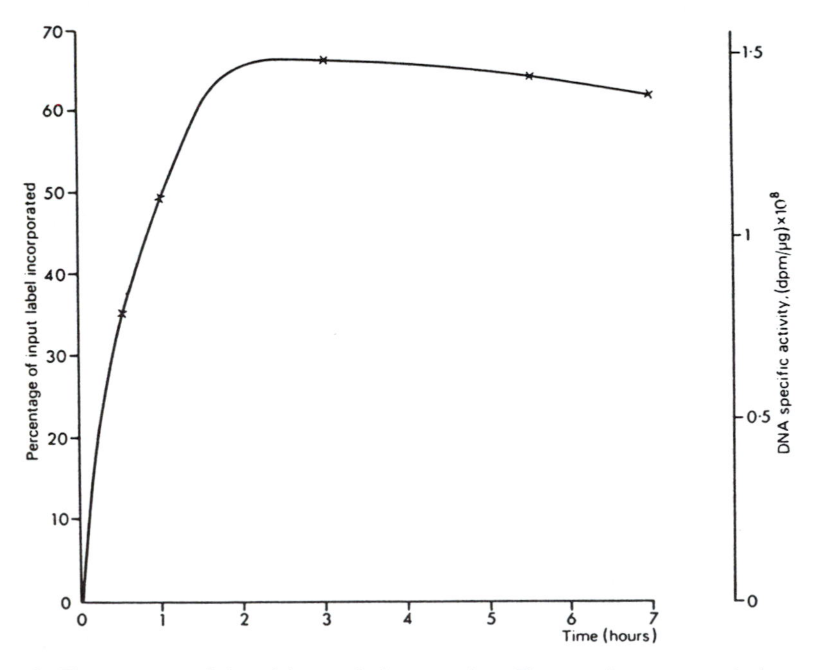

Figure 3. Time course of the nick-translation reaction. The reaction was carried out with 500 ng DNA under standard conditions.

because both DNase I and the 5′–3′ exonuclease activity of DNA pol I will continue to degrade the DNA, causing loss of incorporated label and a progressive reduction in probe size. Both enzymes should be inactivated if the reaction mix is to be stored before use. A chelating agent (EDTA) and/or detergent (SDS) should be used in preference to heat denaturation, as this can cause a transient increase in nuclease activity.

3.1.5 Probe specific activity

The specific activity of a probe prepared by nick-translation depends on the extent of replacement of existing nucleotides in the DNA substrate and on the specific activity of the input labelled nucleotides (8). In theory, complete replacement can be obtained if the labelled nucleotide is present in the same quantity as its equivalent in the unlabelled DNA template. For example, 1 μg DNA contains approximately 750 pmol dCMP (assuming equimolar amounts of the four constituent deoxynucleotides). If 750 pmol [α-^{32}P]dCTP (approximately 300 μCi at 400 Ci/mmol) are added to the reaction, then complete replacement is theoretically possible. In practice, an input of ~1000 pmol [α-^{32}P]dCTP (400 Ci/mmol) is necessary, because levels of incorporation of labelled nucleotides higher than ~75% rarely occur using standard protocols. The maximum specific activity of probe produced under these conditions is 6.6×10^8 d.p.m./μg.

At levels of labelled nucleotide below ~1000 pmol/μg DNA, the specific activity of the final product is generally proportional to the amount of label added. Changing to nucleotides of higher specific activity (for example, 3000 Ci/mmol), without altering the total radioactivity added, will not significantly affect the final specific activity. However, since the effective labelled nucleotide concentration will be lower, the rate of reaction, as monitored by percentage incorporation of radiolabel into DNA, will appear to be faster. This is simply due to a lower final level of replacement in the template.

It is, nevertheless, possible to produce a very high specific activity probe using label of higher specific activity, because a greater amount of label can potentially be added to the reaction. For example, it is possible to use up to 3000 μCi/μg DNA for label at 3000 Ci/mmol giving a final probe specific activity of approximately 5×10^9 d.p.m./μg. Although it is not generally economical to use such high levels of label in a reaction containing microgram amounts of DNA, it is also possible to achieve such specific activities by using less DNA (<100 ng).

The protocol outlined in *Protocol 3* uses a low amount of labelled nucleotide (~33 pmol) which will effectively label substrate over a wide range of concentrations (50–500 ng per reaction, equivalent to 37.5–375 pmol dCMP). With less than 100 ng of DNA, labelling efficiency may begin to decrease. Therefore, to balance efficiency of utilization of label against probe specific activity, it may be found worthwhile to decrease the level of input label to 50 μCi. If label of relatively low specific activity is to be used (e.g. 400 or 800 Ci/mmol),

DNA concentrations of less than 300 ng per reaction will not be labelled efficiently. Under the conditions outlined in *Protocol 3*, 50% incorporation of label should be obtained with 500 ng input DNA giving a final probe specific activity of 2×10^8 d.p.m./μg. A similar level of incorporation with 50 ng input DNA will give a specific activity of 2×10^9 d.p.m./μg.

3.1.6 Use of other radiolabels and multiple labels

The above discussion has centred on the use of ^{32}P-labelled nucleotides. Similar considerations apply for ^{35}S-, ^{3}H-, and ^{125}I-labelled nucleotides, but several specific points are worth noting.

The maximum specific activity available with a ^{35}S-labelled thionucleotide is ~1000 Ci/mmol, so maximal attainable probe specific activities will be somewhat lower than with high specific activity ^{32}P-dNTPs and, in particular, low amounts of substrate will be labelled less efficiently. However, as thionucleotides are resistant to excision, it is possible to carry out reactions for longer periods without significant loss of probe size or incorporated label. Under otherwise identical conditions, ^{35}S-labelled thionucleotides do appear to give slightly higher levels of incorporation than ^{32}P-labelled nucleotides. The maximum specific activity at which ^{3}H-nucleotides are available is only approximately 100 Ci/mmol, so that attainable probe specific activities (approximately 10^7 d.p.m./μg) are a further tenfold lower than those possible with ^{35}S-nucleotides.

It is possible to obtain probes of higher specific activity by using more than one labelled substrate in a reaction. Although this is feasible with ^{32}P-nucleotides, the specific activities attainable with a single label are high enough for virtually all applications. Additionally, the presence of more than one nucleotide at limiting concentration reduces overall labelling efficiency and probe size. Probe stability is also reduced at very high specific activity. With ^{35}S-, and particularly ^{3}H-, labelled nucleotides specific activity gains with multiple labels are more significant.

3.2 Primer extension methods

In common with nick-translation, primer extension methods utilize the ability of DNA polymerases to synthesize a new DNA strand complementary to a template strand, starting from a free 3′-hydroxyl. In this case the latter is provided by a short oligonucleotide primer annealed to the template. Two general approaches are possible. In the first, a mixture of primers of random sequence is used in order to produce a uniformly labelled DNA copy of any sequence. The second method uses a unique primer to restrict labelling to a particular sequence of interest. It is essential to use a polymerase lacking a 5′–3′ exonuclease activity, otherwise degradation of the primer will occur. Both the Klenow fragment of *E. coli* DNA polymerase I, which lacks the 5′–3′ exonuclease activity, and reverse transcriptase have been used successfully in this way.

Klenow fragment was originally produced as a protease cleavage fragment of *E. coli* DNA polymerase I, but it is now widely available as a cloned enzyme. It is used more frequently than reverse transcriptase to label nucleic acids for hybridization. Its major application is with random hexamer primers in the technique of random primer labelling, but unique primers have also been used to produce labelled copies of single-stranded M13 DNA. Both of these approaches will be described below, but they are preceded by a short outline of labelling methods employing reverse transcriptase.

Both random and unique primers (e.g. oligo dT) are used with reverse transcriptase during cDNA synthesis. Radiolabel may be incorporated simply to monitor the reaction or in order to produce cDNA probes (13). These can be used for any of the applications outlined for uniformly labelled probes, but are particularly appropriate for the specific application of subtractive cDNA cloning (14).

Although very high specific activities are achievable using an oligo-dT primer and reverse transcriptase, it is difficult to ensure complete copying of a long message, due to problems of nucleotide limitation. In most cases it is possible to opt for a higher nucleotide concentration and consequent lower probe specific activity if a complete copy is required. For those cases where this is not adequate, or where the RNA to be copied is poly(A)$^-$ (from certain viruses, for example), the use of random hexanucleotides with reverse transcriptase is a possible alternative. The main disadvantage of this approach is that it will lead to the copying of all RNA molecules present and, if no prior purification of mRNA has been carried out, ribosomal and transfer RNA will be present in excess over message. Unique primers are also used with reverse transcriptase in the technique of primer extension mapping (15) by which a labelled primer is annealed to an RNA molecule and is then extended in order to map the 5′-end of the RNA.

3.2.1 Random primer labelling

Hexanucleotides of random sequence, either derived from DNase I digestion of calf thymus DNA or produced by oligonucleotide synthesis, have been used to prepare labelled copies of both DNA and RNA. Feinberg and Vogelstein (11, 16) first described this approach for the labelling of DNA fragments to very high specific activity using Klenow polymerase. The reaction is illustrated diagrammatically in *Figure 4*.

The absence of the 5′–3′ exonuclease activity ensures that labelled nucleotides incorporated by the polymerase are not subsequently removed as monophosphates. This leads to very efficient utilization of labelled precursors and very flexible reaction conditions. The amount of newly synthesized DNA often exceeds the amount of input DNA. It is therefore likely that strand displacement occurs, so that the same region of a given single strand of input DNA may be copied more than once by DNA synthesis primed from different random hexanucleotides.

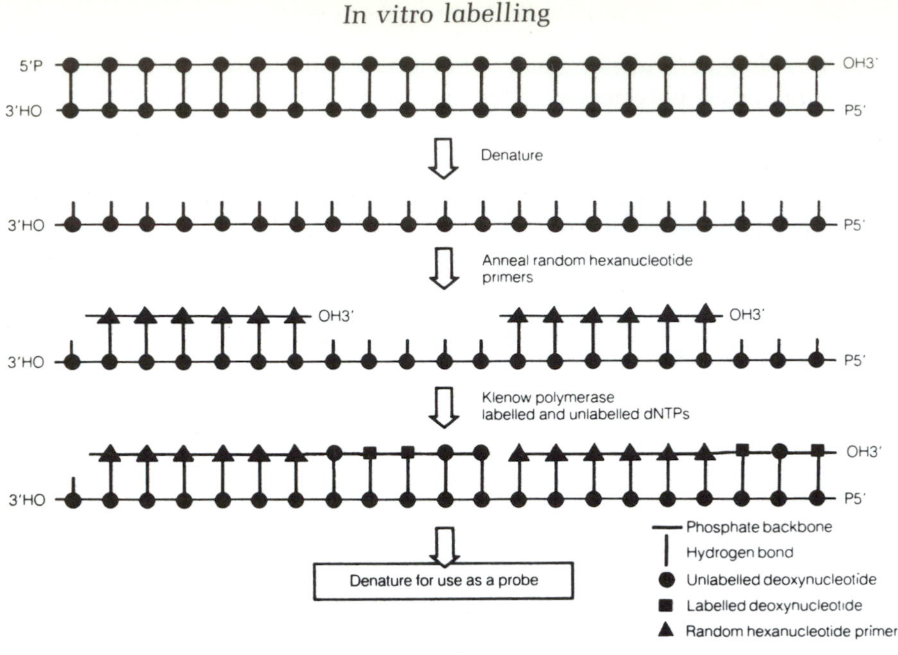

Figure 4. The random primer labelling reaction.

Nature and quantity of substrate

The procedure given in *Protocol 4* is appropriate for labelling between 25 and 250 ng of single-stranded DNA. Using amounts of labelling nucleotide approximately equivalent to the content of the same nucleotide in the DNA template, levels of incorporation as high as 70–80% are routinely achievable with as little as 25 ng DNA. At higher input DNA levels, incorporation will remain high but final specific activity will be lower for a given input label. To maintain specific activity it is necesary to add more label and, because of volume limitations, it may become necessary to scale up the total reaction.

Protocol 4. Preparation of probes by random primer labelling

Materials

10 × labelling buffer (600 mM Tris–HCl, pH 7.8, 100 mM $MgCl_2$, 100 mM 2-mercaptoethanol)
TE buffer, pH 8.0 (10 mM Tris–HCl, pH 8.0, 1 mM EDTA)
300 μM dATP in TE buffer, pH 8.0
300 μM dCTP in TE buffer, pH 8.0
300 μM dGTP in TE buffer, pH 8.0
300 μM dTTP in TE buffer, pH 8.0
[α-^{32}P] dNTP at 3000 Ci/mmol, 110 TBq/mmol
DNA polymerase I (Klenow fragment)

Random hexanucleotide primers dissolved at a concentration of 20 absorbance units/ml in TE buffer containing nuclease-free BSA at 4 mg/ml.

Method

1. Prepare an appropriate nucleotide mix from the 300 μM stocks above by mixing 3.3 μl of three non-radioactive dNTPs, excluding that to be used as label. Alternatively, if the same labelled nucleotide is frequently used, it may be simpler to prepare a 5 × nucleotide mix containing each dNTP at 100 μM except that to be used as label. For example, if [α-^{32}P] dCTP is to be used, the 5 × nucleotide mix contains 100 μM dATP, 100 μM dGTP and 100 μM dTTP.
2. Dissolve the DNA to be labelled in either distilled water or TE buffer to a concentration of 2–25 μg/ml.[a] Denature double-stranded DNA by heating to 95–100 °C for 2 min in a boiling water bath, then chill on ice.
3. Add the following to 10 μl of the nucleotide mix on ice in the order given.

• Denatured DNA to be labelled	25–250 ng
• Water	to a final volume of 50 μl
• 10 × labelling buffer	5 μl
• Primer solution	5 μl
• [α-^{32}P] dNTP	5 μl (50 μCi, 1.85 MBq)
• Klenow polymerase	2 units
• Total volume	50 μl

   Mix gently by slowly pipetting up and down once or twice. Spin briefly (2 sec) in a microcentrifuge to collect contents at the bottom of the tube.
4. Incubate at between 20° and 37 °C for approximately 2 h.
5. Terminate the reaction and remove unincorporated labelled nucleotides by ethanol precipitation or chromatography on Sephadex (see *Protocol 2*).
6. Denature the DNA by heating to 95–100 °C for 5 min before hybridization.

[a] DNA in most restriction enzyme buffers can also be used.

Random priming requires a single-stranded template which may be derived either from cloning in a suitable single-stranded bacteriophage (for example, M13) or more frequently from brief heat denaturation of a double-stranded molecule. If the latter is used, a linear molecule is preferable to avoid rapid renaturation of complementary DNA circles. Random primers, Klenow polymerase, labelled and unlabelled nucleotide precursors are then added in a suitable buffer and the reaction is allowed to proceed.

Random priming has been used successfully to label DNA which is relatively impure. For example, it is often necessary to label specific DNA fragments which have been purified by agarose gel electrophoresis. This procedure is frequently used to separate a cloned insert from vector sequence which may show some cross-hybridization with the target DNA. Feinberg and Vogelstein (16) have demonstrated that it is possible to label DNA fragments in the presence of low gelling temperature agarose without prior purification. A protocol for preparation of DNA fragments in agarose is given in *Protocol 5*. The procedure also efficiently labels DNA from minilysates prepared by either the alkaline (17) or boiling (18) lysis methods

Protocol 5. Labelling of DNA fractionated by electrophoresis in low melting point agarose

The DNA samples produced by the following protocol have been found to be labelled to approximately the same extent and at nearly the same rate as purified DNA. We recommend 5–24 h incubation for optimum labelling.

1. Fractionate restriction endonuclease-digested DNA in a suitable low melting point agarose gel containing 0.5 μg/ml ethidium bromide. Estimate the DNA content of the band by reference to a set of standards on another track. Ideally the band should contain at least 250 ng of DNA so that 25 ng can conveniently be used in the labelling reaction without prior concentration of the DNA.
2. Cut out the desired band cleanly, with the minimum amount of excess agarose, and transfer to a pre-weighed 1.5 ml microcentrifuge tube.
3. Add water at a ratio of 3 ml per gram of gel, and place the tube in a boiling water bath for 7 min to melt the gel and denature the DNA. If the DNA is not used immediately, divide the boiled samples into aliquots and store at −20 °C. Reboil for only 1 min and avoid reboiling any aliquot more than three times.
4. Transfer the tube to a water bath at 37 °C for at least 10 min.
5. Add the volume of DNA/agarose solution that contains 25 ng of DNA to the standard random primer labelling reaction. This volume should not exceed 25 μl in a 50 μl reaction. The reaction may appear to gel during incubation, but polymerization will still proceed if this happens.

Reaction temperature and rate

As the 5′–3′ exonuclease activity of DNA pol I is absent from Klenow polymerase, a wide range of temperatures may be adopted for random primer labelling. Room temperature is frequently employed, but 37 °C may be used for more rapid labelling thus increasing the flexibility in choice of reaction times. Reactions may also be left to proceed overnight if required, although in this case room temperature is preferable to 37 °C in order to minimize contaminating nuclease activity. In general, input DNA is labelled to high specific activity within 2 h at room temperature. However, the apparent reaction rate, as measured by percentage incorporation of radiolabel, depends on a number of factors in addition to temperature. Of particular importance is the chemical concentration of labelling nucleotide. For example, the rate decreases when the concentration of labelled nucleotides is increased at fixed substrate concentration (11) and the advantage of higher temperatures is more significant under these conditions. The concentration of radiolabel given in *Protocol 4* is appropriate for a wide variety of applications, but a change to nucleotide of lower specific activity (for example, if using a ^{35}S-labelled nucleotide) will lead to a slower apparent reaction rate using the same amount of radioactivity. *Figure 5* shows a typical time-course using 25 ng substrate and 30 pmole of labelled precursor (equivalent to approximately 100 μCi of labelled nucleotide at 3000 Ci/mmol). Maximal labelling is achieved within an hour at both 20 °C and 37 °C. Lower levels of substrate or higher concentrations of label will slow this reaction rate.

Specific activity

As with nick-translation, the major determinants of probe specific activity are the specific activity and amount of the label, and the amount of substrate. The reaction temperatures commonly used permit significant strand displacement (see Section 3.1.3) so that, for example, starting from 25 ng template a further 30–50 ng labelled DNA may be synthesized. Both this and the presence of unlabelled template DNA should be taken into account when calculating

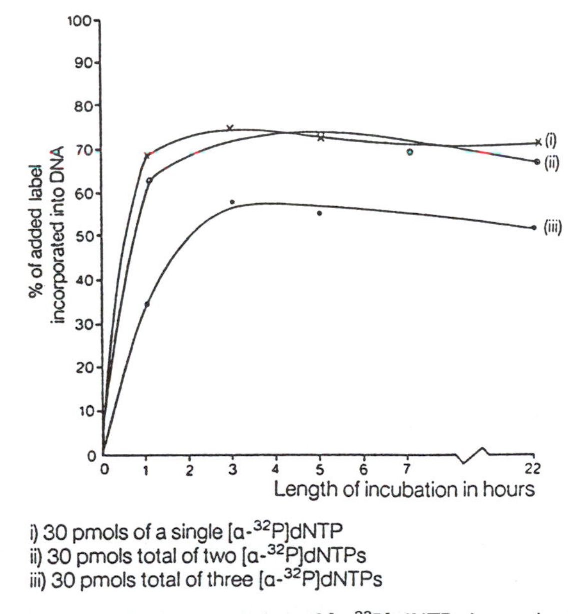

Figure 5. Time-course for the incorporation of [α-^{32}P] dNTPs in random primer labelling. In all cases 25 ng λ DNA were used and reactions were carried out at room temperature (20 °C).

probe specific activity. At the level of input label suggested in *Protocol 4* a specific activity of approximately 2×10^9 d.p.m./μg will be achieved using 25 ng template and levels of incorporation of 70–80% should be obtained. If 250 ng template is used, the probe specific activity will be approximately fivefold lower, although this can be increased by using more label. The addition of more label to a reaction containing 25 ng template DNA will lead to higher specific activities but this is offset by reduced efficiency of utilization. Label at different specific activities can be used and *Table 3* shows some typical results obtained using 25 ng template DNA with a variety of labels. The amounts of label quoted in the table give a good balance between achievable specific activity and economy of utilization. Probe specific activities in excess of 5×10^9 d.p.m./μg are achievable using 200 μCi of an [α-^{32}P]dNTP at 6000 Ci/mmol to label 25 ng of DNA.

Table 3. Probe specific activities obtainable during random primer labelling

Nucleotide	Specific activity (Ci/mmol)	Quantity per reaction (μCi)	(μl)	(pmol)	Specific activity of probe (d.p.m./μg)
[α-^{32}P]dCTP	>400	50	5	125	4.8×10^8
	~800	80	8	100	9.4×10^8
	~3000	50	5	17	1.8×10^9
	~6000	200	20	32	5.3×10^9
[α-^{35}S]dCTPαS	>400	10	1	25	3.1×10^8
	>600	15	1.5	25	4.6×10^8
	>1000	25	2.5	25	7.7×10^8
[1′,2′,2,8-^{3}H]dATP	50–100	10	–	100	1.1×10^8
[^{125}I]dCTP	>1500	376	7.5	25	1.2×10^9

Figures shown are for 25 ng DNA in a reaction volume of 50 μl.
To obtain figures in Bq refer to Appendix 4.

Use of other radiolabels and multiple labels

As with nick-translation, the discussion has concentrated on the use of ^{32}P-labelled nucleotides and many of the considerations discussed in Section 3.1.6 apply equally to random primer labelling. As illustrated in *Table 3*, the specific activities achievable with other nuclides are significantly lower than those achievable with ^{32}P.

We have investigated the use of more than one labelled nucleotide species in some detail. At the same level of input label there has been found to be little advantage in using more than one ^{32}P-labelled nucleotide because, although a slightly higher specific activity may be achieved, the yield of probe and efficiency of utilization of label are both reduced. With ^{35}S, a slight advantage is obtained, but with ^{3}H, where limiting nucleotide concentrations are not used, a significant improvement in specific activity is possible. A maximum of $\sim 1.2 \times 10^8$ d.p.m./μg with three labels and $\sim 2 \times 10^8$ d.p.m./μg with four labels can be obtained compared with $\sim 7.5 \times 10^7$ d.p.m./μg with one label only. In general, percentage incorporation rates are reduced when using two or more labels, and it is advisable to leave the reaction for at least 5 h for maximum incorporation.

Probe size

The two major determinants of probe size appear to be primer concentration (or, more strictly, the ratio of primer to substrate concentration) and the concentration of radiolabelled nucleotide. Hodgson and Fisk (19) have recently demonstrated that, at a fixed level of substrate (200 ng), the average size of nascent single-strand probe DNA is an inverse function of primer concentration:

$$\text{length} = K/\sqrt{\ln}\ Pc$$

where Pc is the primer concentration and K is a constant.

We have investigated the effect of nucleotide concentration using 25 ng linear wild type lambda DNA as substrate and a primer concentration as given in *Protocol 4* (2 absorbance units/ml). Over a range of total input label between 10 and 100 pmol per 50 μl reaction (20–200 pmol for ^{3}H) with between one and three labelling nucleotide species (one and four for ^{3}H), average probe size varies between 200 and 300 bases for ^{32}P, 100 and 150 bases for ^{35}S, and 400 and 500 bases for ^{3}H.

Probe length obtained with tritium-labelled nucleotides may be too great for some *in situ* applications; this can be remedied by prior sonication of the substrate. Exhaustive sonication of DNA yields a population of molecules with a mean size of about 200 base-pairs. Probes produced from such a template using >10 pmol of two [^{3}H]dNTPs average ~95 bases in length. In general it has been found that the variance of sizes obtained by random primer labelling is somewhat wider than with nick-translation. It should also be remembered that, with probes labelled to high specific activity with ^{32}P, radiolysis occurs very rapidly and storage of probes, even for 24 h, will lead to a significant reduction in mean probe length.

3.2.2 Unique primer labelling

The use of Klenow polymerase with unique primers to label DNA was developed out of the M13 dideoxy sequencing methodology (6). Two major approaches have been used. In the first of these, the oligonucleotide primer is annealed to the region 5′ to the multiple cloning site of an M13 vector and, in the second, the primer is annealed to the region 3′ to the multiple cloning site. For both approaches a single-stranded substrate is used. This is normally the (+) strand of any of the M13 vectors; that is, the strand packaged and extruded into the culture medium. A typical reaction requires 50 ng of M13 (+) strand template and 2 ng of primer.

Although it is theoretically possible to carry out unique primer labelling with denatured duplex DNA, in practice the degree of specificity of labelled probe is usually low, most probably due to partial renaturation of the template.

Primer annealed 5′ to the multiple cloning site

This approach, originally described by Hu and Messing (20), uses a hybridization primer which anneals to the region 5′ to the multiple cloning site on the template. DNA synthesis is then initiated and the vector rather than the insert sequence is copied. To avoid read-through into the insert sequence, the reaction is limited by a low concentration of labelled nucleotide. The reaction is terminated by the addition of chelating agent (EDTA) to remove Mg^{2+} and the probe is not denatured before use.

The single-stranded insert acts as a non-radioactive strand-specific probe while the labelled strand complementary to the vector sequence acts as a 'tag'. This method is particularly suited to screening M13 clones for sequences complementary to the insert either by plaque screening or dot blot. Any single-stranded regions of the M13 itself will not hybridize as both they and the target will be (+) strand.

Very high specific activities are theoretically achievable; that is, up to 4.5×10^9 d.p.m./μg DNA using [α-^{32}P]dNTP at 6000 Ci/mmol. This assumes complete copying of vector sequence and excludes the minor contribution of insert sequence. As the vector is generally much larger than the insert in M13, the ability to copy the vector in this way allows a significant amplification of labelling over that achievable were the insert itself to be labelled by more conventional methods. For example, with an insert of approximately 500 bases, the effective maximum specific activity is in excess of 5×10^{10} d.p.m./μg insert, assuming complete copying of the vector.

Primer annealed 3′ to the multiple cloning site

The second approach uses the universal sequencing primer which anneals to the region 3′ to the multiple cloning site on the (+) strand M13. Thus, primer extension produces a labelled copy of any insert sequence. In this case the reaction is not nucleotide-limited, so that a complete copy of the insert is made. As this region is then rendered double-stranded, it is possible to excise the insert using suitable restriction enzymes which cut in the multiple cloning site either side of the insert. The labelled insert fragment can then be gel-purified and denatured prior to use as a probe.

As with the previous procedure, very high specific activity probes can be produced (up to 4.5×10^9 d.p.m./μg DNA with [α-^{32}P]dNTP at 6000 Ci/mmol). This method of labelling clones in M13 has been used in the DNA fingerprint technique of Jeffreys (21). Probes produced in this way are also suitable for mapping studies.

3.3 RNA polymerase-based methods

RNA polymerases catalyse the polymerization of ribonucleoside triphosphates (rNTPs) into RNA using a DNA template. Thus, labelled rNTPs may be incorporated into strand-specific complementary RNA probes that can be used in many of the applications to which DNA probes have traditionally been applied. These include filter and *in situ* hybridization (22), RNase mapping (23), RNA splicing studies (24), *in vitro* translation (25), and as antisense RNA to specifically block gene expression (26). RNA probes have been reported to show higher sensitivity than equivalent nick-translated probes in both Northern blots (27) and *in situ* hybridization. In the latter technique, RNA probes have found particular favour in many laboratories, because anti-sense RNA can be used as a control in the localization of specific mRNA species. A further advantage in filter hybridization is that non-specifically bound RNA probe can be removed by treatment with RNase A, an enzyme that is highly specific for single-stranded RNA (27). This specificity has also led to the development of an alternative to S1 nuclease mapping, called RNase mapping (23), for RNA quantification and for mapping intron-exon boundaries and transcription start and termination points (see Section 4.3.2).

The most relevant contemporary approaches to the synthesis of RNA probes make use of a number of bacteriophage–encoded RNA polymerases (28) that have the unique ability to retain high specificity for their homologous promoters *in vitro*. This enables the specific transcription of template DNA sequences which have been cloned downstream of the appropriate promoter in a suitable vector (see *Figure 6*)

3.3.1 Preparation of the template for transcription

The requirement for a specific cloning step has led to the development of a variety of vectors in which a bacteriophage promoter is present upstream of a multiple cloning site. A summary of some representative vectors is given in *Table 4*.

3.3.2 The transcription reaction

The production of high specific activity RNA probes by the *in vitro* transcription of a suitably prepared recombinant is, in practice, very simple. A procedure suitable for a wide range of applications is given in *Protocol 6*. In the design of the transcription protocol, there are a number of factors to be considered, including probe specific activity, the amount of probe required, the need for full length transcripts and the economy of use of radiolabelled nucleotides. Some of the more important are discussed below.

Protocol 6. Preparation of probes using SP6 RNA polymerase

The protocol given is for labelling with [α-^{32}P] UTP, but similar protocols are appropriate for other nucleotides and radiolabels.

Materials

5 × transcription buffer (0.2 M Tris–HCl, pH 7.5, 30 mM $MgCl_2$, 10 mM spermidine)
Unlabelled NTP mix (2.5 mM each of ATP, CTP, and GTP in 20 mM Tris–HCl, pH 7.5)
0.2 M dithiothreitol (DTT) freshly prepared
Human placental ribonuclease inhibitor (optional)
[α-^{32}P]UTP
SP6 RNA polymerase
Template DNA cut downstream of an SP6 promoter with an appropriate restriction enzyme.

Method

1. Set up the following reaction in a microcentrifuge tube at room temperature.

5 × transcription buffer	4 μl
Unlabelled NTP mix	1.5 μl
0.2 M DTT	1 μl
Ribonuclease inhibitor	20 units (1 μl)
H_2O	to a final volume of 20 μl
Linearized template DNA	2 μg
[α-^{32}P]UTP, 800 Ci/mmol (29.6 TBq/mmol), 20 mCi/ml (740 MBq/ml)	10 μl (equivalent to 12.5 μM)[a]
SP6 RNA polymerase[b]	4–10 units (1 μl)

2. Incubate at 40 °C for 60 min.

3. Unincorporated nucleotides may be removed by ethanol precipitation or Sephadex chromatography (see *Protocol 2*) but, as high levels of incorporation usually result, this is frequently unnecessary. In this case the reaction may be terminated by the addition of 5 μl 0.2 M EDTA, pH 8.0.

[a] A variety of other labelled UTPs can be used, as long as a minimum concentration of 12.5 μM is maintained. Lower concentrations will reduce the proportion of full length transcripts obtained.

[b] Similar reaction conditions are appropriate for T3 and T7 RNA polymerases, except that the template must contain a suitable promoter and the reaction temperature is usually 37 °C.

Nucleotide concentration and choice of radiolabelled nucleotide

In most protocols, it is usual to include the radiolabelled ribonucleotide at a concentration of 12.5 μM; this approximates to the apparent K_m of SP6 RNA polymerase for UTP, CTP, and GTP. ATP is rarely used as the label in reactions with SP6 RNA polymerase since its apparent K_m is higher than those of the other nucleotides. In addition, it is worth noting that GTP is

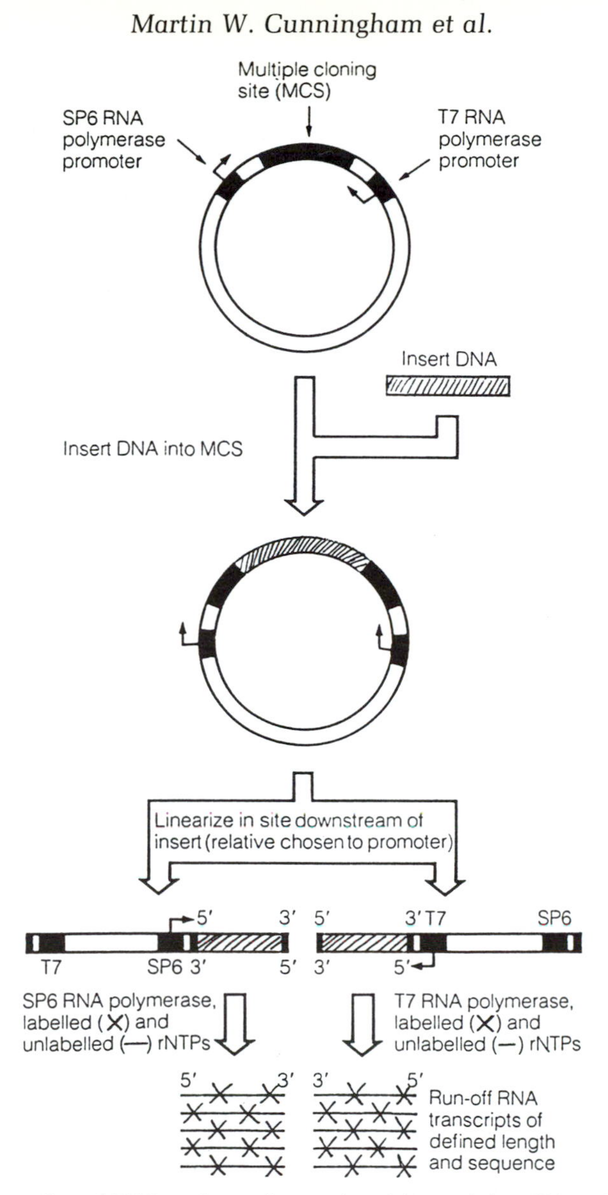

Figure 6. Preparation of RNA probes using a plasmid containing SP6 and T7 RNA polymerase promoters.

usually the initiating nucleotide in the transcription reactions; therefore because the initiation step is rate-limiting, the use of relatively low concentrations of this nucleotide can lead to variations in the overall efficiency of the reaction. In contrast, both CTP and UTP are routinely incorporated with high efficiency under the conditions described (equating to an incorporation level of 80% within 60 min using 2 μg of template DNA).

Table 4. Characteristics of some available 'transcription vectors'

Name	RNA pol enzyme	Comments	Availability
pSP 64	SP6	Standard cloning vectors.	Various suppliers.
pSP 65		Use as a pair for cloning in opposite orientations.	
pAM 18 pAM 19	SP6/T7	Typical of the range of dual promoter vectors allowing transcription of 'sense' or 'anti-sense' RNA from a single vector using two different polymerases.	Amersham, but equivalent vectors available from a number of suppliers.
pTZ 18	T7	Contain M13 origin of replication to allow preparation of ssDNA for sequencing.	Pharmacia and others.
Bluescribe	T7	M13 origin plus dual promoters with colour selection.	Stratagene.
M13T7	T7	M13-derived vector allows convenient preparation of ssDNA for sequencing. No helper phage required. ssDNA is converted *in vitro* to dsDNA for efficient transcription.	Amersham.

Probe specific activity

The major determinant of RNA probe specific activity is the specific activity of the radiolabelled nucleotide used, since it is possible either to remove the template DNA using DNaseI (27) or to avoid competition in the hybridization reaction by not subjecting the reaction mixture to denaturation prior to use as a probe. In general, it is not advisable to use more than one labelled nucleotide in the transcription reaction as the presence of more than one nucleotide at limiting concentration seriously compromises the efficiency of the reaction.

Using the protocol described in *Protocol 6*, a probe specific activity of 1.3×10^9 d.p.m./μg RNA is obtained. The nomogram in *Figure 7* may be used to determine probe specific activities under other conditions, including the use of ^{35}S- or ^{3}H-labelled ribonucleoside triphosphates.

3.4 End-labelling

A wide variety of techniques are available for introducing label at either the 3′- or 5′-ends of linear DNA or RNA. Usually only a single radioactive atom is introduced per molecule. The specific activities achievable by such techniques are thus significantly lower than those obtained by the uniform labelling methods discussed in the previous sections and ^{32}P-labelled nucleotides are therefore most frequently used. Some of the more common methods for

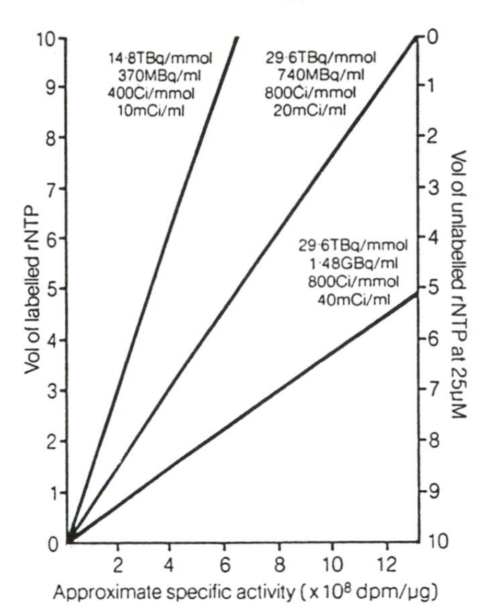

Figure 7. Nomogram for the preparation of RNA probes of known specific activities. The reaction conditions are as quoted in *Table 5*, with a 20 μl reaction volume.

end labelling both DNA and RNA will be described, followed by a general discussion of their applications. T4 DNA polymerase catalysed labelling will also be included in this section, although this enzyme can be used for either end or uniform labelling. The techniques to be discussed can be considered under the categories, 5′-end-labelling, 3′-end-labelling, and end-repair, although the latter is strictly a subset of 3′-end-labelling.

A general point to be remembered when end labelling is that it is useful to calculate the number of picomoles of ends in the reaction that are available for labelling, remembering that a double-stranded molecule has two 5′- and two 3′-ends. Values for the amounts of 3′- or 5′-ends in double-stranded molecules are given in *Table 5*. For RNA or single-stranded oligonucleotides, the picomoles of ends per microgram are half the values given relative to

Table 5. Amounts of 3′- or 5′-ends present in double-stranded DNA

Molecular weight	Base-pairs	pmol of 3′- or 5′-ends per μg of dsDNA
10.0×10^6	15000	0.2
1.0×10^6	1500	2.0
0.1×10^6	150	20.0
0.01×10^6	15	200.0

See Section 3.4 for further discussion.

molecular weight. For example, single-stranded DNA of molecular weight 0.1×10^6 is equivalent to 300 nucleotides and 10 pmol ends per microgram. If a whole restriction enzyme digest is to be end-labelled, then the amount of ends per microgram of DNA may be determined by calculating the number of picomoles of ends per microgram for the undigested DNA, and multiplying this figure by the final number of fragments present (equivalent to the number of cutting sites plus one for an initially linear molecule).

For example, if DNA of molecular weight 2×10^7 is digested with a restriction enzyme with 10 cutting sites, then the number of picomoles of 3′- or 5′-ends per μg of DNA will be

$$0.1 \times 11 = 1.1 \text{ pmol}/\mu\text{g}.$$

Once this has been determined, the appropriate amount of label to be added can be adjusted accordingly, depending on whether the priority is the final specific activity of the DNA (i.e. the proportion of ends labelled) or the efficiency of utilization of label. As the former is normally more important, most published protocols include labelled nucleotide in considerable molar excess to maximize final probe specific activity, particularly under less able labelling conditions. However, it should be remembered that if, for example, the reaction contains 5 pmol 5′-ends and 50 pmol of label, the maximum obtainable level of incorporation is only approximately 10%.

3.4.1 5′-end-labelling with T4 polynucleotide kinase

RNA and DNA may be 5′-end-labelled using T4 polynucleotide kinase. Common uses for this reaction are to label restriction fragments for Maxam and Gilbert sequencing (1) and to label oligonucleotides for use as hybridization probes (29). The enzyme catalyses the reversible transfer of the γ-phosphate of a ribonucleoside 5′-triphosphate donor to the 5′-hydroxyl group of a polynucleotide, oligonucleotide or nucleoside 3′-phosphate (30). [γ-^{32}P]ATP is most frequently used as donor. The radiolabelled γ-phosphate group is transferred to DNA or RNA containing a 5′-hydroxyl terminus. This is termed the ‘forward’ reaction. However, most polynucleotides have a 5′-phosphate group which must be removed using alkaline phosphatase before they can be used as substrates. This step may be avoided by carrying out an ‘exchange’ reaction (31) with T4 polynucleotide kinase (see *Figure 8*). In this case the reaction is driven by excess ADP which causes the enzyme to transfer the terminal 5′-phosphate from DNA to ADP. The DNA is then rephosphorylated by transfer of the labelled γ-phosphate from [γ-^{32}P]ATP. This reaction is more convenient than dephosphorylation followed by the forward reaction, but it usually occurs at a lower efficiency. Procedures based on a number of published protocols (13, 32, 33) are given in *Protocol 7*. Specific activities of 5×10^5d.p.m./pmole of ends and 8×10^5d.p.m./pmole of ends can be achieved with the exchange and forward reactions respectively using blunt-ended DNA fragments.

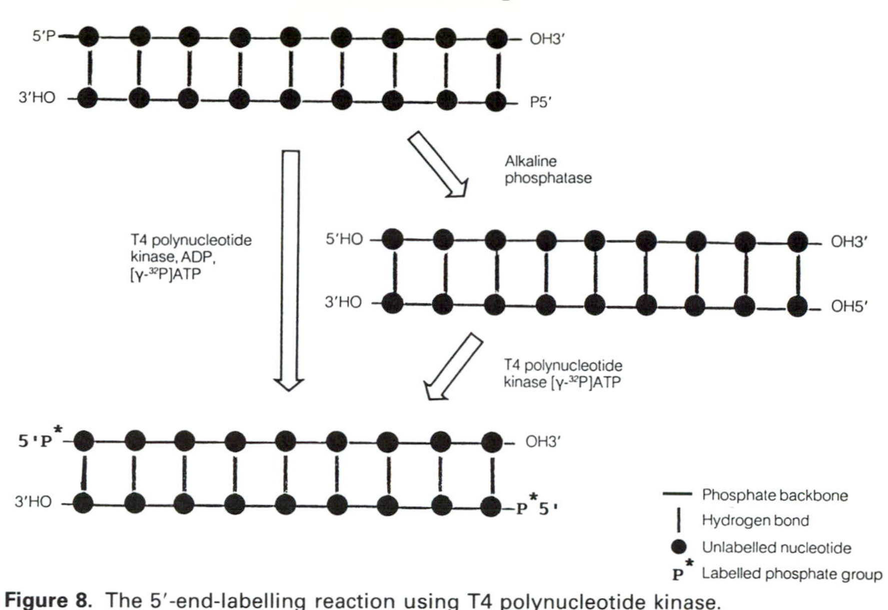

Figure 8. The 5′-end-labelling reaction using T4 polynucleotide kinase.

Protocol 7. 5′-end-labelling using T4 polynucleotide kinase

Forward reaction

Materials

Calf intestinal alkaline phosphatase (CIAP)[a]
10 × CIAP buffer (0.5 M Tris–HCl, pH 9.0; 10 mM $MgCl_2$; 10 mM $ZnCl_2$; 10 mM spermidine)
10 mM Tris, pH 8.0
Phenol saturated with TE buffer
Chloroform or ether
5 M NaCl
Ice-cold ethanol
TE buffer (10 mM Tris, pH 8.0, 1 mM EDTA)
5 × kinase buffer (0.5 M Tris–HCl, pH 7.6; 0.1 M $MgCl_2$; 50 mM DTT; 1 mM spermidine)
Polynucleotide kinase
[γ-^{32}P] ATP at 3000 Ci/mmol (110 TBq/mmol) or 5000 Ci/mmol (185 TBq/mmol) and 10 mCi/ml (370 MBq/ml)

Method

The following protocol includes dephosphorylation (steps 1–2) using calf intestinal alkaline phosphatase (CIAP). For molecules with free 5′-OH groups (e.g. most chemically synthesized oligonucleotides) this is not necessary and the procedure should be followed from step 3.

The procedure starts with 10 pmol ends or 5 pmol duplex DNA (equivalent to approximately 3.3 μg of a 1 kb duplex DNA fragment or 0.065 μg of a 20-mer oligonucleotide). For the dephosphorylation reaction, the amount of enzyme should be increased or decreased depending on the amount of input DNA, but for the kinase reaction, the amount of enzyme given is adequate for between 1 and 50 pmol of ends. However, at higher levels of input DNA, the amount of label recommended will not give maximal labelling of available ends and, if this is required, it will be necessary to increase the amount of label. If necessary, label can be dried down before use.

1. Dissolve the DNA in a small volume of 10 mM Tris, pH 8.0, then set up the following mix in a centrifuge tube on ice.

• DNA	10 pmol ends
• 10 × CIAP buffer	5 μl
• H_2O	to a final volume of 50 μl
• CIAP	0.05 units

   Incubate at 37 °C for 30 min for duplex DNA or at 55 °C for 30 min for RNA. For DNA with blunt ends or recessed 5′ termini, better results may be obtained by use of the higher temperature, or by successive 15-min incubations at 37 °C and 55 °C.

2. Briefly spin, then add an equal volume of buffer saturated phenol. Extract twice with phenol and twice with chloroform or ether. Add 1/10th volume of 5 M NaCl followed by two volumes of cold ethanol for DNA or three volumes for RNA. Precipitate at −20 °C overnight or at −80 °C for 30 min.[b]

3. Redissolve the pellet in 10 μl TE buffer and set up the following mix in a microcentrifuge tube on ice.

• DNA (10 pmol ends)	10 μl
• 10 × kinase buffer	5 μl
• H_2O	to a final volume of 50 μl
• [γ-^{32}P]ATP (3000 Ci/mmol, 110 TBq/mmol)	200 μCi (7.4 MBq, 65 pmol)[c]
• Polynucleotide kinase	10–20 units

   Incubate at 37 °C for 30–45 min.

4. Unincorporated label may be removed by phenol extraction followed by ethanol precipitation, or by chromatography though G-50 Sephadex (see *Protocol 2*). For oligonucleotides, unincorporated label is best removed by gel electrophoresis, HPLC, or thin layer chromatography (34).

5. Denature double-stranded probes by heating to 95–100 °C for 5 min before hybridization.

Exchange reaction

Materials

10 × exchange buffer (0.5 M imidazole–HCl, pH 6.6, 100 mM $MgCl_2$, 50 mM DTT, 3 mM ADP, 1 mM spermidine).[d]
Polynucleotide kinase
[γ-^{32}P]ATP at 3000 Ci/mmol (110 TBq/mmol) or 5000 Ci/mmol (185 TBq/mmol) and 10 mCi/ml (370 MBq/ml)

Method

1. Set up the following mix in a microcentrifuge tube.

DNA to be labelled	10 pmol ends
10 × exchange buffer	5 μl
H_2O	to a final volume of 50 μl
[γ-^{32}P]ATP at (3000 Ci/mmol, 110 TBq/mmol)	200 μCi (7.4 MBq, 65 pmol)
Polynucleotide kinase	10 units

Incubate at 37 °C for 30–60 min.

2. Remove unincorporated label by phenol extraction followed by ethanol precipitation, or by chromatography through G-50 Sephadex (*Protocol 2*).
3. Denature double-stranded probe by heating to 95–100 °C for 5 min before hybridization.

[a] The calf intestinal enzyme is suggested here because it is more readily inactivated. If the bacterial enzyme is used, the same buffer can be employed but the reaction should be performed at 68 °C and it may be found necessary to increase the number of units used to achieve complete dephosphorylation. An additional phenol extraction may be necessary in order to effectively inactivate the enzyme.

[b] If any problems are experienced inactivating the CIAP, which may result in poor labelling, terminate the reaction by addition of EDTA to 1 mM (which chelates the essential Zn^{2+} ions) and SDS to 0.5% (w/v), and heat to 68 °C for 15 min before phenol extraction.

[c] The final concentration of ATP in the reaction should be $\geq$1 μM for optimal incorporation; the quoted conditions give a final concentration of 1.3 μM. It is possible to substitute [γ-^{32}P]ATP of higher specific activity (e.g. 5000 Ci/mmol) to achieve a higher specific activity probe, but this may result in reduced labelling efficiency due to the lowered ATP concentration.

[d] Spermidine is not essential but improves enzyme stability, and hence labelling efficiency in some cases.

The reaction given is appropriate for oligonucleotides but, if they are to be subsequently purified by gel electrophoresis or thin layer chromotography, it is preferable to minimize reaction volume and load directly from the reaction. In this case it will be necessary to dry down the label and resuspend it in 1 or 2 μl H_2O. Final reaction volume should be approximately 10 μl. Effective protocols for 5′-end-labelling of oligonucleotides and subsequent purification have been given in a previous volume in this series (34).

Both single- and double-stranded nucleic acids can be used as substrates. Generally the labelling of blunt or recessed 5′-ends in duplex DNA is less efficient than that of protruding 5′-ends, particularly in the case of the exchange reaction. A short pre-incubation of the DNA at 70 °C followed by quick chilling on ice may improve labelling efficiency in this case. Additionally, increasing final spermidine concentration to 1 mM and the addition of 10% glycerol may improve enzyme stability, allowing somewhat longer (60–90 min) reaction times to be used to maximize labelling efficiency. Removal of the 5′-cap is essential for 5′-end-labelling of mRNA (35). [γ-^{35}S]ATPγS can also be used as donor, although higher enzyme concentrations are required for efficient reaction rates. ^{35}S-labelled oligonucleotide probes are particularly useful for *in situ* hybridization and are resistant to 5′-exonuclease activities.

The major advantages of 5′-end-labelling may be summarized as:

(a) both DNA and RNA can be labelled;
(b) oligonucleotides and small DNA and RNA fragments can be conveniently labelled;
(c) the location of the labelled group is known;
(d) restriction digest fragments may be labelled so that several probes can be prepared at once.

Points (b)–(d) apply equally to 3′-end-labelling with terminal transferase, and points (c) and (d) to all end-labelling techniques.

A major practical disadvantage of 5′-end-labelling with polynucleotide kinase is the necessity for prior dephosphorylation of substrate if the forward reaction is to be used. As with all end-labelling methods, assuming all available ends are labelled, the achievable specific activity depends on the specific activity of labelled nucleotide used.

3.4.2 3′-end-labelling with terminal deoxynucleotidyl transferase

Terminal deoxynucleotidyl transferase (Tdt) can be used in conjunction with ^{32}P, ^{35}S, or ^{3}H-labelled nucleotides to 3′-end-label DNA for a variety of applications. The enzyme adds a series of supplied deoxynucleotides onto the 3′-end of DNA (36) (see *Figure 9*). Single-stranded and double-stranded DNA are substrates for Tdt, although the end-labelling reaction is most efficient with single-stranded DNA of at least three residues. The method is appropriate for end-labelling chemically synthesized oligonucleotides as an alternative to 5′-end-labelling with polynucleotide kinase. The reaction requires the presence of a divalent cation and is therefore inhibited by chelating agents such as EDTA.

In the presence of Mg^{2+} the enzyme catalyses the addition of only one or two nucleotides onto the 3′-ends of DNA which must be single-stranded. Replacement of Mg^{2+} by Co^{2+} results not only in a more efficient reaction with single-stranded primer but also in the addition of more than two residues

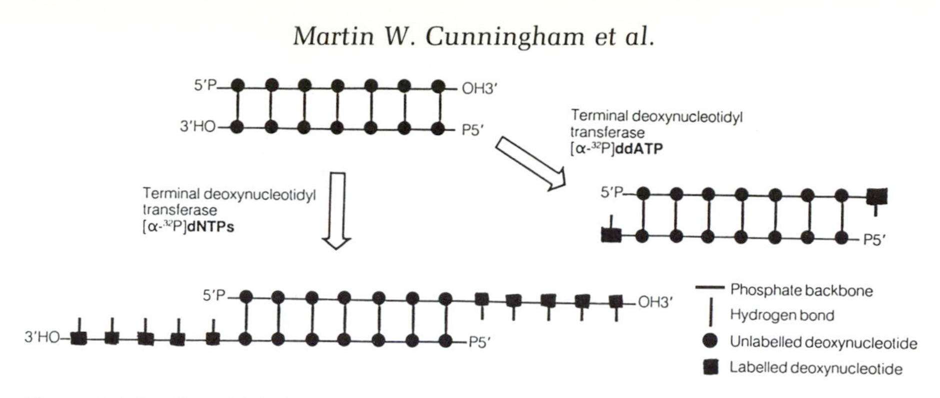

Figure 9. The 3′-end-labelling reaction using terminal deoxynucleotidyl transferase.

and furthermore allows the direct labelling of duplex DNA having protruding, blunt, or recessed 3′-ends (37).

When employing this enzyme to incorporate a single end label, the most reliable approach is to use a dideoxynucleoside 5′-triphosphate (usually [α-^{32}P]ddATP) as precursor (see *Figure 10*); this lacks a 3′-hydroxyl group, so polymerization stops when one residue has been added (38). Before the adoption of this terminator, cordycepin 5′-[α-^{32}P] triphosphate ([α-^{32}P]KTP) (see *Figure 10*) was frequently used. KTP, although also lacking a 3′-hydroxyl group, does have a hydroxyl group at the 2′-position and is a rather poor substrate for terminal transferase. [α-^{32}P]ddATP, with a 2′-deoxy as well as a 3′-deoxy group, is a much more efficient substrate. *Table 6* illustrates some efficiencies obtained using terminal transferase with [α-^{32}P]KTP and [α-^{32}P]ddATP to label 3′-ends, and using T4 polynucleotide kinase and [α-^{32}P]ATP to label 5′-ends. The results demonstrate that, for all three categories of ends, the labelling efficiency using [α-^{32}P]ddATP is three to four times greater than with [α-^{32}P]KTP. They also show that for 3′-protruding and blunt ends, 3′-end-labelling with [α-^{32}P]ddATP is significantly more efficient than 5′-end-labelling of the complementary strand with [γ-^{32}P]ATP. However, the labelling efficiency at blunt ends is also dependent on the sequence at or close to the 3′-end. Fragments with a higher ratio of GC base-pairs label less efficiently than those with a low ratio.

Table 6. Comparison of end-labelling efficiencies with different labels

		Counts incorporated per 10 pmol of DNA ends (d.p.m. × 10^{-6})		
Type of 3′-end	**Type of lambda digest**	**5′-ends labelled with [γ-^{32}P]ATP**	**3′-ends labelled with [α-^{32}P]ddATP**	**[α-^{32}P]KTP**
Protruding	*Pst* I	1.9	16.6	6.0
Recessed	*Hind* III	24.5	12.6	3.5
Blunt	*Alu* I	11.4	61.5	16.5

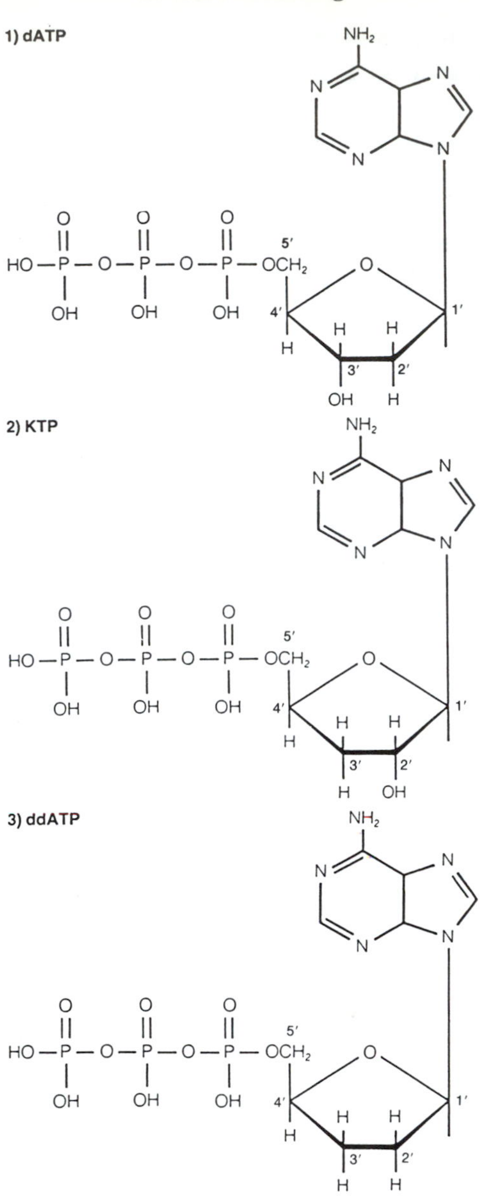

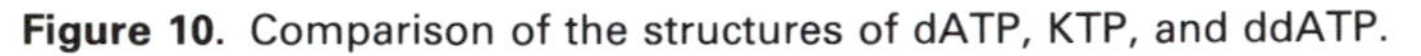

Figure 10. Comparison of the structures of dATP, KTP, and ddATP.

Protocol 8 gives a procedure suitable for the introduction of a single label at protruding, blunt or recessed 3′-ends. Using an *Alu*1 digest of λ, and [α-^{32}P]ddATP at ~3000 Ci/mmol, a DNA specific activity of >8.0×10^6 d.p.m./μg can be achieved. With other DNAs the maximum specific activity that can be achieved will depend on the relative concentrations of 3′-

end and label, and also on the type of 3′-end used. As discussed in Section 3.4, the relative ratio of label to DNA can be altered depending on whether final activity or efficiency of utilization of label are the primary requirements. Under the conditions given, the level of incorporation should be approximately 20%. ^{35}S-labelled thionucleotides are incorporated more slowly, so more enzyme and longer incubation times should be used.

Protocol 8. 3′-end-labelling using terminal deoxynucleotidyl transferase

Materials

10 × Tdt buffer (1.4 M sodium cacodylate, pH 7.2, 10 mM cobalt(II) chloride, 1 mM dithiothreitol).
[α-^{32}P]ddATP (3000 Ci/mmol, 110 TBq/mmol or 5000 Ci/mmol, 185 TBq/mmol)
0.2 M EDTA

Note: Cacodylate is a poisonous arsenic compound and should be handled with care. All contact with the skin should be avoided.

Method

1. Dissolve purified DNA equivalent to 10 pmol of 3′-ends in 20 μl H_2O.[a]
2. Set up the following mix in a microcentrifuge tube on ice.
   - DNA to be labelled (10 pmol ends) — 20 μl
   - 10 × Tdt buffer — 5 μl
   - H_2O — to 50 μl final volume
   - [α-^{32}P]ddATP (3000 Ci/mmol, 110 TBq/mmol or 5000 Ci/mmol, 185 TBq/mmol) — 5 μl (50 μCi, 1.85 MBq) (16 pmol or 10 pmol)
   - Terminal deoxynucleotidyl transferase — 10 units
3. Place the tube in a constant temperature bath at 37 °C for 1 h.
4. Terminate the reaction by addition of 10 μl 0.2 M EDTA. Remove unincorporated nucleotides by ethanol precipitation or Sephadex G-50 chromatography (see *Protocol 2*).
5. Denature double-stranded probe at 95–100 °C for 5 min before hybridization.

[a] If the availability of DNA is the limiting factor (e.g. 1–2 pmol 3′-ends), increase the [α-^{32}P]ddATP to 7.5 μl (75 μCi) and the Tdt to 15 units in step 2.

3.4.3 Other 3′-end-labelling methods

Both poly A polymerase (39) and T4 RNA ligase (40) are used to label the 3′-ends of RNA molecules, usually in preparation for chemical or enzymatic sequencing (see Section 4.2)

3.4.4 End repair catalysed by Klenow polymerase

In the presence of suitable deoxynucleotides, the 5′–3′ polymerase activity of the Klenow fragment can be used to fill in from a recessed 3′-end produced by restriction endonuclease cleavage, using the corresponding 5′-overhang as template (see *Figure 11*). The nucleotides chosen for labelling will depend on the sequence of the 5′-overhang. It is often possible to fill-in and label with only one nucleotide or with several, depending on the length of the overhang, its sequence and the labelled nucleotides supplied. Thus, some variation in specific activity is possible. By careful choice of nucleotide it is also possible to label only one species of end in a mixture of fragments cut with different restriction enzymes. Single fragments produced by a double cut can also be selectively labelled at one end only, thereby producing strand specific probes.

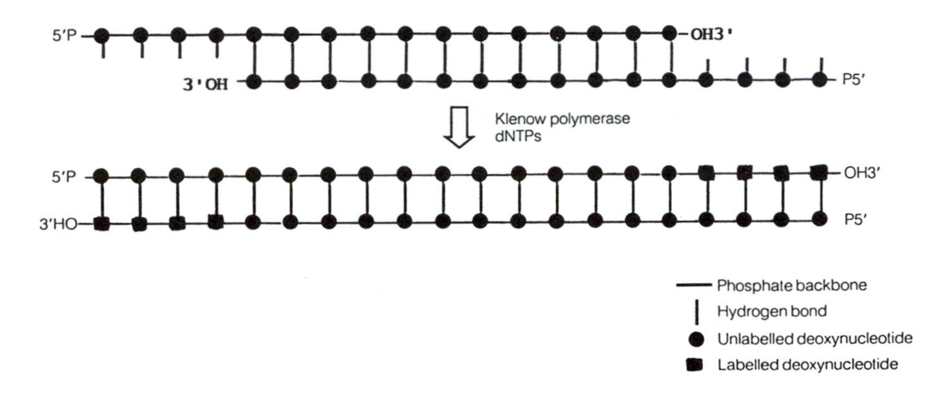

Figure 11. In-filling reaction using Klenow polymerase.

When used for end-labelling, it should be remembered that Klenow polymerase will continue to carry out pyrophosphate exchange (removal of terminal nucleotide by 3′–5′ exonuclease activity followed by repolymerization) when all residues have been filled in. At low nucleotide concentrations, this can cause conversion of all free dNTP corresponding to the terminal nucleotide to dNMP, resulting in loss of the terminal nucleotide. For this reason short reaction times (10–15 min at room temperature) are recommended. Termination by methods other than heat inactivation is also advisable, as raised temperature increases the rate of the exchange reaction.

Klenow polymerase may also be used to label blunt ends, as the 3′–5′ activity of the enzyme is adequate for the removal of terminal 3′ nucleotides, allowing subsequent replacement by a labelled equivalent.

The major advantages of Klenow-catalysed end-labelling are that:

- it can generally be used after restriction digestion with no intermediate purification;

- relatively high specific activities can be achieved;
- selective labelling is possible, facilitating production of strand specific probes.

The major disadvantage is that it cannot be used efficiently for 3′-overhangs.

A method suitable for most situations is given in *Protocol 9*. As with other end labelling methods, the relative amounts of available ends and pmoles of label can be altered depending on requirements. A variety of radionucleotides may be used but, because label density is low, ^{32}P is generally the label of choice.

Protocol 9. End-repair using Klenow polymerase

Materials

10 × end-repair buffer (500 mM Tris–HCl, pH 7.5, 100 mM $MgCl_2$, 10 mM dithiothrietol)
5 × nucleotide mix (containing each nucleotide at 100 μM except those to be used as label) in 10 mM Tris–HC1, pM8.0, 1 mM EDTA
[α-^{32}P] dNTP(s) at 3000 Ci/mmol (110 TBq/mmol), 10 mCi/ml (370 MBq/ml)
Klenow polymerase
0.2 M EDTA
Chase solution (1 mM each of dATP, dCTP, dGTP, dTTP).

Method

1. Set up the following reaction in a microcentrifuge tube on ice.
   - DNA to be labelled — 1 μg (equivalent to 1 pmol ends for a 3 kb fragment)
   - H_2O — to a final volume of 20 μl
   - 10 × end-repair buffer — 2 μl[a]
   - 5 × nucleotide mix — 2 μl
   - [α-^{32}P]dNTP(s) at 3000 Ci/mmol (110 TBq/mmol) — 1 μ (10 μCi, 370 kBq; equivalent to 3 pmol[b])
   - Klenow polymerase — 2 units

   Incubate at room temperature for 15 min.[c]
2. Terminate reaction by addition of 5 μl 0.2 M EDTA. Unincorporated nucleotides may be removed by ethanol precipitation or chromatography on Sephadex G-50 (see *Protocol 2*).
3. Denature double-stranded probe at 95–100 °C for 5 min before hybridization.

[a] It is also possible to label DNA in most restriction enzyme buffers without prior purification, as the primary requirements for Klenow polymerase are Mg^{2+} ions and a roughly neutral pH.

[b] The use of high specific activity nucleotides is shown above, in order to provide maximal probe specific activity. However, it is also possible to use low specific activity label (400 or 800 Ci/mmol (14.8 or 29.6 TBq/mmol)). In this case much less label is required (approximately

1 μCi, 37 kBq, of 400 Ci/mmol, 14.8 TBq/mmol, in the above reaction), but probe specific activity is correspondingly less. The use of more than one labelled nucleotide in the reaction is likely to lead to reduced efficiency of incorporation as enzyme is limited by more than one nucleotide. As discussed for previous end labelling techniques, a variety of DNA/label ratios are possible, depending on whether probe specific activity or utilization of label is the chief priority.

[c] If it is important that all end-labelled molecules are of the same length when using 3′-recessed DNA fragments, it is advisable to carry out a cold chase step after labelling. Add 2 μl of chase solution and incubate for a further 5 min at room temperature before addition of EDTA.

Although Klenow polymerase is most frequently used, DNA polymerase I holoenzyme is still sometimes used for filling in restriction fragments with 3′-recessed ends. In this case it is necesary to inhibit the 5′–3′ exonuclease activity using high salt (100 mM NaCl) and low temperature (4 °C). The enzyme should not be heat-inactivated as this may lead to transient activation of the 5′–3′ exonuclease.

3.4.5 Labelling with T4 DNA polymerase

T4 DNA polymerase has similar properties to the Klenow fragment in that it possesses a 5′–3′ polymerase and a 3′–5′ exonuclease activity (41). However, the exonuclease activity of the T4 enzyme is about 250-fold greater than that of the Klenow fragment. This has the practical advantage that it is possible to control the relative levels of the two activities by omitting or providing deoxynucleoside triphosphates. The enzyme can be used to label DNA in two distinct ways: 3′-end-labelling of DNA or labelling by replacement synthesis.

3′-end-labelling of DNA

Using T4 DNA polymerase it is possible to control the precise position of labelling by selective addition of deoxynucleotides. Either [α-^{32}P] or [α-^{35}S] deoxynucleoside triphosphates can be used in this reaction. Three approaches are outlined below.

Recessed 3′-end. In this case the polymerase activity will synthesize a strand complementary to the 5′-overhang. It is thus only necessary to include a radiolabelled nucleotide complementary to the first residue in the overhang to achieve effective labelling. If all four deoxynucleotides are added the product will be a blunt-ended molecule labelled at or near the terminus. The Klenow fragment of DNA polymerase I will also carry out this reaction.

Blunt end. The enzyme is used in the presence of only one labelled deoxynucleotide. Nucleotides are removed from the 3′-end by the 3′–5′ exonuclease activity until a residue identical to the radiolabelled nucleotide is reached. When this nucleotide is removed, the 5′–3′ polymerase activity of the enzyme inserts a labelled nucleotide in its place. A continuous series of exchange reactions now occurs at this point, producing a molecule labelled at its 3′-end. If the nucleotide used is identical to the initial 3′-residue of the molecule, a labelled blunt-ended molecule will be produced. If any other labelled nucleotide is chosen, then the molecule will be left with a 5′-overhang.

Protruding 3′-end. The 3′-overhang will be removed by the powerful 3′–5′ exonuclease activity until the molecule is blunt-ended, from which point the reaction will proceed as above.

These reactions are most widely used to label restriction enzyme fragments. It is often possible to carry out the initial digestion in T4 polymerase assay buffer to simplify the procedure. However, not all restriction enzymes are active in this buffer and it is usually advisable to carry out a small-scale pilot reaction first. It may then be found necessary to include a phenol/chloroform extraction and an ethanol precipitation between the two stages.

Labelling by replacement synthesis

DNA can be labelled to high specific activity in a two stage reaction. Initially, long recessed 3′-termini are produced by incubation of the DNA with T4 polymerase in the absence of added dNTPs. When these are provided in the second stage, the polymerase activity uses the DNA as a primer-template for the re-synthesis of sequences complementary to the long 5′-overhangs. Such molecules are frequently used as hybridization probes.

The main advantages of the method are:

- no hairpin structures occur, in contrast to nick-translation;
- either end of the molecule can be isolated following restriction enzyme digestion and gel purification for use as a strand-specific probe.

The main disadvantage is that the first stage must not be allowed to proceed to the centre of the molecule. Reaction conditions are therefore generally chosen to favour incorporation of label near the ends. Hence, uniform labelling is not achieved and, in a mixture of restriction fragments, the extent to which all molecules can be labelled is dictated by the size of the smallest fragment present.

Protocols for both the end labelling and the replacement synthesis reactions are given in *Protocol 10*. They are closely based on the adaptation of the original method given by Maniatis (32).

Protocol 10. Labelling with T4 DNA polymerase

End-labelling reaction

Materials

10 × T4 DNA polymerase buffer (0.33 M tris-acetate, pH 7.9, 0.66 M potassium acetate, 0.10 M magnesium acetate, 5 mM DTT, 1 mg/ml BSA)

10 × nucleotide mix (containing three nucleotides at 1 mM, but excluding that to be used as label) in 10 mM Tris–HCl, pH 8.0, 1 mM EDTA

[α-^{32}P]dNTP at 3000 Ci/mmol (110 TBq/mmol), 10 mCi/ml (370 MBq/ml)
T4 DNA polymerase
Chase solution (1 mM each of dATP, dCTP, dGTP, and dTTP)

Method

1. Set up the following reaction in a microcentrifuge tube on ice.

DNA to be labelled	1 μg (1 pmol ends for a 3 kb fragment)
H_2O	to a final volume of 20 μl
10 × T4 polymerase buffer	2 μl[a]
10 × nucleotide mix	2 μl
[α-^{32}P]dNTP	2 μl (20 μCi, 740 kBq; equivalent to 7 pmol)[b]
T4 DNA polymerase	2–3 units

   Incubate at 37 °C for 5 min.
2. Add 1 μl of chase solution. Incubate at 37 °C for a further 10 min.[c]
3. Terminate reaction by addition of 5 μl 0.2 M EDTA, then heat to 70 °C for 5 min.
4. Unincorporated nucleotides may be removed by phenol extraction or by Sephadex G-50 chromatography (see *Protocol 2*).

Labelling by replacement synthesis

Materials

10 × T4 DNA polymerase buffer (see above)
T4 DNA polymerase
10 × nucleotide mix (containing three nucleotides at 1 mM, excluding that to be used as label) in 10 mM Tris–HCl, pH 8.0, 1 mM EDTA
Chase solution (1 mM each of dATP, dCTP, dGTP and dTTP)
[α-^{32}P]dNTP at 400 Ci/mmol (14.8 TBq/mmol], 10 mCi/ml (370 MBq/ml).

Method

1. Set up the following reaction on ice in a microcentrifuge tube.

DNA to be labelled	1 μg (1 pmol ends for a 3 kb fragment)
H_2O	to a final volume of 20 μl
10 × T4 polymerase buffer	2 μl
T4 DNA polymerase	1 unit

2. Incubate at 37 °C for the required length of time. Under the above conditions, approximately 15 residues are excised from each 3′-end per minute.
3. Add 2 μl of 10 × nucleotide mix.

4. Add the fourth dNTP as [α-^{32}P]dNTP (400 Ci/mmol, 14.8 TBq/mmol)[d] in an amount at least equivalent to the moles of nucleotide excised from the DNA by exonuclease. Incubate at 37 °C for 1 h.
5. Add 2 μl of chase solution and incubate at 37 °C for a further 15 min.
6. Terminate reaction by addition of 5 μl 0.2 M EDTA, then heat to 70 °C for 5 min.
7. Unincorporated nucleotides may be removed by ethanol precipitation or by Sephadex G-50 chromatography (see *Protocol 2*).

[a] It is often possible to label DNA directly after restriction enzyme digestion without additional purification. If this is desirable, the restriction enzyme digest should be carried out in T4 DNA polymerase buffer in a volume of approximately 20 μl, before addition of nucleotide mix, label, and T4 DNA polymerase. As not all enzymes function in this buffer, it is advisable to test the enzyme in this buffer initially. If the enzyme does not work, then it will be necessary to phenol extract and precipitate the DNA before setting up the labelling reaction.
[b] It is also possible to use low specific activity label (see *Protocol 9*, note b). As discussed for previous end-labelling techniques, a variety of DNA/label ratios are possible, depending on whether probe-specific activity or utilization of label is the chief priority.
[c] As the 3′–5′-exonuclease of the T4 DNA polymerase is more active than that of Klenow polymerase, this 'chase' step is essential to ensure fragments of equal length.
[d] It is possible to use high specific activity label (3000 or 6000 Ci/mmol, 110 and 220 TBq/mmol) to maximize probe specific activity, but, to avoid further net exonuclease activity and hence poor incorporation, it may be necessary to add a large amount of label which may need drying down. The specific activity of the sequences will be approximately 7×10^8 d.p.m./μg if a single [α-^{32}P]dNTP at 400 Ci/mmol (14.8 TBq/mmol) is used.

3.4.6 Applications of end-labelled DNA

The applications of end-labelled molecules are somewhat different from those described for uniformly labelled nucleic acids. Although molecules end-labelled with ^{32}P can be used in the detection of mammalian single-copy genes, more rapid results will be achieved if a uniform labelling method is employed. Oligonucleotides 5′-end-labelled with ^{32}P have been used for both single-copy gene detection (29) and the detection of single-base mismatches in gel blots or in plaque/colony lifts (42). 5′-end-labelling with T4 polynucleotide kinase is the most common means of introducing label into oligonucleotides but alternative methods have been developed in recent years (see Section 3.5).

End-labelling is used for the most part in situations where direct detection of the labelled molecule (for example, by autoradiography) is to take place and where an initial intact molecule is desirable. For example, restriction fragments with recessed 3′-ends can be readily end-labelled with Klenow polymerase and used either as gel markers or for further restriction mapping studies. More detailed mapping by the nuclease S1 technique (15) can also be carried out using an end-labelled restriction fragment.

A major application for end-labelled nucleic acids is sequencing; both enzymatic (43) and chemical (44) methods of RNA sequencing, and the

Maxam and Gilbert method for DNA sequencing (1) utilize end-labelled molecules (see Section 4.2). The option of labelling either the 3′- or the 5′-terminus allows the sequence to be read from either end of the molecule. A disadvantage of this technique with double-stranded DNA is that two 3′- or 5′-ends will be labelled. To avoid overlapping sequences, it is necessary to remove one labelled fragment by either a strand separation gel or restriction digestion followed by gel purification. An alternative is to subclone into a vector that allows only one end of a restriction fragment to be labelled; for example, by using an enzyme that cuts at indeterminate residues within its recognition sequences so that a different sequence can be present at either end of an excised insert (45).

In summary, *Tables 7* and *8* outline the major properties of the various uniform and end-labelling methods.

Table 7. Major properties of uniform labelling methods

	Nick-translation	Random primer	Phage polymerase
Template	dsDNA	ssDNA (and denatured dsDNA)	dsDNA
Optimal form	linear or circle	linear	linear
Amount of template	0.5–1 μg	25 ng	1–2 μg
Reaction time	~2 h	~2 h	~1 h
Efficiency of incorporation	~60%	~75%	~75%
Nature of probe	DNA	DNA	RNA
Competing strand	yes	yes	no
Amount of probe	0.5–1 μg	~50 ng	~250 ng
Potential specific activity of probe (d.p.m./μg)	2×10^9	5×10^9	5×10^9
Probe length	~500 bp (medium variance)	~200 bp (high variance)	defined
Insert specific	no	no	yes
Requirement for subcloning	no	no	yes
Labelling in agarose	(yes)[a]	yes	no

Values quoted are for commonly used protocols.
[a] Although effective labelling in agarose has been observed, its efficiency is variable.

3.5 Labelling of oligonucleotides

The use of relatively short nucleotide sequences (such as chemically synthesized oligonucleotides) as hybridization probes has increased in recent years, particularly for the detection of point mutations, but also as general probes for mapping. A variety of methods have been developed for labelling short nucleotide sequences both to improve the convenience of the labelling reaction and to increase probe specific activity. Some of these approaches will be briefly summarized.

Traditionally, chemically synthesized oligonucleotides have been either 5′-end-labelled with polynucleotide kinase or, less frequently, 3′-end-labelled with terminal transferase. With improvements in the convenience of chemical synthesis of oligonucleotides, it has been possible to develop methods allow-

Table 8. Major properties of end-labelling methods

End	Enzymes	Applications
5′	T4 PNK	Oligonucleotide and RNA probes for blots DNA probes for S1 mapping Maxam and Gilbert DNA sequencing
3′	Tdt	DNA probes for S1 mapping Maxam and Gilbert DNA sequencing
3′	Poly(A)pol	RNA sequencing
3′	T4 RNA ligase	RNA sequencing
3′	Klenow	DNA probes for S1 mapping Maxam and Gilbert DNA sequencing Gel markers
3′	T4 DNA pol	DNA probes for S1 mapping Maxam and Gilbert DNA sequencing Gel markers DNA probes for blots (replacement synthesis)

ing the incorporation of more labelled residues into an oligonucleotide probe. One approach is to anneal two oligonucleotides complementary over part of their sequence, then to fill-in the single-stranded regions using Klenow polymerase in the presence of labelled nucleotides. This primer extension procedure has been described in a previous volume of this series (34). Another similar approach is to anneal two partially complementary oligonucleotides and, in this case, the double-stranded region constitutes a promoter for T7 RNA polymerase. Transcription then takes place across a single-stranded region, producing virtually full-length run-off transcripts, although a large amount of short oligonucleotides (2–6 nucleotides) are also produced (46). In a similar approach involving an additional DNA polymerase I catalysed step, a double-stranded sequence is produced. Transcription of this can yield probes of a specific activity $>10^9$ d.p.m./μg (47). It is also possible to modify standard phage polymerase reactions to produce short probes either by careful choice of restriction cut (if a convenient site is available) or by limiting nucleotide concentration.

For short probes, removal of unincorporated label by ethanol precipitation gives poor recovery, and alternative methods such as purification by gel electrophoresis, thin layer chromatography, HPLC or Sephadex G-25 chromatography are recommended. A procedure is given for gel purification of oligonucleotide probes in a previous volume of this series (34).

4. Choice of labelling strategy

The choice of labelling method and radioisotope is largely dependent on the way in which the labelled nucleic acid is to be used, and on the detection methods available. In this section, we focus on three major areas of application; filter hybridization, nucleic acid sequencing and nucleic-acid mapping (*in situ* hybridization is covered in Chapter 7).

4.1 Filter hybridization

Although the underlying theory of filter hybridization is complex, in practical terms there are four main probe-related variables which can affect the quality of the results obtained. These are: type of probe (DNA or RNA), probe specific activity, probe concentration, and probe length.

In addition, there are a number of general factors which may be manipulated, including: hybridization time, presence of rate enhancers, presence of helix destabilizers, presence of blocking agents, and hybridization temperature.

4.1.1 Probe-related factors

Traditionally, DNA probes have been used in filter hybridization. However, RNA probes are now preferred by some researchers, and there are some reports of increased sensitivity when using RNA probes (27). This is most probably due to the avoidance of probe reannealing and to the fact that unlabelled template may be removed or rendered ineffective in hybridization. In addition, under certain conditions, RNA:RNA and RNA:DNA hybrids are more stable than DNA:DNA hybrids. A further advantage is that non-specifically bound RNA can be removed by treatment with RNase A, which is highly specific for single-stranded RNA.

Although it has become usual when using RNA probes to employ helix destabilizing agents such as formamide, we have found that it is often possible to obtain equally sensitive results in their absence. Since formamide has the general effect of reducing the rate of hybridization, we prefer to avoid its use. However, we have also noted that, when using certain SP6-generated transcripts as probes, optimal signal-to-noise ratios are obtained at temperatures which differ from those predicted by hybridization theory. In addition, there are variations from probe to probe. It is therefore advisable when using such a probe for the first time to determine empirically the optimum hybridization temperature. In some instances we have found it necessary to include formamide in the hybridization buffer in order to reduce the optimal hybridization temperatures to convenient levels. Investigation into the reasons for this behaviour are in progress.

The sensitivity of detection of radiolabelled probes can be improved by increasing specific activity. However, the same effect may be obtained by increasing probe concentration, provided the duration of hybridization is chosen carefully and if the general factors are controlled correctly. For example, we have achieved an equivalent level of detection in Southern blot hybridization using either a ^{32}P-labelled, nick-translated probe at a specific activity of 10^8 d.p.m./μg and a concentration of 10 ng/ml, or a random primer labelled probe at 10^9 d.p.m./μg and 1 ng/ml under standard hybridization conditions.

Furthermore, a nick-translated probe at 10^8 d.p.m./μg and 100 ng/ml gives an equivalent signal-to-noise ratio following a 1 h hybridization as the same probe at 10 ng/ml hybridized overnight, using a proprietary hybridization buffer optimized for control of the general hybridization factors mentioned above (*Figure 12*). Similar results may be obtained using random primer labelled DNA probes and SP6 generated RNA probes (data not shown).

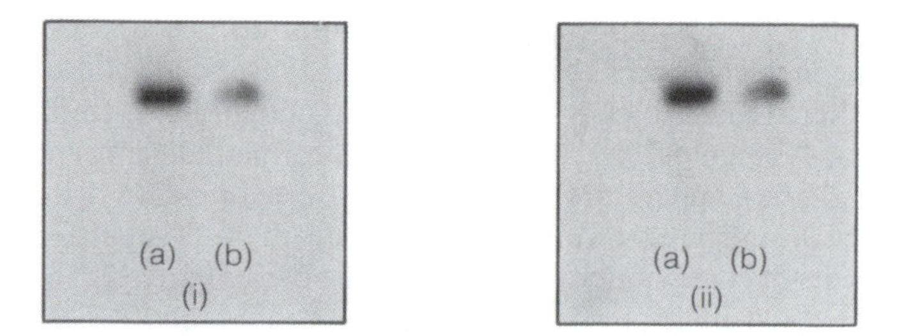

Figure 12. Effect of probe concentration and specific activity on hybridization signal. (a) 10 μg and (b) 2 μg tracks of *Hin*dIII-digested human genomic DNA blotted and hybridized with a *Raf* 1 probe, nick translated with [α-^{32}P] dCTP to a specific activity of 10^8 d.p.m./μg. (i) Overnight hybridization at 10 ng probe/ml, (ii) 1 h hybridization at 100 ng probe/ml. Autoradiography: 16 h exposure on Hyperfilm™ at −70 °C with intensifying screens.

In theory then, the researcher is free to choose any appropriate combination of probe type, specific activity, probe concentration, and hybridization time which suits his/her experimental requirements (or perhaps budget and/or working hours). There are, however, a number of other factors to consider. One of the most important of these is probe length. In our hands, nick-translated or random primer labelled probes give optimal signal-to-noise ratios only over a relatively narrow range of probe lengths from 150 to 500 nucleotides. This is particularly noticeable when performing rapid hybridization reactions at high probe concentrations up to 500 ng/ml. Clearly then, the ratio of DNase I to DNA polymerase in nick translation, or the concentration of primers in random primer labelling is of importance. For this reason, it is convenient to use one of the commercially available labelling kits that are pre-optimized for use in these procedures.

4.1.2 General factors

Non-probe-related factors are of equal importance in generating data of high quality. At Amersham, a large number of hybridization buffers have been analysed. The following general observations may be made.

Rate enhancers such as dextran sulphate or polyethylene glycol are beneficial only if combined with optimal amounts of blocking agents such as SDS, non-homologous DNA, and/or Denhardt's reagent (48). Their effects may be due to two factors: promotion of network formation by overlapping sequences (49) and increased effective probe concentration by volume exclusion (50). The magnitude of rate enhancement can be up to 100-fold in the case of DNA probes and about threefold for single-stranded RNA probes. The enhancement is more pronounced with increasing probe length and is not seen at all when oligonucleotide probes are used. It should be noted that the use of dextran sulphate or other polymers can lead to lowered signal-to-noise ratios, especially with short or single-stranded RNA probes, unless the length of hybridization time is optimized.

The use of helix destabilizing agents in relation to hybridization of RNA probes has been discussed in Section 4.1.1 above. Formamide is also commonly used in Northern (RNA) blot hybridization in combination with either DNA or RNA probes. As a guide, each 1% increase in formamide concentration will lower the T_m (the temperature at which half the hybrids are dissociated) by 0.61 °C. Since the introduction of Denhardt's reagent (48) a number of effective blocking agents or combinations of agents have been employed. These include heparin (51) dried milk powder, high concentrations of SDS, and sodium pyrophosphate. Not surprisingly, a wide range of hybridization buffers and methods are in use. However, *Protocol 11* gives a procedure that can be used in most instances. For rapid (1 h) hybridization, a system is commercially available from Amersham.

Protocol 11. A general procedure for hybridization

1. DNA blot hybridization

   Make up a pre-hybridization solution as follows (for 25 ml)

Solution	*Volume*	*Final concentration*
20 × SSPE[a]	6.25 ml	5 × SSPE
100 × Denhardt's solution[b]	1.25 ml	5 × Denhardt's
10% SDS	1.25 ml	0.5%
50% Dextran sulphate	5.00 ml	10%

   Make up to 25 ml with sterile H_2O. Add to membrane in a plastic sandwich box (5 ml per 100 cm^2).
2. Denature 0.5 ml of a 1 mg/ml solution of sonicated (10 × 30s at full power) non-homologous DNA by heating to 100 °C for 5 min. Chill on ice and add to pre-hybridization solution.
3. Pre-hybridize in a shaking water bath at 65 °C for 1 h.
4. Denature labelled probe (unless using an RNA or single-stranded DNA probe) by heating to 100 °C for 5 min. Add probe to pre-hybridization solution so that it does not directly contact the membrane. Do not exceed a probe concentration of 20 ng/ml.
5. Continue incubation for at least 12 h.
6. Incubate filters with 50 ml 2 × SSPE, 0.1% (w/v) SDS at room temperature for 10 min. Repeat.

7. Replace with 50 ml 1 × SSPE, 0.1% (w/v) SDS. Incubate at 65 °C for 15 min.
8. Replace with 50 ml 0.1 × SSPE, 0.1% (w/v) SDS. Incubate at 65 °C for 15 min. Repeat.
9. Remove filter, wrap in Saran Wrap and carry out autoradiography as described in Chapter 4.

[a] 20 × SSPE = 3.6 M NaCl, 0.2 M sodium phosphate, pH 7.4, 0.02 m EDTA.
[b] 100 × Denhardt's = 2% (w/v) BSA, 2% (w/v) Ficoll, 2% (w/v) polyvinylpyrrolidone.

Northern blots, or RNA probes may be used in this procedure. In some instances, however (see text), it may be beneficial to include formamide at a final concentration of 50% in the hybridization step.

4.2 Nucleic acid sequencing

As discussed in Section 2.1.2, RNA and DNA sequencing methods rely on the electrophoretic resolution of sets of single-stranded fragments with one defined end in common. In Maxam and Gilbert sequencing, the defined end is determined by end-labelling homogeneous fragments, while in dideoxy sequencing it is at the 5′-end of the sequencing primer. The two different approaches therefore allow quite different labelling strategies. Label is normally incorporated uniformly during dideoxy sequencing allowing the use of ^{35}S-labelled thionucleotides for increased resolution (see Section 2.2). Similar methods based on the use of bacteriophage RNA polymerases and 3′-deoxynucleotide terminators have been published (52). It is also possible to employ an end labelling approach by using either labelled terminators or end-labelled primers (53). In this case, as with Maxam and Gilbert sequencing, label density is lower so that ^{32}P is generally the label of choice.

Maxam and Gilbert sequencing (1) requires the DNA to be labelled at only one end of a single strand. This material is usually obtained first by labelling both strands of a duplex at either the 3′- or 5′-ends, and then removing one labelled end by restriction enzyme digestion. 5′-end-labelling is achieved using T4 polynucleotide kinase and [γ-^{32}P]ATP (3.4.1). The choice of 3′-end-labelling technique is dependent upon the nature of the end. Blunt or protruding 3′-ends are labelled most efficiently using terminal deoxynucleotidyl transferase (see Section 3.4.2). Recessed 3′-ends should be labelled by incorporation of ^{32}P-labelled nucleotide using Klenow polymerase (see Section 3.4.4). Specialized vectors, which permit the direct labelling of only one end of one strand of an insert by use of Klenow polymerase, have been developed (45).

Sequencing of RNA by either chemical (44) or enzymatic (43) methods generally requires an end-labelled substrate. 5′-end-labelling may be carried out using T4 polynucleotide kinase, following removal of the 5′ cap (35). 3′-end-labelling presents the researcher with a choice between use of poly(A)polymerase (39) or T4 RNA ligase (40). The former catalyses the sequential addition of A residues using ATP as precursor. In order to limit this reaction to addition of a single residue cordycepin-5′-triphosphate may be used (see Section 3.4.2). However, this analogue is incorporated at low efficiency. An alternative is to limit the ATP concentration.

Use of T4 RNA ligase, together with a radioactively labelled ribonucleoside 3′,5′-bisphosphate such as [5′-^{32}P]pCp, limits incorporation to a single labelled nucleotide per RNA molecule. This reaction is efficient even in the presence of low concentrations of termini and may be carried out at low temperature to minimise RNA degradation.

4.3 Mapping of nucleic acids

4.3.1 DNA mapping

The structural analysis of DNA at a relatively gross level is achieved by mapping of restriction sites. With cloned DNA the information is obtained by direct detection of ethidium bromide fluorescence or, alternatively, by autoradiography of gel-fractionated restriction fragments. The fragments may be directly end-labelled, for example at the 5′-end using T4 polynucleotide kinase, or at the 3′-end by Klenow in-filling. They may also be annealed to an oligonucleotide, again 5′-end-labelled, as in the technique of cos mapping (54). With DNA that is not cloned, but is present in a heterogeneous population, Southern blot analysis is used.

4.3.2 RNA mapping

For mapping structural features of transcripts, variations on three techniques are in general use.

Nuclease S1 mapping (15)

In nuclease S1 mapping the RNA is hybridized to one strand of a previously characterized restriction fragment of genomic DNA, part of which is complementary to the transcript. Subsequent digestion with nuclease S1 degrades DNA that is not hybridized. A discrete segment remains that is equal in length to the section initially hybridized to the transcript. Subsequent gel analysis allows information to be gained about the position of splice sites and termini.

In order to detect the DNA, an end-labelled fragment is normally used. Although it would be possible to use a uniformly labelled DNA fragment to improve sensitivity, both commonly employed methods (nick-translation and random primer labelling) are likely to leave nicks. These would both interfere with accurate hybridization and allow nuclease S1 digestion within the hybridized segment. However, a uniformly labelled fragment can be prepared by unique primer extension on a single-stranded template or by replacement synthesis using T4 DNA polymerase. When mapping the 5′-terminus of an RNA or of an exon, a restriction fragment 5′-end-labelled with T4 polynucleotide kinase is used, and when mapping equivalent 3′-termini, the probe is also 3′-end-labelled by Klenow in-filling. A more detailed description of S1 mapping has been given in a previous volume in this series (15).

RNase mapping
The rationale behind RNase mapping is similar to that of S1 mapping except that a uniformly labelled RNA produced by transcription using a bacteriophage RNA polymerase is used. Single-stranded regions of the probe are then digested by ribonuclease A. The concentration of this enzyme is somewhat less critical than that of nuclease S1 which, when present in excess, can also digest the hybrid. It is possible to identify both the number and size of exons, and hence of introns, and to define their relative positions, using progressively shorter transcripts. These are produced by restriction digestion of the SP6 or T7 recombinant at sites progressively closer to the promoter (27).

Primer extension mapping
In this technique, a radioactively labelled probe derived from within the gene (usually close to the 5′-end), is annealed to the transcript and extended using reverse transcriptase. After nuclease S1 digestion, the size of the labelled fragment allows the position of the 5′-terminus to be located. In theory, any site along the RNA molecule can be used, but better results are usually obtained if the primer is annealed close to the 5′-end. As is the case with nuclease S1 mapping, either a 5′- or a 3′-end-labelled probe can be used, although the former should be used if the sequence of the primer extension product is to be determined. For a more detailed description of primer extension mapping see ref. 15.

In both S1 and primer extension mapping, it is possible to use uniformly labelled probes to improve sensitivity. However, it is essential to use these quickly before significant strand scission due to radioactive decay has occurred. Therefore, it is generally better to use an end-labelled probe to avoid this problem.

5. Conclusion

We have summarized the variety of radiolabels available for use with nucleic acids, and described the ways in which they can be introduced into nucleic-acid molecules. Section 4 has described how the requirements for sensitivity and resolution influence the choice of labelling method. In this section, we have attempted to summarize the choice of label and labelling method available for any given application (excluding *in situ* hybridization, see Chapter 7) in *Table 9*.

For most hybridization applications, any of the three major uniform labelling methods can be used successfully. The final choice can then be made by considering factors not necessarily dependent upon the ultimate sensitivity and resolution required. For example, if a probe is to be used repeatedly over a long period, or is required for expression studies, it can be worthwhile to subclone into a transcription vector and synthesise an RNA probe using phage polymerase methods. Alternatively, if a probe is to be used in several hybridization experiments over a short period, or a high probe concentration is required, nick-translation may be the method of choice. If a wide variety of probes available in small quantities are to be screened, then random primer labelling may be preferred, particularly if insert-specific labelling is required. Personal preferences, for instance for RNA probes rather than DNA, are also perfectly valid.

Table 9. Recommended labels and labelling methods for nucleic acids

Labelling method Application	**Required sensitivity**	**Recommended label and labelling method**
Filter hybridization		
High sensitivity, for example single-copy mammalian gene	1–5 pg	^{32}P, any uniform labelling method
Intermediate sensitivity, for example single copy *Drosophila* gene	10–50 pg	^{32}P, any uniform labelling method
Low sensitivity, for example single copy *E. coli* gene and colony/plaque hybridization	50 pg–5 ng	^{32}P, any uniform or end-labelling method
Mapping of nucleic acids		
Nuclease S1 mapping	10 pg–1 ng	^{32}P, any end-labelling method
Primer extension mapping	10 pg–1 ng	^{32}P, 5′-end-labelling
RNase mapping	10 pg–1 ng	^{32}P, RNA polymerase based
Restriction mapping	1–100 ng	^{32}P, nick-translation or end-labelling
Nucleid-acid sequencing		
Sanger dideoxy sequencing	10–50 pg	^{35}S, unique primer
Maxam and Gilbert sequencing	10–50 pg	^{32}P, any end-labelling method

In contrast to the hybridization methods discussed above, some of the analytical techniques for which end-labelled nucleic acids are used dictate the methods of labelling to a greater extent. However, some element of choice usually remains. For example, Maxam and Gilbert sequencing may require either 3′ or 5′-end-labelling depending on the desired direction of sequencing, but a variety of methods are still possible. In most other cases, the procedure that is employed determines the method of labelling to a very large extent. These are generally procedures such as cDNA synthesis in which a radiolabel is used simply to monitor the progress of the reaction. Only Sanger dideoxy sequencing, which can be included in this category, has been discussed in any detail in this chapter.

The choice of the optimum radionuclide for a particular application can be as difficult as that of labelling method. *Table 9* also summarizes what we would consider to be the best radiolabel for the applications discussed in terms of the resolution and sensitivity required. However, other labels can frequently be used effectively, and we hope that the above discussion will be of help in deciding which radiolabel and labelling method to use for an experiment.

Acknowledgements

We should like to thank Alan Hamilton, John McCombe, and Mike Evans for their critical reading of this manuscript and our colleagues Andrew Bertera, Carol Hawes, Phil Hedge, Mahomed Jassat, Alan Syms, Chris Spencer, and Christine Taylor for their various contributions towards the information contained in this chapter. Many thanks also to Sylvia Garry for typing the manuscript and all its revisions.

B. Proteins

GRAHAM S. BAILEY

6. Introduction

Radioactive labelled peptides, proteins and glycoproteins, produced by *in vitro* labelling, are used extensively in many areas of biochemistry, pharmacology and medicine. For example, they are frequently employed as tracer molecules in quantitative determinations such as measurement of hormone and hormone-receptor concentrations, and kinetic and equilibrium studies of both agonist and antagonist binding to receptors. Those studies require the accurate determination of very low amounts of the labelled protein. Such very low amounts can be accurately measured by using a tracer molecule labelled to a high specific radioactivity.

In theory several different radioisotopes could be employed to label a particular protein *in vitro*, for example ^{14}C, ^{3}H, ^{125}I, ^{131}I. However, in practice, the vast majority of peptides, proteins and glycoproteins are labelled with ^{125}I. The detectable count rate produced by one gram atom of ^{125}I is approximately 33 000 times and 100 times greater than that produced by one gram atom of ^{14}C and ^{3}H respectively. Furthermore, ^{125}I-labelled proteins can be counted directly in a γ counter whereas molecules labelled with the β emitters ^{14}C and ^{3}H, can only be counted after suitable preparation in a liquid scintillation counting system.

It can be noted that ^{131}I is also a γ emitter and that the theoretical specific activity of ^{131}I at 100% isotopic abundance is about seven times greater than that of ^{125}I. However, commercially available ^{131}I is at only 20% isotopic abundance compared to greater than 90% for commercial ^{125}I. Furthermore, ^{131}I has a shorter half-life (8 days) and lower counting efficiency (30%) than ^{125}I ($t\frac{1}{2}$ 60 days, counting efficiency 80%). An additional problem is that ^{131}I

is more hazardous to use than ^{125}I because its γ rays are more penetrating. Thus, for most situations ^{125}I is the preferred radioisotope for the *in vitro* labelling of proteins.

7. Radioiodination

7.1 General principles

Several different methods are available for the direct radioiodination of peptides, proteins and glycoprotein (see *Table 10*). Essentially, they differ in the nature of the oxidizing agent that is used to convert the commercially-available radioactive iodide [$^{125}I^-$] into the reactive species which is believed to be hydrated, positively charged, iodine radical [$I^+(H_2O)$] in the chloramine T procedure (67). This reactive species reacts most readily with tyrosine residues of the protein but substitution into other residues, particularly histidine, can occur in some circumstances. In the main, monoiodo-tyrosine residues are produced using relatively low levels of radioactive iodine. Indeed, diiodotyrosine residues which can be formed using relatively high levels of iodine should be avoided due to greater instability of the labelled product.

Table 10. Methods of radioiodination

	Method	Reference Number
(a)	*Direct oxidative iodination*	
	Iodine monochloride	55
	Chloramine T	56
	Electrolysis	57
	Lactoperoxidase	58
	Chlorine	59
	Sodium hypochlorite	60
	1,3,4,6-Tetrachloro-3α, 6α-diphenyl glycoluril [Iodogen™]	61
	N-bromosuccinimide	62
	Protag-125™	63
(b)	*Indirect conjugation iodination*	
	N-Succinimidyl-3-(4-hydroxy 5[^{125}I] iodophenyl)propionate [Bolton and Hunter reagent]	64
	Methyl-*p*-hydroxybenzimidate [Wood's reagent]	65
	Aniline	66

It is generally agreed that on average the incorporation of one iodine atom per protein molecule produces the optimum labelled molecule both in terms of specific radioactivity and stability. Obviously, the incorporation of one radioactive iodine atom per protein molecule will result in the minimum possible change in the protein. Nevertheless, it has to be borne in mind that the iodine atom is approximately the same size as a benzene nucleus and thus the addition of even one iodine atom may have a profound effect on the structure and physiochemical properties of the labelled molecule, especially if the latter is a small biologically active peptide. It is always best to compare the labelled and unlabelled molecules in terms of their behaviour on electrophoresis, gel filtration, ion-exchange chromatography, and of their binding reactions with receptors or antibody molecules, as far as is practical (68, 69).

In conjugation labelling, a suitable molecule containing a phenol or imidazole group is first radioiodinated by one of the direct oxidative procedures. The iodinated product is then coupled to the protein to be labelled. The indirect procedure enables peptides which lack tyrosine residues to be labelled and is unlikely to damage the peptide as the latter does not come into contact with the oxidant. However, as conjugation labelling consists of two stages overall iodination yields tend to be lower than for the direct methods, resulting in lower specific radioactivities. Also, indirect labelling is technically more demanding.

In the next sections four well-established methods of radioiodination are described. They have been chosen because of their versatility and widespread use.

7.2 Radioiodination using chloramine T as oxidant

7.2.1 Introduction

The procedure (*Protocol 12*), introduced by Hunter and Greenwood (56), is probably the single most widely used method for radioiodination of peptides and proteins. It can be recommended as the first method to be tried for the radioiodination of a previously unlabelled protein since it is a relatively simple and inexpensive procedure. Chloramine T (the sodium salt of the *N*-monochloro derivative of *p*-toluene sulphonamide) in aqueous solution breaks down to produce hypochlorous acid which oxidizes the radioactive iodide. The optimum pH for substitution of the cationic iodine into tyrosine residues appears to be pH 7.5 and a strong buffer is required to maintain the reaction mixture at that pH. When sufficient incorporation of ^{125}I into the protein has taken place, a reductant is added to stop further reaction. There are many variations to the general procedure differing in relative amounts of protein, iodine, chloramine T and reductant, and in nature of the reductant, and in time of the reaction. One such variation is described in the next section. The best set of conditions for any particular protein have to be established by trial and error.

Protocol 12. Practical procedure for radioiodination using chloramine T as oxidant

1. Prepare the following reagents:

Reagent	*Composition*
Buffer A	0.5 M sodium phosphate, pH 7.5
Buffer B	0.05 M sodium phosphate, pH 7.5

Buffer C	0.01 M sodium phosphate, pH 7.5 containing 1 M NaCl, 100 mg/100 ml bovine serum albumin and 1 g/100 ml KI.
Chloramine T	40 mg/100 ml of buffer B, prepared fresh, just prior to use
Sodium metabisulphite	4 mg/100 ml of buffer C
Protein to be iodinated	2 mg/ml in buffer B

2. Add 25 μl of chloramine T solution to a small plastic test-tube containing 5 μl of protein to be labelled and 50 μl buffer A.
3. Carefully add 5 μl of Na ^{125}I (100 mCi/ml, 3.7 GBq/ml).
4. Mix the contents gently by tapping the tube, ensuring that all droplets are collected together at the bottom of the tube. Poor mixing is probably the commonest cause of a low yield of labelled protein by this method.
5. Allow the reaction to proceed for up to 10 min.
6. Add 50 μl of sodium metabisulphite solution and thoroughly mix the contents of the tube.
7. Separate the radioactive labelled protein from the remaining radioactive iodide (see Section 3).

7.2.2 Variations to the practical procedure

As previously stated, there are many variations in practical details that can be made for the chloramine T procedure. It is not possible here to list all of those variations but some important general points will be made.

(a) Usually the method is carried out at room temperature. However, if the protein is very labile it may be better to carry out the reaction with all reagents cooled in ice water.

(b) It is best to use the smallest possible amount of chloramine T since high amounts of the reagent have been found to damage certain proteins, for example human protactin (70). However, it should be noted that chloramine T will also oxidize sulphhydryl groups of proteins and that sufficient oxidant must be present for that process as well as oxidation of iodide for the successful radioiodination of proteins of high reducing potential; for example, keyhole limpet haemocyanin (71). The optimum amount of chloramine T can only be established by experiment with each individual protein.

(c) Likewise, the optimum time of reaction can be obtained by trial and error. If very short times, for example 15 sec, are to be employed adequate initial mixing becomes of paramount importance.

(d) It is best to use the smallest possible amount of sodium metabisulphite as reductant to avoid potential damage to the labelled protein. Alternative reducing agents such as L-cysteine may be preferred (72), or perhaps better still, L-methionine.

(e) This type of reaction can also be carried out with an insoluble derivative of a related oxidant [sodium salt of *N*-chloro-benzene sulphonamide] with the reported advantage that the iodination can be stopped without the need to add a reducing agent (73). The insolubilized oxidant is commercially available (Iodo-Beads™, Pierce Chemical Company).

7.3 Radioiodination using Iodogen as oxidant

This procedure (*Protocol 13*) uses an oxidant [1,3,4,6-tetrachloro-3α, 6α-diphenyl glycoluril] that is only very slightly soluble in aqueous media so that any damage to the protein during labelling, brought about by the oxidant, is likely to be very small. The oxidant is dissolved in a suitable organic solvent and is deposited as a thin solid coat on the walls of a small glass or plastic tube. The radioiodination is carried out in the Iodogen-coated tube and is terminated simply by removal of the incubation mixture. Nustad and co-workers have carried out a detailed study of the radioiodination of a number of proteins using the Iodogen procedure (74). They reported that iodination efficiency was independent of pH in the range from pH 6 to pH 8.5 and that maximum incorporation of radioactive iodine occurred at a molar ratio of Iodogen to protein of 8 or more. Using a molar ration of Na ^{125}I to protein in the range 1 to 1.2, up to 1 mol radioactive iodine was incorporated per mol protein over a period of around 10 min. A typical procedure is given in *Protocol 13*.

Protocol 13. Practical procedure for radioiodination using Iodogen as oxidant

1. Dissolve 1 mg Iodo-Gen™ (Pierce Chemical Company) in 10 ml dichloromethane. The solution is stable for several weeks if kept in the dark.
2. Add 10 μl of Iodogen solution to a small plastic tube. Flush the tube with nitrogen gas until the solvent has evaporated. The tube can be sealed and stored at −20 °C until required.
3. Cool the tube in iced water and add in succession, 10 μl of 0.5 M sodium phosphate buffer, pH 7.5, 10 μl of protein (1 mg/ml) to be labelled and 5 μl of Na ^{125}I (100 mCi/ml, 3.7 GBq/ml).
4. Gently shake the tube and allow the reaction to proceed for 10 min.
5. Stop the iodination by transferring the reaction mixture from the coated tube to an uncoated plastic tube containing 0.25 ml of 0.01 M sodium phosphate buffer, pH 7.5, containing 1 M NaCl.

6. Leave for 10 min and then add 0.25 ml of 0.01 M sodium phosphate buffer, pH 7.5, containing 1 mg/ml bovine serum albumin and 10 mg/ml KI.
7. Separate the radioactively labelled protein from the residual radioactive iodide (see Section 3).

7.4 Radioiodination using lactoperoxidase

The method, as originally published by Marchalonis (58), used lactoperoxidase at pH 7.3 in the presence of a low concentration of hydrogen peroxide to radioiodinate a number of different proteins. The iodination was stopped by the addition of cysteine or mercaptoethanol. Since that time many variations have been suggested including differences in amounts of lactoperoxidase and hydrogen peroxide, and in frequency of addition of the latter reagent. Furthermore, the optimum pH for radioiodination by this procedure has to be found by experiment, and is governed by the nature and stability of the protein or glycoprotein to be labelled (75, 76). Also, it is possible to generate the hydrogen peroxide *in situ* by the action of glucose oxidase on glucose (77) and to use solid-phase lactoperoxidase (78). Such a system is commercially available [Enzymobead^{TM} Lactoperoxidase-Glucose Oxidase System, Bio-Rad Laboratories]. It should be noted that sodium azide, which is frequently added to solutions as an antibacterial agent, is a potent inhibitor of lactoperoxidase, and thus all solution must be azide-free for this method of radioiodination (*Protocol 14*).

7.5 Radioiodination using the Bolton and Hunter reagent

This is an indirect method in which an acylating reagent [*N*-succinimidyl-3-(4-hydroxyphenyl)propionate] (I) is first radioiodinated by the chloramine T procedure (see *Figure 13*). Rapid iodination is essential as the acylating reagent in aqueous media is quickly degraded by hydrolysis. The iodinated reagent (Bolton and Hunter reagent prepared as described in *Protocol 15*) (II) is extracted into an organic solvent from where it can be deposited as a solid layer at the bottom of a small conical tube. It is then reacted with the protein to be labelled. Conjugation occurs mainly with ϵ NH_2 groups of lysine residues (III) (see *Figure 13*) to produce the labelled protein (IV). After a reaction time of up to 30 minutes at 4 °C, an excess of glycine is added to conjugate with any residual acylating reagent. For the experiment protocol see *Protocol 16*.

Protocol 14. Practical procedure for radioiodination using lactoperoxidase

1. Prepare stock solutions of 10 mg/ml lactoperoxidase and 5 mg/ml hydrogen peroxide in 0.05 M sodium phosphate buffer, pH 7.5.[a]

2. Immediately prior to use dilute the stock solutions with the forementioned buffer to give 20 μg/ml lactoperoxidase and 10 μg/ml hydrogen peroxide.
3. Add 10 μl of diluted hydrogen peroxide solution to a small plastic tube containing 45 μl of 0.5 M sodium phosphate buffer, pH 7.5, 5 μl protein (2 mg/ml) to be labelled, 5 μl of diluted lactoperoxidase and 5 μl of Na ^{125}I (100 mCi/ml, 3.7 GBq/ml).
4. Gently shake the tube and allow the reaction to proceed for 20 min during which time two further additions of 10 μl diluted hydrogen peroxide are made.
5. Stop the iodination by adding 0.8 ml of 0.05 M sodium phosphate buffer, pH 7.5, containing 0.1% sodium azide.
6. Leave for 5 min and then add 0.8 ml of 0.05 M sodium phosphate buffer, pH 7.5, containing 1 mg/ml bovine serum albumin and 10 mg/ml KI.
7. Separate the radioactive labelled protein from the remaining radioactive iodide (see Section 3).

[a] In this protocol it is assumed that pilot experiments have established that pH 7.5 is the optimum pH for the radioiodination of the protein concerned.

Protocol 15. Preparation of [^{125}I] Bolton and Hunter[a] reagent

Note: it is essential that the entire procedure is carried out as rapidly as possible (less than 1 min).

1. prepare a 40 μg/ml solution of *N*-succinimidyl-3-(4-hydroxyphenyl) propionate in dry, re-distilled benzene. The solid reagent is available from a number of commercial suppliers including, Sigma Chemical Company, Fluka AG, ICN Biomedicals Inc.
2. Put a 10 μl aliquot of that solution into a small, stoppered, conical glass tube.
3. Remove the solvent by evaporation using a gentle stream of dry nitrogen.
4. Add 20 μl Na ^{125}I (100 mCi/ml, 3.7 GBq/ml) followed by 10 μl of chloramine T solution (5 mg/ml in 0.5 M sodium phosphate buffer, pH 7.5). Gently mix by tapping the tube.
5. Immediately add 10 μl sodium metabisulphite solution (12 mg/ml in 0.05 M sodium phosphate buffer, pH 7.5) and mix.
6. Add 10 μl KI solution (20 mg/ml in 0.05 M sodium phosphate buffer, pH 7.5), followed by 5 μl dimethylformamide (DMF) and 250 μl benzene.
7. Stopper the tube and gently vortex to enable the labelled reagent to go into the benzene layer.
8. Remove the benzene layer and put it into another conical glass tube.

9. Wash the remaining aqueous layer two times with 250 μl of benzene and add the benzene washings to the second glass tube.

[a] Ready radioiodinated Bolton and Hunter reagent is available from a number of commercial suppliers including Amersham International, New England Nuclear, ICN Biomedicals, thus avoiding the need to personally carry out the above procedure.

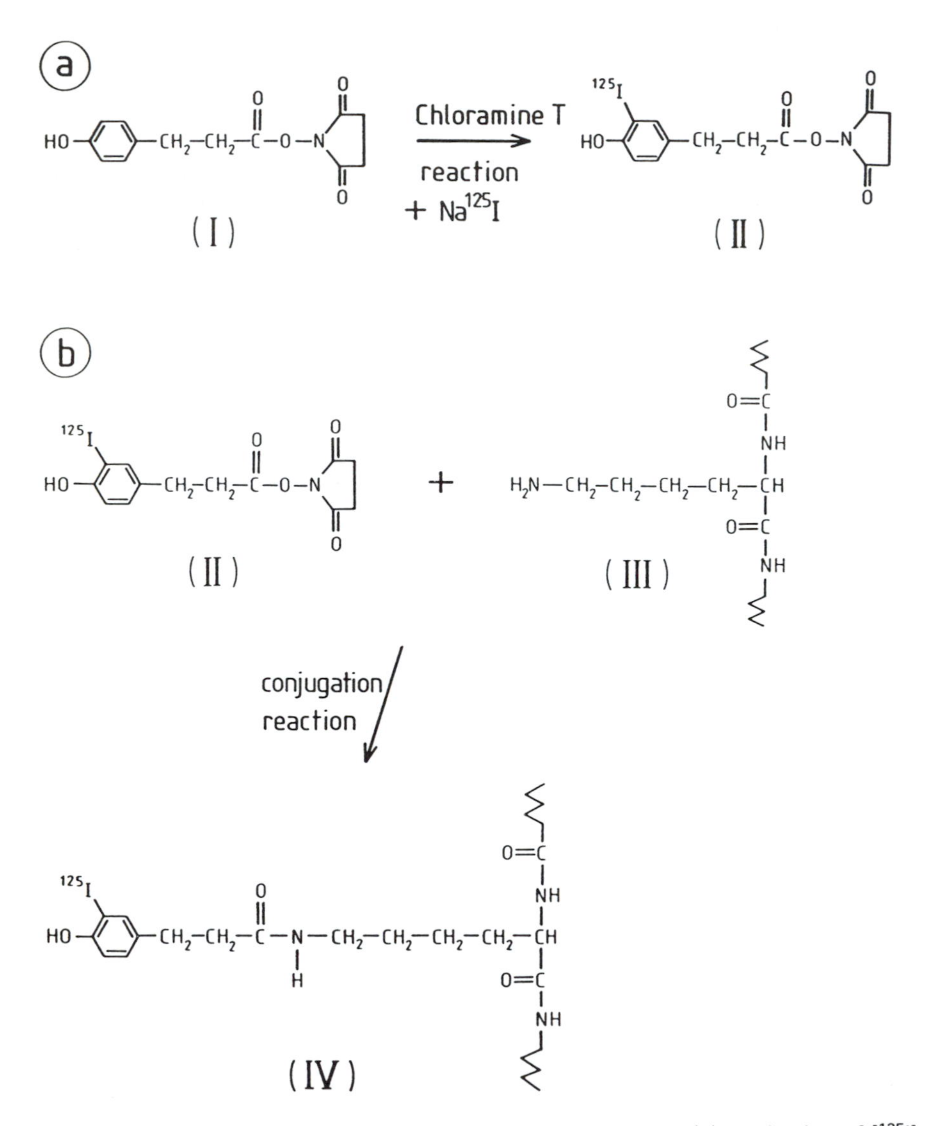

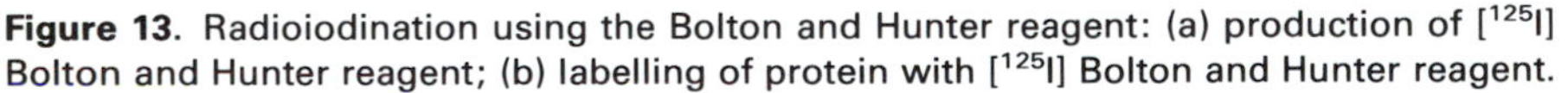
Figure 13. Radioiodination using the Bolton and Hunter reagent: (a) production of [^{125}I] Bolton and Hunter reagent; (b) labelling of protein with [^{125}I] Bolton and Hunter reagent.

Protocol 16. Practical procedure for radioiodination using Bolter and Hunter reagent

1. Evaporate to dryness the benzene-DMF solution of [^{125}I] Bolton and Hunter reagent (prepared in solution according to *Protocol 15*) or 0.2 ml commercial reagent (5 mCi/ml, 185 MBq/ml).
2. Cool all reagents in iced-water.
3. Add 10 μl protein solution (0.5 mg/ml in 0.1 M sodium borate buffer, pH 8.5) to the dry reagent.
4. Gently shake the tube and allow the conjugation to proceed for 15 min at 4 °C.
5. Add 0.5 ml 0.2 M glycine in 0.1 M sodium borate buffer, pH 8.5. Mix and leave for 5 min at 4 °C.
6. Add 0.5 ml and 0.05 M sodium phosphate buffer containing 2.5 mg/ml gelatin.
7. Separate the radioactively labelled protein from the other products (see Section 3).

8. Purification of the radioactively labelled protein

Most procedures that employ radioactively labelled proteins require that the labelled protein is free of unreacted radioactive iodide. Gel filtration is the most widely used procedure for the purification of labelled proteins and glycoproteins, as it affords a simple, rapid, and efficient method of separation. However, high performance liquid chromatography (HPLC) provides a better means of separation for peptides and small proteins (79). Separation also allows an estimate to be made of the yield of the radioiodination procedure. Knowing the yield, the specific radioactivity of the labelled protein can then be calculated.

9. Calculation of yield and specific radioactivity

An example is shown below using results from a typical radioiodination.

Amount of rat submandibular gland kallikrein used = 10 μg.
Amount of ^{125}I added = 500 μCi (18.5 MBq).

Chloramine T procedure as described in *Protocol 12*. Separation on 1 × 60 cm column of Sephadex G-75 resin under gravity flow collecting 0.6 ml fractions. 5 μl aliquots of the collection fractions and of the incubation mixture applied to the column were taken and counted in a γ-counter. The first peak to be eluted contained the labelled enzyme (shown by precipitation with specific antiserum) and the second peak unreacted iodide.

Counts in 5 μl incubation mixture prior to gel filtration = 1 441 643 counts/10 sec.

Counts in 5 μl aliquots of fractions of iodide peak = 507 221 counts/10 sec.

Thus counts in 5 μl aliquots of fractions of kallikrein
= original counts − iodide counts
= 1 441 643–507 221 counts/10 sec
= 934 422 counts/10 sec.

Yield of radioiodination procedure = % incorporation of ^{125}I into kallikrein

$$= \frac{934\,422 \times 100}{1\,441\,643} = 64.8\%.$$

Amount of radioactivity incorporated into kallikrein
= % incorporation × original radioactivity
= 64.8% × 500 μCi
= 324 μCi (12 MBq)

$$\text{Specify radioactivity} = \frac{\text{amount of radioactivity}}{\text{amount of protein}}$$

$$= \frac{324\ \mu\text{Ci}}{10\ \mu\text{g}}$$

$$= 32.4\ \mu\text{Ci}/\mu\text{g}\ (1.2\ \text{MBq}/\mu\text{g}).$$

10. Labelling of proteins with ^{14}C and ^{3}H

Although radioactive-labelling of proteins is frequently carried out using radioiodination, there are situations where labelling with the much longer-lived radioisotopes, ^{14}C or ^{3}H, is more appropriate. For example, ^{14}C-labelled proteins can be used as substrates in sensitive assays of proteolytic enzymes (80, 81). Widely used methods of labelling proteins with ^{14}C and ^{3}H are described in the following two sections.

10.1 ^{14}C-labelling of proteins by reductive methylation

The basis of the method is that a free amino group of the protein forms a Schiff's base with ^{14}C-formaldehyde and is then reduced by sodium cyanoborohydride.

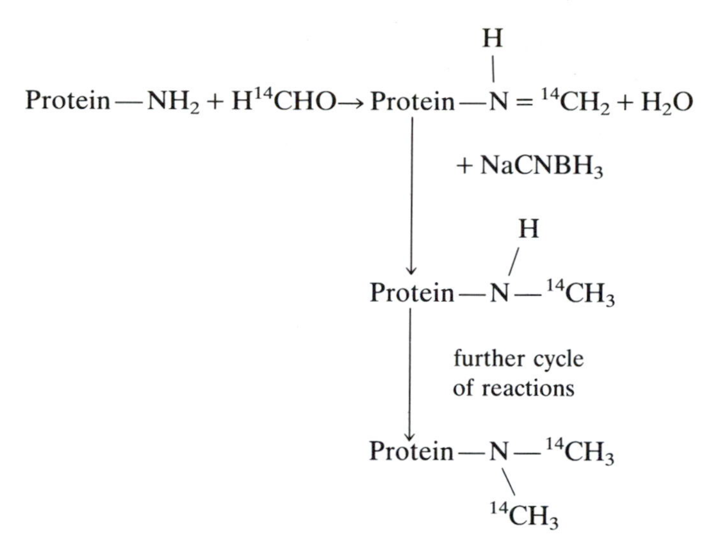

In that way the protein is labelled via its free α amino group and ϵ amino group of lysine residues. Other reductants such as sodium borohydride can be employed but, in contrast to sodium cyanoborohydryde, it cannot be used at neutral pH because of hydrolysis and it can also denature proteins (82). As pointed out by Dottamiuo-Martin and Ravel, a wide range of reaction conditions can be employed and the optimal conditions for the alkylation of a specific protein have to be established by trial and error (83). A representative procedure (83) is recorded in *Protocol 17*.

Protocol 17. Practical procedure for ^{14}C-labelling by reductive methylation

1. Add 10 μl of [^{14}C]formaldehyde (30 mCi/mmol, 1.11 GBq/mmol) to 25 μl of protein solution (1 mg/ml in 0.04 M potassium phosphate buffer, pH 7.0).
2. Immediately add 10 μl of a freshly prepared solution of sodium cyanoborohydride (6 mg/ml in the phosphate buffer).
3. Incubate at 25 °C for 1 h with periodic shaking.
4. Increase the volume of the reaction mixture to 250 μl by adding phosphate buffer.
5. Purify the labelled protein by extensive dialysis (16 h) at 4 °C against phosphate buffer.

10.2 ^{3}H-labelling of proteins with *N*-succinimidyl [2,3-^{3}H] propionate

This is a very mild method of labelling and produces labelled proteins with high retention of biological activity and stability (84). The labelling reagent reacts with free amino groups of the protein and in particular with the ϵNH_2 groups of lysine residues (31).

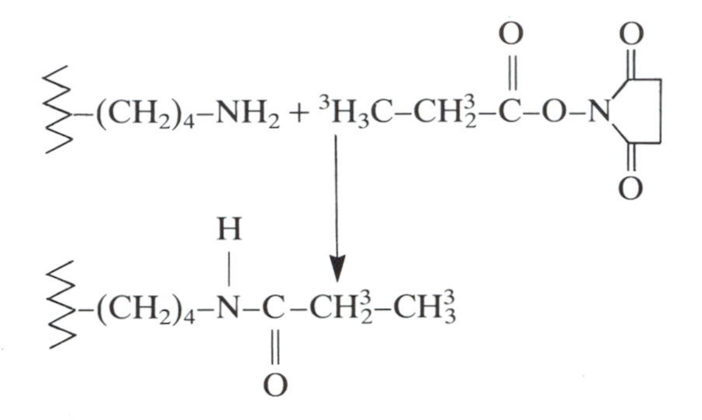

Numerous conditions can be employed (85) and a representative procedure (86) is presented in *Protocol 18.*

Protocol 18. Practical procedure for ^3H-labelling

1. Deposit 6.5 nmol of *N*-succinimidyl [2,3-^3H] propionate (37 Ci/mmol) at the bottom of a small plastic tube by evaporation of the solvent under a stream of nitrogen.
2. Add 160 μl of protein solution (0.25 mg/ml in 0.05 M Tris–HCl buffer, pH 7.5, containing 0.05 M KCl and 0.5 mM dithiothreitol) to the coated tube.
3. Leave for 60 min at 250 °C.
4. Separate the labelled protein from unreacted reagent by gel filtration on a column of BioGel P30.

References

1. Maxam, A. M. and Gilbert, W. (1980). In *Methods in Enzymology* (eds. L. Grossman and K. Moldave), Vol. 65, p. 499. Academic Press, London and New York.
2. Laskey, R. A. (1984). *Review 23*, Amersham International plc.
3. Biggin, M. D., Gibson, T. J., and Hong, G. F. (1983). *Proc. Natl. Acad. Sci. (USA)* **80,** 3963.
4. Southern, E. M. (1975). *J. Mol. Biol.* **98,** 503.
5. Alwine, JC., Kemp, D. J., and Stark, G. R. (1977). *Proc. Natl. Acad. Sci (USA)* **74,** 5350.
6. Sanger, F. (1981). *Science* **214,** 1205.

7. Rigby, P. W. J., Dieckmann, M., Rhodes, C., and Berg, P. (1977). *J. Mol. Biol.* **113,** 237.
8. *Technical Bulletin* 80/3, Amersham International plc.
9. Moore, S. (1981). In *The Enzymes* (ed. P. D. Boyer), Vol. XIV, p. 281. Academic Press, London and New York.
10. Lehman, I. R. (1981). In *The Enzymes* (ed. P. D. Boyer), Vol. XIV, p. 16. Academic Press, London and New York.
11. Feinberg, A. P. and Vogelstein, B. (1983). *Anal. Biochem.* **132,** 6.
12. Mann, D. A., Gurtler, V., and Reed, K. C. *Anal. Biochem.* (submitted).
13. Arrand, J. E. (1985). In *Nucleic Acid Hybridisation: A Practical Approach* (eds. B. D. Hames and S. J. Higgins), p. 17. IRL Press, Oxford and Washington, DC.
14. Gray, P. W. (1982). *Nature, Lond.* **295,** 503.
15. Williams, J. G. and Mason, P. J. (1985). In *Nucleic Acid Hybridisation: A Practical Approach* (eds. B. D. Hames and S. J. Higgins), p. 139. IRL Press, Oxford and Washington, DC.
16. Feinberg, A. P. and Vogelstein, B. (1984). *Anal. Biochem.* **137,** 266.
17. Birnboim, M. C. and Doly, J. (1979). *Nucl. Acids Res.* **7,** 1513.
18. Holmes, D. S. and Quigley, M. (1981). *Anal. Biochem.* **114,** 193.
19. Hodgson, C. P. and Fisk, R. Z. (1987). *Nucl. Acids Res.* **15,** 6295.
20. Hu, N. and Messing, J. (1982). *Gene* **17,** 271.
21. Jeffreys, A. J., Wilson, V., and Thein, S. L. (1985). *Nature, Lond.* **314,** 67.
22. Cox, K. H., De Leon, D. V., Angerer, L. M., and Angerer, R. C. (1984). *Dev. Biol.* **101,** 458.
23. Zinn, K., diMaio, D., and Maniatis, T. (1984). *Cell* **34,** 865.
24. Green, M. R., Maniatis, T., and Melton, D. A. (1983). *Cell* **32,** 681.
25. Krieg, P. A. and Melton, D. A. (1984). *Nucl. Acids Res.* **12,** 7057.
26. Melton, D. A. (1985). *Proc. Natl. Acad. Sci. (USA)* **82,** 144.
27. Melton, D., Krieg, P. A., Rebagliati, M. R., Maniatis, T., Zinn, K., and Green, M. R. (1984). *Nucl. Acids Res.* **12,** 7035.
28. Little, P. F. R. and Jackson, I. J. (1987). In *DNA Cloning*, Vol. III: *A Practical Approach* (ed. D. M. Glover), p. 1. IRL Press, Oxford and Washington, DC.
29. Connor, B. J., Reyes, A. A., Morin, C., Itakura, K., Teplitz, R. L., and Wallace, R. B. (1983). *Proc. Natl. Acad. Sci (USA)* **80,** 278.
30. Richardson, C. C. (1981). In *The Enzymes* (ed. P. D. Boyer), Vol. XIV, p. 299. Academic Press, London and New York.
31. Berkner, K. L. and Folk, W. R. (1977). *J. Biol. Chem.* **252,** 3176.
32. Maniatis, T., Fritsch, E. F, and Sambrook, J. (1982). *Molecular Cloning – A Laboratory Manual.* Cold Spring Harbor Laboratory, NY.
33. Dillon, J-A. R., Nasim, A., and Nestmann, E. R. (1985). *Recombinant DNA Methodology*. John Wiley, New York.
34. Thein, S. L. and Wallace, R. B. (1986). In *Human Genetic Diseases: A Practical Approach* (ed. K. E. Davies), p. 33. IRL Press, Oxford and Washington, DC.
35. Shinski, H., Miwa, M., Kato, K., Noguchi, M., Matsushima, T., and Sugimura, T. (1976). *Biochemistry* **15,** 2185.
36. Bollum, F. J. (1974). In *The Enzymes* (ed. P. D. Boyer), Vol. X, p. 145. Academic Press, London and New York.
37. Roychoudhury, R., Jay, E., and Wu, R. (1976). *Nucl. Acids Res.* **3,** 863.
38. Yousaf, S. I., Carrol, A. R., and Clarke, B. E. (1984). *Gene* **27,** 309.

39. Sippel, A. (1973). *Eur. J. Biochem.* **37,** 31.
40. England, T., Gumport, R., and Uhlenbeck, O. (1977). *Proc. Natl. Acad. Sci. (USA)* **74,** 4839.
41. Lehman, I. R. (1981). In *The Enzymes* (ed. P. D. Boyer), Vol. XIV, p. 51. Academic Press, London and New York.
42. Zoller, M. J. and Smith, M. (1983). In *Methods in Enzymology* (eds. R. Wu, L. Grossman and K. Moldave), Vol. 100, p. 468. Academic Press, London and New York.
43. Donnis-Keller, H. (1979). *Nucl. Acids Res.* **7,** 179.
44. Peattie, D. (1983). In *Methods of DNA and RNA Sequencing* (ed. S. M. Weissman), p. 261. Praeger.
45. Volckaert, G., De Vleeschouwer, E., Blocker, H., and Frank, R. (1984). *Gene Anal. Techniques* **1,** 52.
46. Milligan, J. F., Groebe, D. R., Witherell, G. W., and Uhlenbeck, O. C. (1987). *Nucl. Acids Res.* **15,** 8783.
47. Wolfl, S., Quaas, R., Hahn, U., and Wittig, B. (1987). *Nucl. Acids Res.* **15,** 858.
48. Denhardt, D. (1966). *Biochem. Biophys. Res. Commun.* **28,** 64.
49. Wahl, G. M., Stern, M., and Stark, G. R. (1979). *Proc. Natl. Acad. Sci. (USA)* **76,** 3683.
50. Wetmur, J. G. (1975). *Biopolymers* **14,** 2517.
51. Singh, C. and Jones, K. W. (1984). *Nucl. Acids Res.* **12,** 5627.
52. Axelrod, V. D. and Kramer, F. R. (1985). *Biochemistry* **24,** 5716.
53. McGraw III, R-A. (1984). *Anal. Biochem.* **143,** 267.
54. Rackwitz, M-R, Zehetner, G., Frischauf, A. M., and Lehrach, H. (1984). *Gene* **30,** 195.
55. McFarlane, A. S. (1958). *Nature, Lond.* **182,** 53.
56. Hunter, W. M. and Greenwood, F. C. (1962). *Nature, Lond.* **194,** 495.
57. Rosa, U., Scassellati, G. A., Pennisi, F., Riccioni, N., Gianoni, P., and Giordani, R. (1964). *Biochim. Biophys. Acta* **86,** 519.
58. Marchalonis, J. J. (1969). *Biochem. J.* **113,** 299.
59. Butt, W. R. (1972). *J. Endocrinol.* **55,** 453.
60. Redshaw, M. R. and Lynch, S. S. (1974). *J. Endocrinol.* **60,** 527.
61. Fraker, P. J. and Speck, J. C. (1978). *Biochem. Biophys. Res. Comm.* **80,** 849.
62. Reay, P. (1982). *Ann. Clin. Biochem.* **19,** 129.
63. Guenther, J. G. and Ramsden, H. (1984). *Int. Biotech. Lab.* **2,** 38.
64. Bolton, A. E. and Hunter, W. M. (1973). *Biochem. J.* **133,** 529.
65. Wood, F. T., Wu, M. M., and Gerhart, J. C. (1975). *Anal. Biochem.* **69,** 339.
66. Hayes, C. E. and Goldstein, I. J. (1975). *Anal. Biochem.* **67,** 580.
67. Sonoda, S. and Schlamowitz, M. (1970). *Immunochemistry* **7,** 885.
68. Hunter, W. M. (1983). In *Immunoassays for Clinical Chemistry: A Workshop Meeting* (eds. W. M. Hunter and J. C. T. Corrie), p. 263. Churchill Livingstone, Edinburgh.
69. Nustad, K., Gautvik, K., and Ørstavik, T. (1979). *Adv. Exp. Med. Biol.* **120A,** 225.
70. Sherman, L. A., Harwig, S., and Hayne, O. A. (1974). *Int. J. Appl. Rad. Isotopes* **25,** 81.
71. McConahey, P. J. and Dixon, F. J. (1980). In *Methods in Enzymology*, Vol. 70, p. 210. Academic Press, London and New York.

72. Brown, N. S., Abbot, S. R., and Corrie, J. E. T. (1983). In *Immunoassays for Clinical Chemistry: A Workshop Meeting* (eds. W. M. Hunter and J. E. T. Corrie), p. 267. Churchill Livingstone, Edinburgh.
73. Markwell, M. A. K. (1982). *Anal. Biochem.* **125,** 427.
74. Paus, E., Bromer, O., and Nustad, K. (1982). In *RIA and Related Procedures in Medicine*, p. 161. International Atomic Energy Agency, Vienna.
75. Miyachi, Y., Vaitukaitis, J. L., Nieschlag, E., and Lipsett, M. B. (1972). *J. Clin. Endocrin.* **34,** 23.
76. Holohan, K. N., Murphy, R. F., Buchanan, K. D., and Elmore, D. T. (1973) *Clin. Chim. Acta* **45,** 153.
77. Murphy, M. J. (1976). *Biochem. J.* **159,** 287.
78. Karonen, S. L., Mörsby, P., Siren, M., and Seuderlug, U. (1975). *Anal. Biochem.* **67,** 1.
79. Lazure, C., Dennis, M., Rochement, J., Seidah, N. G., and Chretien, M. (1982). *Anal. Biochem.* **125,** 406.
80. Roth, J. S., Losty, T., and Wierbicki, E. (1971). *Anal. Biochem.* **42,** 214.
81. Donnelly, W. J., Barry, J. G., and Richardson, T. (1980). *Biochim. Biophys. Acta* **626,** 117.
82. Crestfield, A. M. and Moore, S. (1963). *J. Biol. Chem.* **238,** 622.
83. Dottavio-Martin, D. and Ravel, J. M. (1978). *Anal. Biochem.* **87,** 562.
84. Dolly, J. O., Nockles, E. A. V., Lo, M. M. S., and Barnard, E. A. (1981). *Biochem J.* **193,** 919.
85. Tang, Y. S., Davis, A. M., and Kitcher, J. P., (1983). *J. Lab. Comp. Radiopharm.* **20,** 277.
86. Caras, I. W., Friedlander, E. J., and Block, K. (1980). *J. Biol. Chem.* **255,** 3575.

7

Subcellular localization of biological molecules

RICHARD CUMMING and RACHEL FALLON

1. Introduction

1.1 Compartmentalization within the cell

The complexity of higher organisms is due in part to the wide variety of tissues and cells which have become specialized for different functions. Cells can often be differentiated from each other on the basis of size, shape, fine structural differences and on the basis of phenotypic molecular markers. Techniques such as tissue culture can help to elucidate the function of individual cell types by growing them and manipulating them under controlled conditions. A further degree of complexity comes from the organization of structures within the cell; at the simplest level we can divide the cell into nucleus and cytoplasm, while sophisticated techniques allow distinction between many different structures such as mitochondria, lyzosomes, ribosomes, etc. It is now widely appreciated that these structures perform varied cellular functions and that different molecules are localized within different subcellular compartments (e.g. neuronal compartmentation of cytoskeletal proteins) (1). In addition to the well-known movement of nucleic acids between the nucleus and the cytoplasm, it is now known, for example, that translocation of specific molecules from one region of the cell to another appears to represent the mechanism of action of molecules including steroids, cyclic nucleotides and receptors.

Organelles including mitochondria, nuclei, and cytoskeletal structures such as microtubules and intermediate filaments can also be biochemically separated by fractionation. However, certain subcellular structures, for example endosomes, cannot be routinely observed in electron micrographs and require labelling with markers for identification. Furthermore, biochemical fractions such as microsomes and synaptosomes represent preparations that are closely related but not identical to subcellular structures as they occur *in vivo*.

A commonly used technique is cell fractionation based on tissue homogenization followed by differential centrifugation or density gradient centrifugation. Differential centrifugation exploits the difference in size and density of different organelles; these have different sedimentation coefficients and can be separated by ultracentrifugal techniques (for a review of practical centrifugation, see ref. 2).

Care must be taken to ensure that subcellular fractions are as pure as possible and the use of enzyme markers, for example acid phosphatase for lyzosomes, *N*-acetylglucosamine galactosyl transferase for the Golgi complex, together with confirmatory electron microscopy is essential. Recent developments in biochemical subcellular fractionation have included the use of free flow electrophoresis, flow cytometry for sorting organelles and the use of more selective biochemical/immunochemical markers of subcellular compartments. Diffusion between subcellular compartments is always a concern but will naturally vary depending on the organelle and the molecule being identified.

A general scheme that has been used for separation of different subcellular fractions from brain is shown in *Protocol 1*.

Protocol 1. General scheme for separation of subcellular fractions from brain using density gradient centrifugation

1. Homogenize brain tissue in ~10 volumes of 0.32 M sucrose using a glass homogenizer.
2. Centrifuge at 1000 *g* for 10 min; remove the supernatant and centrifuge this at 17 000 *g* for 60 min. Resuspend the pellet in 0.32 M sucrose (approximately 2–3 ml per gram of starting tissue).
3. Carefully layer over a density gradient (made up at least 1 h before use) consisting of layers of 10 ml of 1.2 M sucrose and 10 ml of 0.8 M sucrose; spin for 2 h at 53 000 *g*. Layering of solutions of different sucrose concentrations is achieved by slowly pipetting down the side of the centrifuge tube.
   The material separates into three layers in the tube with small myelin fragments at the top, synaptosomes in the middle and mitochondria at the bottom. These may then be separated and analysed or purified further.

Note: All solutions and manipulations must be maintained at 4 °C.

2. Subcellular localization of proteins

2.1 Introduction

A multitude of techniques exist for identifying molecules within biochemical fractions, but many of them lack specificity and sensitivity. For this reason we have chosen to focus on procedures using antibodies and nucleic-acid probes which are now available for detecting biological molecules with high specificity and high sensitivity. In order to determine the presence of a particular molecule in a cell or tissue homogenate or subcellular fraction the preparation is commonly separated on a gel (polyacrylamide for proteins, agarose for nucleic acids) transferred to a membrane and then detected using the probe and detection system. Although techniques such as radiommunoassay and dot blotting enable the presence and quantity of a molecule to be determined they do not allow the molecular size to be determined or the relationship to other separated molecules to be ascertained. Membrane transfer is frequently used as it avoids the problems of spurious reactions within gel matrices. Two examples are given here: (a) *'Western blotting'* for identification of specific protein(s) in a mixture described below, and (b) *'Northern blotting'* for identification of specific mRNAs in a mixture described in Section 3.9.1.(a).

An alternative technique known as immunoprecipitation which does not require membrane transfer is discussed in Section 2.3

2.2 'Western' blotting

For further details on gel separation and blotting methods the reader is referred to refs. 3, 4, 5, and 6.

2.2.1 Protein separation

In order to identify a particular protein in a complex mixture using Western (electro-) blotting, the first step is polypeptide separation using gel electrophoresis under denaturing conditions using SDS (sodium dodecyl sulphate)—polyacrylamide gel electrophoresis (SDS-PAGE) (2D gel separation utilizing isoelectric focusing can also be used but is beyond the scope of this chapter, see ref. 7).

Protocol 2. Protein separation using polyacrylamide gels

1. Solubilize samples (protein concentration ~ 2 mg/ml) by mixing 1:1 with a dissociation buffer containing 2.5% SDS, 2% 2-mercaptoethanol, 4 mM EDTA, 20% sucrose, 0.25 M Tris–HCl, pH 6.8. Samples should be boiled immediately to prevent proteolysis of susceptible polypeptides.
2. Set up slab gels (160 × 160 × 2mm) consisting of a stacking gel (3.5% acrylamide, 0.09% bis-acrylamide, 0.1% SDS, 2 mM EDTA, 0.125 M Tris–HCl, pH 6.8) overlaid on a separating gel. The latter can be made to an acrylamide concentration ranging between 5.5 and 12.5% depending on the molecular weight of the polypeptides of interest to be studied. The ratio of acrylamide:bis-acrylamide in the separating gel is 37:1. The acrylamide is made up in 0.1% SDS, 2 mM EDTA, 0.375 M Tris–HCl, pH 8.9. Polymerize the gels by adding solid ammonium persulphate to 0.04% (w/w) followed by *N*,*N*,*N*′,*N*′-tetramethylethylene diamine to 0.08% (w/v) immediately before pouring.

3. Apply samples to wells in the stacking gel (formed by using a plastic 'comb') and overlay with electrode buffer (6 g Tris, 28.8 g glycine, 1 g SDS in 1 litre of distilled water, pH 8.3). Electrophorese at 40 mA until the refractive front reaches the bottom of the gel. Coomassie Blue dye may be included to check the process of separation. Alternatively, add coloured molecular weight markers (available commercially) as an additional separate sample.
4. Use the gels directly for electroblotting or stain for total protein overnight at room temperature with 0.025% (w/v) Coomassie Blue R in 50% (v/v) methanol, 5% (v/v) acetic acid. Destain in 7.5% (v/v) acetic acid and 5% (v/v) methanol at room temperature. For higher sensitivity, silver staining gel methods can be used.

2.2.2 Electroblotting

A tank for electroblotting may be constructed out of Perspex (24 × 16 × 9 cm) and fitted with platinum wire frame electrodes (see *Figure 1*). Tanks are also commercially available (e.g. Bio-Rad Transblot cell, Pharmacia gel destaining tank).

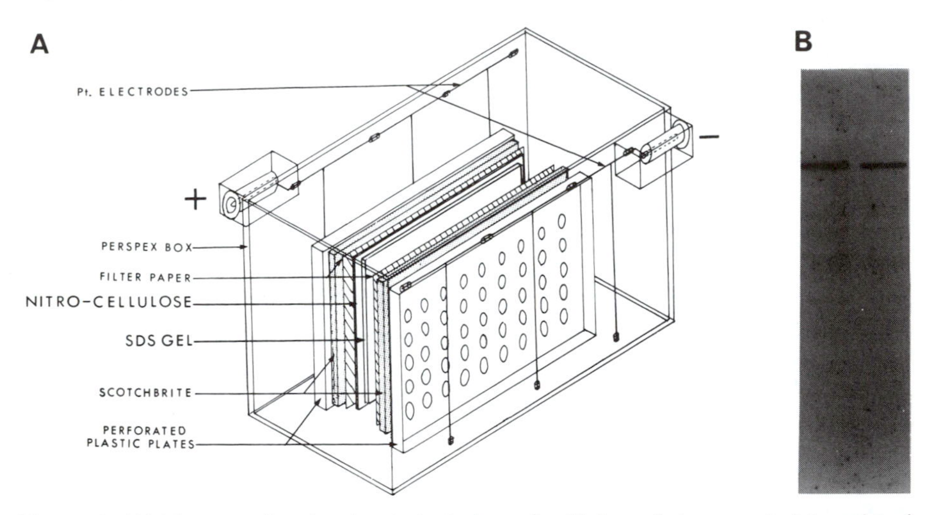

Figure 1. (A) Diagram showing 'exploded' view of gel/nitrocellulose sandwich and tank for SDS-PAGE blotting; this illustrates the basis of 'home-made' and commercial equipment. (B) Western blot of C6 rat glial cell lysate probed with a monoclonal antibody to protein kinase C and detected using [^{125}I] protein G. A single band corresponding to a molecular weight of 79 kd is clearly observed.

Protocol 3. Electroblotting

1. Sandwich the gel with nitrocellulose paper using domestic nylon scouring pads, blotting paper and plastic racks (pipette tip holders); cut the nitro-

cellulose paper from large sheets. Hold the sandwich together with rubber bands. Use plastic gloves when handling the gel and nitrocellulose paper and also for assembly.

2. Make up the blotting buffer for SDS-PAGE transfer as follows: 15.15 g Tris, 72.1 g glycine, 1000 ml of methanol in 5 litres of distilled water (adding 0.1% SDS into the buffer will help in the transfer of high molecular weight proteins).
3. Place the components of the sandwich and the gel itself into buffer for a few minutes prior to assembly. The gel and nitrocellulose paper must be carefully apposed to ensure that air bubbles are not trapped; also take care when handling the fragile gels especially those with a low percentage of acrylamide.

   Cut the nitrocellulose paper slightly larger than the gel and mark the top left-hand corner of the gel and the nitrocellulose paper using, for example, a scissor cut. Submerge the sandwich completely and allow it to rest on its long axis on the base of the tank. Place weights on top if necessary to ensure it does not tip over. Put the loose-fitting lid on the tank and connect the electrodes to a stabilized power supply. For blotting of SDS-polyacrylamide gels the nitrocellulose paper *must* be at the anode (+); electroblot at 10 V, 35 mA for approximately 16 h.
4. After transfer, disconnect the electrodes and carefully separate the components of the sandwich. Cut the nitrocellulose paper to exactly the size of the gel before separation, remembering to mark the top left-hand corner again. Remove the gel and stain with Coomassie Blue to detect any residual polypeptides on the gel that have failed to transfer.
5. Wash the nitrocellulose paper thoroughly in phosphate-buffered saline (PBS: 10 mM sodium phosphate, pH 7.2, 130 mM sodium chloride) and process as a sheet (for comparing different protein samples with the same antibody) or cut into tracks to compare different antibodies. This is conveniently carried out by lining up the nitrocellulose paper with the gel 'comb'. When handling nitrocellulose strips flat-bladed forceps should be used to prevent marking the surface.

   Note: The nitrocellulose paper must be kept moist at all stages after electroblotting.
6. To determine the fidelity of polypeptide transfer, cut out selected strips and stain with the dye Ponceau S (0.2%; Sigma) for approximately 5 min. After staining wash out the dye (if required) with PBS and then probe the blots with antibody. If necessary bands can be marked with pencil lines for re-identification after immunostaining.

Note: For separation and blotting of polypeptides from mini gels (94 mm × 84 mm) a variety of systems are now commercially available, e.g. Bio-Rad mini-protean II electrophoresis and transfer system.

2.2.3 Immunostaining of nitrocellulose replicas

Procedures commonly used for the detection of individual proteins include non-radioactive techniques, isotopically labelled reagents (generally ^{125}I) or biotin-streptavidin techniques, for example using ^{35}S-streptavidin. A protocol based on the use of ^{125}I detection reagents is described below:

Protocol 4. Procedure for staining nitrocellulose replicas

1. Block the nitrocellulose (sheet or strips) in Tris buffered saline (TBS: 50mM Tris–HCl, pH 7.4, 130 mM sodium chloride) containing 1% w/v gelatine, 0.05% (v/v) Tween 20 at 37 °C for 1 h.
2. Incubate primary antibodies at selected dilutions in 0.05% (w/v) Tween/TBS for several hours at room temperature. Take care to ensure that cross-contamination does not occur between strips of nitrocellulose incubated with different reagents; suitable vessels include 10 ml plastic tubes on a rotating support, shallow plastic dishes and silicone rubber moulds.
3. After thorough washing of the nitrocellulose in TBS, detect the antibody using ^{125}I-labelled anti-immunoglobulins or ^{125}I protein A/protein G [final concentration 2.4 MBq/ml Bq/ml)] for 1 h at room temperature diluted in 0.05% (w/v) Tween/TBS. Wash thoroughly (until counts in buffer read background levels), allow blots to air-dry, then seal with Saran Wrap (Dow) to prevent adhesion of the blots to the autoradiography film. Run controls including the omission of primary antibodies. After appropriate exposure using X-ray film (see Chapter 4) develop the film according to the manufacturer's instructions. An example showing the identification of protein kinase C in a cell extract using Western blotting and ^{125}I protein G is shown in *Figure 1*.

   In addition to confirming that a specific molecule is present in the mixture or subcellular fraction these methods have the ability to determine the precise size of a protein and to note whether or not related molecules are present, e.g. proteolytic fragments that may show immunological cross-reactivity to the protein under investigation.

2.3 Immunoprecipitation

In this technique the antigen is isolated from a radiolabelled mixture by specific precipitation with antibody and analysed by polyacrylamide gel electrophoresis followed by autoradiography. This allows detection of the antigen, characterization of its molecular weight (especially useful in biosynthetic studies) and identification of other proteins closely associated with it. In addition, the method of radiolabelling the initial mixture may be varied and information obtained on the structure and orientation of the antigen. Generally, the proteins in a mixture are radiolabelled with iodine (see Chapter 5) or alternatively newly synthesized proteins are biosynthetically labelled with ^{35}S amino acids.

The procedure is usually applied to cell extracts and is carried out in the presence of detergents. The most common means of precipitating the antibody involves binding it to *Staphylococcus aureus* bacteria or Protein A–Sepharose. Unfortunately, not all classes or subclasses of immunoglobin bind tightly to protein A and for those that do not a second antibody must be added to induce precipitation, or alternatively Protein G may be used. Non-specific binding is decreased by 'pre-clearing', the labelled extract with non-immune IgG, immune complexes or, more simply, *S. aureus* bacteria. A control precipitation with pre-immune sera should always be carried out.

The procedure described in *Protocol 5* is given as an example of an immunoprecipitation technique that has been successfully used for the identification of receptors, cytoskeletal proteins, and enzymes.

Protocol 5. Protocol for immunoprecipitation

1. Suspend cells in 5 ml of methionine-free DMEM (Dulbecco's modified Eagles's medium Flow laboratories) containing ^{35}S methionine (3.7 MBq, 100 μCi per ml). Incubate for 4 h at 37 °C. Wash cells in PBS (see *Table 1*), spin at 2000 *g* for 5 min, then resuspend in 1 ml of ice-cold solubilization buffer and incubate for 30 min while vortex mixing at 5 min intervals. Centrifuge lysate at 12 500 *g* for 15 min and collect the supernatant; store on ice.

   Solubilization buffer:

• Tris–HCl 50 mM, pH 7.4	
• Sodium chloride	150 mM
• EDTA	5 mM
• Sodium fluoride	1 mM
• PMSF (phenylmethylsulphonyl fluoride)	2 mM
• Triton X-100	1%

2. Dispense convenient volumes of lysate into Eppendorf tubes. Incubate with monoclonal antibody for 2–4 h at room temperature or overnight at 4 °C (ideally the antibody should be in excess; the precise quantity required can be established in a pilot titration experiment).
3. Separate the immune complexes by adding Pansorbin (insoluble protein A, Calbiochem) to give a final 1:10 dilution, and incubate for 60 min at room temperature with constant agitation.
4. Centrifuge at 12 500 *g* for 10 min. Resuspend the pellet in PBS/Tween 20

(0.1%) then respin as before. Discard the wash solution then resuspend the pellet with PBS, respin at 12 500 *g* for 10 min. Discard the supernatant and resuspend the pellet in 50–100 μl of sample buffer (Tris–HCl, 125 mM, pH 6.75; SDS, 4%; glycerol, 20%; 2-mercaptoethanol, 0.1 M; Pyronin Y, 2 mg/100 ml). Boil for 2 min, spin at 12 500 *g* for 5 min to pellet the immunoabsorbant; the supernatant can be used directly or stored at −20 °C.

5. Load samples of the supernatant on to SDS-polyacrylamide gels and run until the tracking dye has traversed the gel length.
6. Dry the gel on a suitable slab dryer then expose the dried gel to autoradiography film (see Chapter 4).

Note: If high levels of non-specific protein interactions are observed, the lysate may be pre-incubated with Sansorbin (Calbiochem) followed by centrifugation between steps 2 and 3.

2.4 Histological techniques

Subcellular fractionation techniques cannot always isolate particular parts of a cell. Furthermore, they have the disadvantage that they will not allow subtle differences in molecular location to be observed. The way that this information may be obtained is to use techniques to visualize the location of molecules within microscopic structures *in situ*. While light microscopy (LM), with a maximum resolution of 0.2 μm, will enable distinction of some subcellular organelles, electron microscopic methods are needed to resolve all subcellular structures with high resolution.

For reasons of brevity we will not discuss autoradiography using *in vivo* administered labelled compounds such as ^{3}H thymidine or ^{3}H amino acids (for nucleic-acid synthesis and protein synthesis respectively) or the use of labelled precursors. Aspects of these techniques are discussed in Chapter 4 and 5.

In this chapter we will discuss *immunocytochemistry* (ICC) which utilizes highly specific antibody probes and *in situ hybridization* (ISH) which uses nucleic-acid probes. ICC is a powerful technique that can be used to identify the position of closely related molecules at different cellular and subcellular locations (see for example, ref. 10). One of the many applications of ISH is the demonstration of precise sites of protein synthesis within the cell by the use of nucleic acid probes for the visualization of specific mRNA transcripts. The principles underlying both techniques are similar although since ISH is a more recent and predominantly radioactive technique this will be covered in more detail.

2.4.1 Immunocytochemistry

(a) Cell/tissue preparation

In order to precisely locate a molecule of interest it is essential that the morphology of the cells is well preserved. Fixation is used to maintain structural integrity while also preventing diffusion of molecules. The choice of fixation method is dependent upon the antigen of interest and the technique employed (9). For optimal results in animals, perfusion fixation is the method of choice. Rapid freezing of tissue may be used to 'fix' cells and chemical fixatives may be applied in the liquid and also vapour phases on freeze-dried tissue. In order to rapidly fix cells it is important that a chemical fixative has rapid penetration to all cells in a tissue and all subcellular sites in order to prevent molecular diffusion. In order for the antibody to gain access to subcellular compartments in a cell, sectioning (e.g. using a microtome or cryostat) or cell membrane permeabilization (using detergents) are used and enzyme treatment may be required. Subcellular preparations may also be fixed and processed for molecular localization (10).

(b) Selection of antibody

Polyclonal and monoclonal antibodies (as well as fragments) may be used for ICC with the latter being preferred for high specificity due to recognition of a single epitope. It should not be assumed that an antibody that has been shown to be specific by criteria such as radioimmunoassay is suitable for ICC (8). Techniques such as Western blotting may also be used to confirm specificity but again may not be directly analogous to ICC since blotting techniques separate molecules prior to detection and may present them in a different configuration to that on a tissue section.

Radioimmunocytochemistry

Although immunocytochemistry is predominantly a non-radioactive technique, one of the main advantages of utilizing radiolabelled antibodies is that quantitation is made simpler by the ability to count silver grains in an emulsion. Quantitation is more complex for non-isotopic techniques particularly at the light microscopic level. ^{125}I-labelled antibodies have been widely used in many areas of immunology due to their high energy of emission but due to the path length the radiation they have poor resolution compared with a label such as ^{3}H and therefore are not widely suitable for subcellular localization studies, particularly at the electron microscopic level. Labelled antibodies have certain advantages over labelled ligands: for example, radiolabelled antibodies to receptors can detect occupied and unoccupied receptor. Several papers have been published on the use of isotopically labelled antibodies for use in immunocytochemistry either using labelled polyclonal antibodies, labelled monoclonal antibodies or using radiolabelled biotin for detection. The later technique will be described in detail here as it is the most generally applicable and offers good resolution since it uses the isotope tritium (^{3}H). The advantages of the technique are:

- ability to perform semiquantitation by grain counting at LM and EM although this requires statistical analysis;

- potential use in combination with non-radioactive methods.

Disadvantages include:

- sensitivity does not appear to be as good as with non-radioactive methods probably because of the low penetration of β particles from the ^{3}H source;
- the resolution is still not as good at the EM level as for non-radioactive markers such as colloidal gold, and the development time is naturally longer for the radio immunocytochemical method.

Although the isotopic technique has a number of disadvantages it is likely that with reports of simple *in vitro* labelling of monoclonal antibodies using ^{3}H, that the technique may become more widely used (9). A procedure is provided in *Protocol 3*.

Protocol 6. Light microscopic radioimmunocytochemistry using biotinylated protein A and (strept)avidin—[3H]biotin complex (9)

1. Fix tissue in 4% paraformaldehyde solution in 0.1 M sodium phosphate buffer, pH 7.4 containing lysine (3.4 g/litre) and sodium periodate (0.55 g/litre) at room temperature. Remove tissue, post-fix for 2–4 h, wash overnight in phosphate buffer with 30% sucrose and section at 30 μm (freezing microtome).
2. Incubate in optimal dilution of primary antibody (to obtain maximum specific signal: background) at 4 °C with continuous agitation (also use pre-absorbed or non-immune antibodies in control experiments).
3. Wash in phosphate buffer for 30 min.
4. Incubate in 1 μg biotinylated protein A/ml of incubation mixture for 60 min at 37 °C.
5. Wash in phosphate buffer for 30 min.
6. Incubate in a solution containing 10 μg of avidin/streptavidin and 370 kBq (10 μCi) of [^{3}H]biotin in 1 ml 0.1 M Tris buffer pH 7.4 pre-mixed for 20 min beforehand. Incubate for 60 min at 37 °C.
7. Wash in phosphate buffer and prepare for autoradiography using nuclear track emulsion.

Note: Carry out all incubations in phosphate buffer with 1% bovine serum albumin and 0.3% Triton X-100.

---

3. Subcellular localization of nucleic acids

In 1871 Miecher isolated cell nuclei from white blood cells using a combination of dilute acid and proteases. He was able to demonstrate acid precipitable 'nuclein' within the isolated nuclei. Since this discovery, experiments detailing subcellular localization of DNA and RNA molecules have been and continue to be invaluable in the elucidation of function and mechanisms of action.

Early work included the use of histological markers or stains to demonstrate total DNA or RNA within whole cells. Such staining procedures remain vitally important today for example in the analysis of chromosomes, cell sorting, and screening (11).

In 1958, Meselson, Stahl, and Vinograd developed the important technique of density gradient centrifiguation in caesium chloride solutions for separating nucleic acids. This technique is still crucial to recombinant DNA technology; for example, to purify bacterial plasmid DNA from chromosomal DNA or RNA (12).

One of the most powerful early techniques for localizing nucleic acids in cells involved the use of radioisotopes that can be detected with high sensitivity using autoradiography and microscopy (13); for example, autoradiography of cells incubated with the radioactive DNA precursor ^{3}H thymidine demonstrates localization of DNA within the eukaryotic cell nucleus. More recently, with the advent of recombinant DNA technology in the early 1970s, radioisotopes have played and continue to play in a pivotal role as detection systems for nucleic acid hybridization and analysis both in solution, on solid supports and *in situ*.

3.1 Identification of nucleic-acid molecules isolated from subcellular fractions

A number of *in vitro* techniques for nucleic acid detection are appropriate to the identification of nucleic acids isolated from subcellular fractions. Their use is summarized below and detailed protocols can be found in previous volumes (refs 14 and 15) in this series.

(a) *Southern blotting*

DNA is purified from a cell homogenate and digested with appropriate restriction enzymes. The DNA fragments are separated according to size using agarose gel electrophoresis. The DNA fragments are then transferred to a suitable membrane support usually by capillary action. The membrane is probed subsequently using a sequence specific radiolabelled probe. This technique produces information with respect to location, size, and quantity of a given target sequence.

(b) *Northern blotting*

RNA is purified from a cell homogenate. The RNA molecules are separated according to size using denaturing agarose gel electrophoresis. The RNA molecules are then transferred to a suitable membrane support and probed in a similar way to the Southern blot described in (i). This technique produces information with respect to size, quantity and integrity of target RNA. A detailed protocol for this technique is given in Section 3.9.1 of this chapter.

(c) *Dot blotting*

In this procedure, purified DNA or RNA is immobilized on a solid support by direct application. The membrane is then probed with a sequence-specific radiolabelled probe. This technique is quantitative and useful for rapid screening of multiple samples but does not give size information and cross-hybridization of the probe with other nucleic acids in the sample can give false results.

(d) *S1 Nuclease mapping and primer extension*

These techniques are used to identify the termini or splice sites of RNA molecules.

RNA is hybridized to an end-labelled DNA probe (usually a sequenced restriction fragment of a genomic clone) that is complementary to the RNA over part of its length and overlaps a site of interest. After hybridization, S1 nuclease can be used to digest away unhybridized single-stranded DNA and RNA. The size of the labelled DNA strand in the protected hybrid is then determined using denaturing gel electrophoresis. This length then corresponds to the distance between the labelled end of the DNA restriction fragment and the RNA terminus or splice site. In addition, if the DNA probe is used in excess, the hybridization goes to completion such that quantitation of the RNA species is possible.

Primer extension is often used to confirm S1 mapping data and vice versa. Similar information is generated with respect to the start of transcription in a particular DNA sequence. Reverse transcriptase is used to extend the DNA probe to the precise 5′ terminus of the RNA molecule. The extension product is sized using denaturing gel electrophoresis. In addition, sequence data can be generated for the site of interest by incorporating dideoxynucleotides using the Klenow fragment of DNA polymerase I.

(e) *Study of nucleic acids using the electron microscope*

The electron microscope (EM) has greatly facilitated increased understanding of the properties and localization of nucleic acids. A range of protocols are available (15) covering the application of EM to the study of nucleic acid hybridization, the interaction of proteins with nucleic acids, visualization of transcriptional complexes and chromatin. The use of radioisotopes has played a significant role in EM work to date although non-radioactive markers such as colloidal gold are increasingly used for improved resolution.

3.2 Localization of nucleic acids using *in situ* hybridization

In situ hybridization (ISH) refers to the detection of nucleic acid sequences within single cells or on chromosomes using labelled sequence-specific probes. This is an extremely powerful subcellular technique that complements the current range of nucleic acid technologies in terms of the information generated. This method will be discussed in detail.

In situ hybridization supplies spatial information with respect to the nucleic acid sequence of interest; for example, whether a particular cell type is expressing a transcript of interest, or the precise location of a gene on a chromosome. Furthermore, extreme sensitivity is possible using ISH since detection of target molecules is confined to a single cell and visualized by virtue of the microscope. Extraction of nucleic acids followed by detection in solution or on a solid support can result in a dilution of the target sequence when only a few cells from a mixed population contain the sequence of interest. This result in loss of sensitivity following detection.

In situ hybridization is carried out using the following basic steps.

- Preparation and fixation of the cell or tissue sample to retain cellular morphology and target nucleic acid in a form compatible with hybridization.
- Pre-hybridization of the sample. This step serves to establish denaturing conditions, block non-specific binding of probe and facilitate probe entry where necessary.
- Hybridization of the nucleic acid probe with target nucleic acid in the tissue sample under the correct conditions of salt, temperature, probe concentration and blocking agents.
- Removal of labelled probe which has not specifically hybridized to target nucleic acid.
- Detection of the labelled probe *in situ*.
- Visualization of the labelled probe using the light or electron microscope.

Since this technology was first described by Gall and Pardue in 1969 (16), radioactive labelling and detection of the probe continues to be the most widely used approach *Figures 2* and *3*). Radioactive detection currently has advantages over non-radioactive detection in terms of high sensitivity, for example detection of unique gene sequences and the prospect of quantitation of signal by grain counting an autoradiograph. The future of this technology however, lies with non-radioactive detection by virtue of a shorter turn-around time of results (for example, 24–48 h compared with 4–6 weeks using ^{3}H-labelled probes), higher resolution resulting in better spatial information (17) and the possibility of using non-hazardous reagents. Currently the non-radioactive procedures are generally perceived as being less sensitive and less amenable to quantitation than radioactive counterparts.

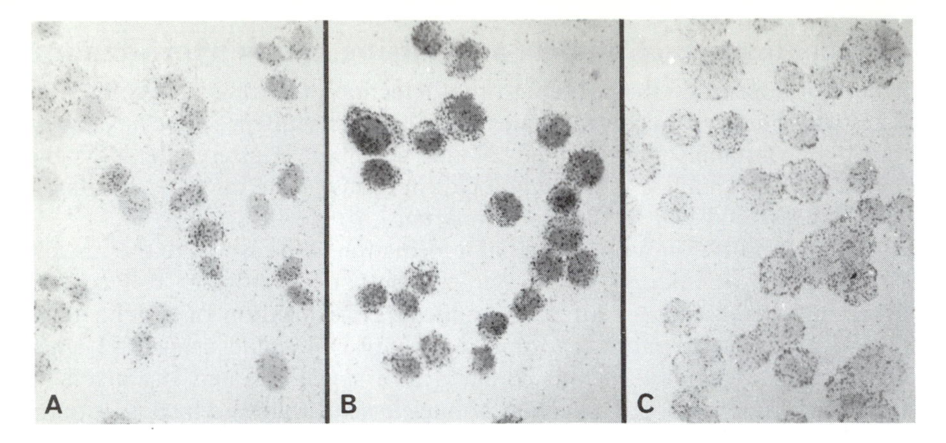

Figure 2. Autoradiograph of cultured chicken embryo cells which have been cytospun on to glass slides and hybridized with a chicken β-actin RNA probe. The autoradiographs demonstrate the degree of resolution produced by different radioactive labels. The cells are shown in bright field illumination: (A) cells hybridized with a ^{32}P-labelled RNA probe (specific activity 1.3×10^9d.p.m./μg) exposed for four days; (B) cells hybridized with a ^{35}S-labelled RNA probe (specific activity 6.6×10^8d.p.m./μg) exposed for six days; (C) cells hybridized with a ^{3}H-labelled RNA probe (specific activity 8.6×10^7d.p.m./μg) exposed for 14 days.

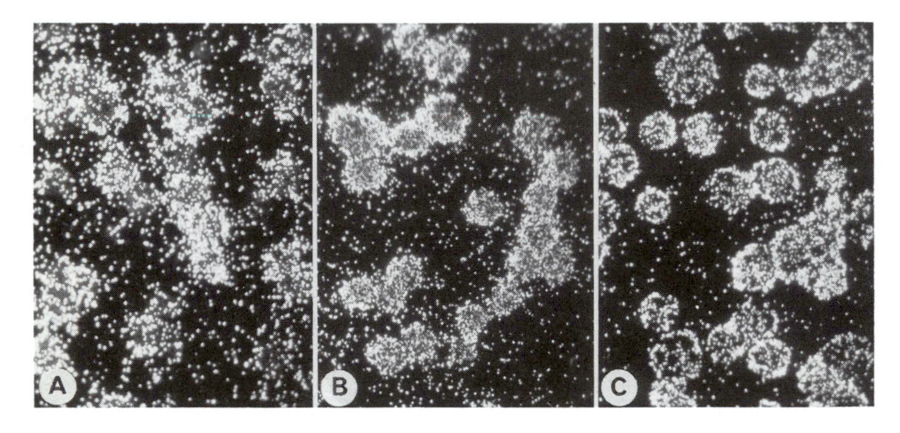

Figure 3. Identical autoradiographs detailed in *Figure 2* shown in dark field illumination.

3.2.1 Applications of *in situ* hybridization

The applications of ISH fall into four main areas detailed below.

i. Gene assignment

In situ hybridization to the DNA of condensed metaphase chromosomes is an important application of this technology. Many genes have now been assigned to particular chromosomes and the position of the gene on mapped according to histological staining procedures. In addition the importance of translocations in disease states emphasizes the potential of this approach for diagnosis (17). Detailed protocols and references covering ISH to metaphase chromosomes are available (18, 19, 20).

ii. Detection of gene sequences and transcripts in interphase cells
Gene sequences have been detected in interphase cells showing a highly ordered arrangement of chromosomes in interphase nuclei with implications for the phenotype of particular organisms (21). In addition, specific transcripts have been localized within certain areas of the cell cytoplasm (22). This has required the high resolution afforded by non-radioactive markers (17).

iii. Detection of RNA transcripts in whole cells or tissue sections
In situ hybridization has been used to study expression at the cellular level providing data which is not available from other techniques; for example, the expression of a particular gene with respect to cell type during development (23) or establishing the site of gene expression which may be different from the location of protein molecules, for example in the central nervous system (CNS) (24).

iv. Detection of intracellular viral nucleic acids
Cells infected with virus contain both viral genomic nucleic acid molecules and viral RNA transcripts. It is possible to detect either or both of these types of molecules using ISH. There are now many publications detailing the detection of viral sequences by ISH (25, 26, 27). In addition it is possible, using strand-specific probes to identify viral nucleic acid replicative intermediates (28).

The application of ISH to detection of virally infected cells is likely to be important for diagnosis where there is a latent viral infection, and when both the morphology of the tissue specimen is required in conjunction with the diagnosis of viral infection, for example detection of HPV in cervical biopsies (29).

Unfortunately, there is a lack of consensus in the literature with respect to ISH methodology and the worker wishing to use the technique is faced with a plethora of protocols. We attempt to provide a reasoned approach to the protocols for the detection of nucleic acids within whole cells or tissue sections.

3.3 Safety considerations when using *in situ* hybridization (30)

This section highlights additional precautions to be taken when carrying out ISH. It is not meant to be a comprehensive guide to laboratory safety.

Clearly, before planning any ISH experiment, set laboratory procedures should already be in place for current radioactive licensing requirements, tissue handling facilities and Genetic Manipulation Advisory Group (GMAG) approval for genetic manipulation work involving construction of appropriate probes.

3.3.1 Use of radioisotopes for *in situ* hybridization

A range of isotopes are useful for ISH, these include ^{32}P, ^{35}S, ^{3}H, and ^{125}I. It is important to have an understanding of the different properties and handling procedures for each prior to use (see Appendix 2).

It is recommended that ISH experiments are carried out in a contained area, close to a sink designated for radioactive liquid waste disposal. This is particularly useful when disposing of the stringency wash solutions. Solid waste such as contaminated tips, cover-slips, etc., should be carefully stored in a watertight container prior to bulk disposal. The contained area should comprise a crevase free tray containing several layers of readily disposable absorbant paper (NB Keep tissues and detergent close by to mop up any spills).

The hybridization and washing chambers should be constructed of material which is readily decontaminated by soaking or wiping with strong detergent; for example, glass or plastic vessels.

Contamination of the hybridization chamber can be minimized by carefully drying around each slide and tissue section so that 'hot' hybridization buffer does not travel by capillary action to the underside of the slide and then to contact points within the chamber.

Finger dosimeters are recommended for this type of work: Personal dose received from ^{32}P of ^{125}I through handling slides should be minimized by remote handling using forceps.

^{32}P and ^{125}I can be detected at the bench using hand-held Geiger and γ counters respectively. These monitors should be available at all times during the experiment when these isotopes are used. The Geiger counter is extremely inefficient at detecting ^{35}S and does not detect ^{3}H (see Appendix 3).

The amounts of solid and liquid waste should be carefully recorded. After each experiment, monitor the area with a hand held monitor if possible or take swabs and count in scintillant using the appropriate channels on a scintillation counter. Always swab down the bench area with a suitable detergent such as Decon 90 after each experiment.

3.3.2 Tissue handling and cell culture

- Handling fresh tissue or cells of human or animal origin is intrinsic to ISH. In the unfixed state, tissues should be considered as potentially hazardous.
- All procedures likely to generate aerosols from biological material must be performed with approved containment, i.e. combined class I/III cabinets for tissue culture.
- All biological waste must be made safe by autoclaving before disposal.

Bench surfaces should be wiped down with a suitable freshly prepared disinfectant.

- A record of cell lines, tissue samples, etc., should be kept providing details of the source and storage conditions.
- Tissue and cells should be fixed immediately to achieve the best retention of nucleic acid and structural morphology for ISH. This also minimizes the handling time required for unfixed preparations substantially.

Clearly, these extra considerations for ISH must be carried out together with a general background of good laboratory discipline in handling *all* potentially hazardous material including solvents, oncogenic DNA, etc.

3.4 Preparation of target cells or tissues

Methods of collection and fixation of tissues or cells are an important and sometimes overlooked preparatory step for ISH. The aim is to preserve the tissue/cell morphology whilst ensuring maximal retention of target nucleic acid in a form which is still accessible to hybridization with a complementary probe.

When the target nucleic acid of interest is RNA, special care should be taken. It is essential to fix tissue or cells with a minimum delay (30 min is a safe limit in most cases) following death or harvest of cells. The collection should be carried out using sterile instruments to minimize contact of the tissues with ribonucleases. In addition, the operator should wear disposable gloves.

A whole variety of fixatives have been used successfully for ISH by different researchers (26, 34). It is generally believed that strongly crosslinking fixatives may prevent the penetration of probes whilst precipitating fixatives might be insufficient for RNA retention. Sometimes, 4% paraformaldehyde has been singled out as the best fixative for ISH (35). In our experience, each tissue type requires a different optimal fixative for optimal hybridization efficiency (36). We recommend that the user optimizes the fixation procedure for each tissue type under investigation by screening with a panel of fixatives (see *Table 1*). Where retrospective studies are undertaken, it may be necessary to optimize permeabilization conditions. Both frozen and more recently paraffin embedded tissue have been used successfully for ISH.

3.4.1 Frozen tissue blocks

Protocol 7. Preparation of frozen tissue blocks for *in situ* hybridization

1. Cut the tissue into small pieces ($1 \times 1 \times 0.5$ cm) and fix by immersion in the appropriate fixative. NB Animal brain tissue should be perfused with fixative.

2. Cryoprotect the tissue blocks by rinsing in 15% sucrose in 1 × PBS (see Table 1) 4 × 1 h. Store excess tissue blocks at 4 °C in 15% sucrose in PBS containing 0.01% sodium azide.
3. Orient the tissue block in a suitable freezing compound then snap freeze it in isoprene pre-cooled in liquid nitrogen.
4. Cut 10 μm cryostat sections and mount on to poly-L-lysine hydrobromide (PLL) coated slides. Dry the slides for at least 4 h in an oven set at 37 °C prior to ISM.
5. The sections may be used immediately or stored in the presence of silica gel at −70 °C for at least seven months without loss of target nucleic acid.

Table 1. A range of tissue fixation conditions

		Fixation time	
	Fixatives	**Cultured cells**	**Tissue block**
1.	Ethanol/acetic acid (3:1 v/v) + ethanol	15 min RT[a]	30 min RT
	Ethanol/acetic acid (95%:5% v/v)	5 min RT	10 min RT
2.	Methanol/acetone (1:2 v/v)	15 min RT	30 min RT
3.	Methanol/acetone (1:1 v/v)	4 min −20 °C	20 min −20 °C
4.	Bouin's 75% picric acid + 25% formalin + 1%	4 min −20 °C	20 min −20 °C
5.	acetic acid	30 min RT	1 h RT
6.	4% Paraformaldehyde (PF)[b] in PBS pH 7.2[c]		
	2% Glutaraldehyde in 1 × PBS pH 7.2	5–30 min RT	1 h RT
7.	Paraformaldehyde-lysine-periodate (PLP)[d]	30 min RT	1 h RT
8.	4% PF/Methanol (1:9)	15 min RT	1 h RT
9.	4% glutaraldehyde in 20% ethylene glycol	30 min RT	1 h RT
10.		5 min 4 °C	

[a] Room temperature.
[b] Prepare paraformaldehyde solution in a fume-hood. Do not breath dust or fumes. Dissolve 12 g of EM grade paraformaldehyde in 300 ml 1 × PBS pH 7.2. Heat the solution to 50–60 °C (do not boil) and stir rapidly until the paraformaldehyde is completely dissolved.
[c] Phosphate buffered saline: 10 mM sodium phosphate pH 7.2, 130 mM sodium chloride.
[d] Paraformaldehyde lysine periodate: 4% EM grade paraformaldehyde, 23.2 mM lysine, 2.58 mM sodium periodate in 0.1 M sodium phosphate pH 7.2.

3.4.2 Cultured cells

Both adherent and non-adherent cultured cells can by cytocentrifuged onto glass slides. These preparations contain rounded cells such that the cytoplasm forms a thin rim around the nucleus.

Protocol 8. Preparation of cultured cells for *in situ* hybridization

1. Rinse adherent cells with isotonic PBS. Detach the cells from the flask using 0.5% trypsin, 0.02% EDTA solution (5 min at 37 °C).
2. Wash the trypsinized cells in medium containing 10% serum (serum inhibits trypsin).

3. Resuspend cells at 1×10^5/ml.
4. Deposit 2000 cells on each slide previously coated with poly-L-lysine hydrobromide using a cytocentrifuge (Shandon).
5. Allow the cells to air dry 5 min.
6. Fix the cells by immersion in the appropriate fixative.
7. Rinse the cytospin preparations in $1 \times$ PBS (2×3 min).
8. Rinse the cytospin preparations in distilled water (2×5 min).
9. Dry the preparations at 37 °C for at least 4 h.
10. The cells may be used immediately for ISH or stored in the presence of silica gel at −70 °C for at least 7 months without loss of target nucleic acid.

Alternatively, adherent cells can be grown directly on slides or coverslips and fixed *in situ*. This is useful when observing cytoplasmic distribution of RNA since the cells are spread out over the solid support. Seed sterile multiwell slides or glass cover-slips with 2000–3000 cells each and culture under appropriate conditions for at least 24 h. Rinse the preparations with three changes of isotonic PBS then follow steps 7–10.

3.4.3 Preparation of glass slides and cover-slips

Loss of tissue sections or cells during the ISH procedure has presented itself as a problem. Several methods of coating or 'subbing' the slides have been developed (31, 32). Poly-L-lysine hydrobromide (150 000–300 000 molecular weight) is efficient in retaining tissue sections and cytospin preparations of cultured cells *Protocol 9*.

Protocol 9. Procedure for coating glass with poly-L-lysine hydrobromide

1. Render the slides clean and grease free by immersion in chromic acid for ten minutes.
2. Rinse the slides *thoroughly* in running tap water until all traces of chromic acid have disappeared.
3. Rinse the slides thoroughly in double-distilled de-ionized water.
4. Immerse the slides in 100% ethanol.
5. Air dry the slides.
6. Wrap the slides in aluminium foil and bake at 250 °C for at least 4 h. The slides are, subsequently, sterile and RNase free.
7. When cooled, coat the slides with a fresh 1 μg/ul solution of poly-L-lysine

hydrobromide either by immersion or by smearing 5 μl of solution over each slide using the edge of another sterile slide.

8. Allow the slides to air dry.
9. Re-wrap the slides in foil and store at room temperature prior to use.

Glass cover-slips can be siliconized to prevent adherence of nucleic-acid probe and cytological material (33).

Protocol 10. Procedure for siliconizing glass cover-slips

1. Immerse the glass cover-slips in a 5% solution of dichlorodimethylsilane in chloroform.
2. Rinse the cover-slips thoroughly in double-distilled de-ionized water.
3. Bake the cover-slips for at least 4 h at 250 °C to complete the process and render them RNase-free prior to use.

3.4.4 Procedures which minimize the possibility of contaminating preparations with RNases

Ribonucleic acid molecules are rapidly degraded by RNases. Since both RNA probes and target RNA molecules are susceptible to degradation, it is important to take the following precautions to avoid contamination with RNases.

- Always wear disposable gloves.
- Use sterile disposable plasticware or baked glassware for solution preparation (bake glassware at 250 °C for at least 4 h to render it RNase free).
- Prepare solutions to include 0.1% diethylpyrocarbonate (DEPC), which destroys RNase activity. Excess DEPC must be converted to carbon dioxide and ethanol by autoclaving treated solutions.
- Use exogenous inhibitors of RNases in solutions which cannot be autoclaved (for example, placental ribonuclease inhibitor or vanadyl ribonucleoside complexes).
- Use baked spatulas to avoid the introduction of RNases into chemical stocks.

3.5 Preparation of radiolabelled probe

3.5.1 Choice of radioisotope

^{32}P, ^{35}S, ^{3}H, and ^{125}I labelled probes have been used for ISH. Unfortunately, there has to be a compromise between autoradiographic resolution and sensitivity which is dependent on the energy of particles emitted from the radioisotope. For example, this means that the low energy β particles emitted by tritum afford the best autoradiographic resolution but exposure times can be of the order of 4–6 weeks. At the other end of the scale, the β particles from ^{32}P are high energy, affording low resolution but much shorter exposure times of the order of a few days. ^{34}S and ^{125}I fall in the middle of this range in terms of resolutions and sensitivity.

The choice of radioisotope rather depends on the resolution required by the investigator; for example, ^{3}H-labelled probes generally offer intracellular localization whereas a ^{32}P-labelled probe would denote presence of a transcript over a cell or group of cells. Researchers often choose ^{35}S or ^{125}I as a compromise, although autoradiographic backgrounds are sometimes a problem with these isotopes.

3.5.2 Choice of probe

There is now a considerable body of evidence (37) to show that single-stranded RNA probes are more sensitive (10–15-fold) when used for ISH than double-stranded DNA probes. DNA probes may reanneal before hybridization within a cell is accomplished. In addition, non-specifically bound RNA probes can be removed from the tissue preparation after hybridization by digesting with appropriate concentrations of RNase to improve the signal-to-noise. (NB New batches of RNase should be titrated prior to use.)

Although many researchers continue to use double-stranded DNA probes for ISH, investigators requiring the highest possible sensitivity should consider inserting probe sequences into commercially available dual promotor vectors which will allow preparation of strand specific probes.

Oligonucleotide probes also have the advantages of being single-stranded but are generally less sensitive since less target nucleic acid is covered. There are, however, situations where short oligonucleotide probes are necessary to define sequence specificity; for example, when probing for a family of related genes or precursors (38).

3.5.3 Alkaline hydrolysis of RNA probes

Synthesis of labelled RNA probes is detailed in Chapter 6 of this volume.

For efficient hybridization in cytological preparations it is generally agreed that probes should be between 50 and 250 bases in length. This can be achieved by alkaline hydrolysis of the full-length transcripts using the protocol described below.

Protocol 11. Alkaline hydrolysis of RNA probes

1. Calculate the hydrolysis time using the following formula (13)

$$t = \frac{L_0 \, L_f}{K \, L_0 \, L_f}$$

L_0 = Initial probe length in kilobases
L_f = Final probe length in kilobases
K = The rate constant for hydrolysis (0.11 kb/min)

2. Prepare the hydrolysis mix:

$NaHCO_3$	0.4 M	20 μl
$NaCO_3$	0.6 M	20 μl
H_2O		150 μl
Probe dissolved in sterile water		10 μl
	Total =	200 μl

NB Containing 10 mM DTT if ^{35}S-labelled probes are used.

3. Incubate the hydrolysis mix at 60 °C for the appropriate time
4. Stop the reaction by adding:

Sodium acetate 3 M	6.6 μl
Glacial acetic acid	1.3 μl
	207.9 μl

5. Precipitate the hydrolysed probe with 2.5 volumes of ethanol.
6. Pellet the probe in a microfuge for 5 min.
7. Dry the pellet.

(At this stage, it is possible to check the original and final lengths of the transcript by electrophoresis in 2% agarose formaldehyde gels followed by autoradiography. The above formula gives reproducible results provided all solutions are RNase free.)

8. Dissolve the probe in hybridization buffer (see Section 6 for preparation) at 0.5 ng/μl for use in the hybridization.

3.6 Hybridization

Protocol 12. Pre-treatment of tissues prior to hybridization

Equipment

1. Humid chambers are required for ISH because the hybridizations are carried out in small volumes typically 10–20 μl. Plastic air-tight sandwich boxes are ideal. Cover the bottom of the sandwich box with a thin layer of the incubation buffer. It is important that this solution has the same salt concentration as the medium currently on each slide to prevent distillation effects. Support slides above the liquid for example using pipettes taped to the bottom of the chamber.
2. Coat an orbital shaker with readily disposable absorbant material and agitate stringency washes at the lowest speed.
3. Set a hot-air oven to the correct hybridization temperature before starting the experiment.

NB Pre-warm solutions as appropriate (e.g. the hybridization solution) prior to use.

Treatment of cells/tissue prior to hybridization

1. Remove slides from storage at −70 °C and allow them to come to room temperature in the presence of silica gel. Although the preparations are fixed at this stage it is important not to introduce RNases when using an RNA probe of defined length. In addition random fragmentation of target nucleic acids may result in altered stringency.
2. Permeabilize the cells by immersing the slides in 0.3% Triton X-100, 1 × PBS pH 7.2 for 15 min.
3. Rinse the slides in 1 × PBS pH 7.2, 2 × 3 min.
4. Improve access of the probe to cellular nucleic acid by limited pronase digestion. Pre-digest pronase to render it RNase-free and use at ~1 μg/ml.

NB. Pronase differs from batch to batch, it is therefore necessary to titrate each batch for optimal results (see *Protocol 13*).

Incubate all slides in the optimal pronase concentration at 37 °C for 30 min.

5. Stop the action of pronase by immersing the slides in 0.1 M glycine, 1 × PBS pH 7.2 for 5 min.
6. Post fix the tissues for 3 min to retain target nucleic acid.
7. Rinse the slides in 1 × PBS pH 7.2, 2 × 3 min.
8. Reduce the non-specific binding of the probe by acetylating the tissue sections.

Immerse the slides in 0.25% acetic anhydride, 0.1 M trithanolamine pH 8.0 for 10 min. (NB Add the acetic anhydride to 0.1 M trethanolamine as it is stirring. Immediately place the slides in the staining jar and begin the incubation.)

Protocol 13. Autodigestion and titration of pronase

1. Prepare a 40 mg/ml pronase solution in sterile distilled water.
2. Incubate at 37 °C for 4 h.
3. Dispense the solution in 0.25 ml aliquots and store at −70 °C.
4. Dilute the pronase stock solution in 50 mM Tris–HCl pH 7.5, 5 mM EDTA for use.

NB For each batch of pronase, test a range of concentrations, e.g. 20 μg/ml, 40 μg/ml, 60 μg/ml and 80μg/ml. Choose the highest pronase concentration consistent with good morphology.

3.6.3 Pre-hybridization/hybridization procedure

Hybridizations are usually performed in formamide to prevent exposure of the tissues to high temperatures during the hybridization which may affect the morphology. It is worth noting that the melting temperature of RNA:RNA duplexes is reduced by 0.35 °C for each percentage point of formimide. Hybridization is optimal at 25 °C below the melting temperature, T_m (13). Ideally the T_m for each probe should be determined using Northern or dot blots. Generally, when RNA probes are used to detect RNA, a hybridization temperature of 50–55 °C in 50% formamide is used.

A pre-hybridization step is generally incorporated at the hybridization temperature.

Protocol 14. Procedure for pre-hybridization and hybridization

1. Prepare a solution of 50% de-ionized formamide and 5 × SSPE (20 × SSPE: 0.2 M sodium phosphate pH 7.4, 3.0 M sodium chloride, 0.02 M ethylene diaminotetracetic acid) pre-warmed to the correct temperature. Immerse the slides and incubate for at least 15 min.
2. Prepare the hybridization mixture containing 50% de-ionized formamide, 5 × SSPE, 10% dextran sulphate, 0.25% bovine serum albumin 0.25% Ficoll 400, 0.25% PVP 360, 0.5% SDS, 100 μg/ml freshly denatured salmon sperm DNA (sonicated to an average size of 200 base-pairs).

   Use this buffer to dissolve the hydrolysed probe at 0.5 ng/μl.
3. Drain the slides from pre-hybridization buffer, apply 10–20μl hybridization buffer containing probe to each slide.
4. For hybridization to double-stranded DNA in tissue/cell preparations heat the slides in a mechanical convection oven at 100–150 °C for 10 min (39).
5. Cover the sections/cells with siliconized RNase free cover-slips (see *Protocol 10*) where necessary. Cytospin preparations are usually well covered by 10 μl hybridization mix without the need for a cover-slip.
6. Incubate in the hybridization chamber at the appropriate temperature (typically 52 °C for RNA probes) for 2 h, or overnight.

3.6.4 Stringency washes

Stringency washes remove non-specifically bound probe from the cell/tissue preparations. Stringency is increased by raising the temperature and/or decreasing the salt concentration of particular washes. As with the hybridization temperature, stringency conditions should be evaluated for each probe using Northern or dot blots. The following protocol outlines the basic procedure.

NB. It is important when using ^{35}S or ^{125}I labelled probes to include 10 mM

DTT or 1000 μM potassium iodide in the stringency washes respectively, to reduce background signal.

Protocol 15. Procedure for stringency washes

1. Following hybridization, remove the slides from the hybridization chamber and remove the cover-slip and hybridization buffer by dipping each slide into a beaker of 2 × SSPE. (See *Protocol 14*).
2. Place the slides in a staining jar containing 2 × SSPE and 0.1% SDS. Wash the slides 4 × 5 min at room temperature with gentle shaking on an orbital mixer.
3. Wash the slides in 0.1 × SSPE, 0.1% SDS pre-warmed to the hybridization temperature 2 × 10 min with gentle shaking.
4. RNA hybrids are relatively resistant to digestion by pancreatic RNase A. It is therefore possible to use low concentrations of RNase to remove non-specifically bound RNA probes.

Immerse the slides in a pre-warmed solution of RNase (10 μg/ml RNase A, 2 × SSPE) at 37 °C for 15 min.

5. Dehydrate the tissues/cells in graded ethanols (70%, 95%, 2 × 100%) containing 0.3 M ammonium acetate. The ammonium acetate is present to maintain the integrity of the hybrids (40).

3.7 Autoradiography (41)

Both X-ray film and nuclear track emulsion coated cover-slips have been used to detect *in situ* hybridization by apposing the tissues/cells against the film. X-ray film is useful for screening many samples or optimizing conditions for hybridization. However, these are not the most common methods of autoradiography. The method described here allows the worker to view the tissue or cells simultaneously with the exposed emulsion overlayed.

Autoradiography requires a completely light-tight darkroom. It is an advantage to have a double-door system or a rotating door fitted so that the user can enter and leave the darkroom without exposing the emulsion during the drying period. If there is no double door, the slides can be dried in a light-tight well-ventilated box.

The dark room should be fitted with a safe light containing a 25-watt bulb and a Kodak Wratten series II filter when using Kodak NTB2 emulsion. (NB Different safe lighting conditions may be required with nuclear track emulsions from different manufacturers.) The safe light should be positioned at least 3 ft away from the working area.

The dark room should also be equipped with a water bath set at 43–45 °C.

Nuclear track emulsions have a gelatin base and are therefore solid at room temperature. Since repeated melting of the emulsion raises the background level of grains, the emulsion should be aliquoted as detailed below.

3.7.1 Aliquoting new emulsion

Protocol 16. Procedure for aliquoting new emulsion

Before adopting safe light conditions

1. Switch on the water bath set at 43–45 °C.
2. Ensure that the temperature in the room is 18–20 °C and that the humidity is ~60% (low humidity can cause stress grain formation during drying of emulsions (40).
3. Place an appropriate volume of distilled water containing 0.6 M ammonium acetate in a flask in the water bath. Leave the water to come to temperature.
4. Check the arrangement of apparatus. Extinguish all lights. Switch to the safelight when ready to proceed.

After adopting safe light conditions

5. Place the new container of emulsion in the water bath. Allow the emulsion to melt for 30 min.
6. Carefully pour the molten emulsion into the flask containing an equal volume of pre-warmed 0.6 M ammonium acetate. *Gently* mix the emulsion by turning the flask several times (too much agitation of the emulsion also raises background grains).
7. Aliquot the emulsion, e.g. into 10 ml amounts in glass scintillation vials.
8. Store aliquots in a light-tight box at 4 °C in a refrigerator *away* from radioisotopes.

3.7.2 Dipping slides in nuclear track emulsion

Protocol 17. Procedure for dipping slides in nuclear track emulsion

1. Carry out steps (1) and (2) from *Protocol 16*.

   In addition, place a metal or glass plate on to a tray containing ice. This is used to facilitate gelling of the emulsion on the slide.
2. Place a suitable dipping chamber which will accommodate one slide at a time to warm in the water dish. We have used a glass dipping chamber constructed using blown glass.

3. Sort slides into sets for different exposure times. Under safe light conditions, melt an aliquot of pre-diluted emulsion at 43–45 °C for 10–15 min.
4. Pour the emulsion slowly down the side of the chamber ensuring that air bubbles do not form.
5. Check for the absence of air bubbles by repeated dipping of blank glass slides, i.e. hold the slide between thumb and forefinger, dip the slide slowly into the emulsion and withdraw. Drain the slide briefly on a pad of absorbant tissues and examine the slide for air bubbles under the safe light. Continue to dip slides until no air bubbles remain.
6. Dip the experimental slides carefully as described. By withdrawing the different slides at as near a constant rate as possible from the emulsion, the uniformity of the emulsion coat between slides is more reproducible (41). After draining each slide briefly on a pad of absorbant tissue, wipe the back of the slide with a tissue and quickly place the slides on to the pre-cooled glass or metal plate.
7. Allow the emulsion to 'gel' for 10 min.
8. Remove the slides from the cooled plate and allow them to dry at room temperature for at least 2 h.
9. Pack the slides into light-tight boxes containing silica gel to maintain dryness. Seal the boxes with black tape and store at 4 °C for appropriate exposure times.

3.7.3 Developing autoradiographs

When nuclear track emulsion is 'wet', it is extremely fragile. Wrinkling and loss of emulsion from the slides will occur if the emulsion is subjected to different temperatures during the development or staining procedure. All solutions should therefore be equilibrated to room temperature (15–20 °C) in the dark room prior to development.

Protocol 18. Procedure for developing autoradiographs

1. Remove the box containing slides from the refigerator. Allow the slides to warm up to room temperature in the presence of desiccant.
2. Under safe lighting conditions, remove the slides from the box and immerse developer (e.g. Kodak D-19 developer, 4 min). Do not agitate the slides.
3. Stop development in distilled water, 5 min.
4. Fix the emulsion, e.g. using Kodak F24 fixer, 5 min.
5. Rinse the slides in distilled water for 5 min.
6. At this stage the lights can be turned on.

3.7.4 Histological staining

Ensure that the staining solutions are at the same temperature as the developing solutions. The emulsion layer can result in variable and imprecise staining; however, counterstaining the tissue/cells through the photographic emulsion does produce acceptable results. Silver grains can be lost through exposure to acidic solutions during staining procedures, e.g. for some haematoxylins and differentiation with acid alcohol.

Protocol 19. A procedure for counterstaining with Pyronin Y

1. Prepare a 2.5% solution of Pyronin Y in distilled water.
2. Immerse the slides in Pyronin Y for 30 sec.
3. Rinse in distilled water for a few seconds. This removes pyronin stain and therefore the correct time should be determined by trial and error.
4. Dehydrate in acetone for 1 min.
5. Rinse in equal parts of acetone and CNP™.
6. Clear in pure CNP™.
7. Mount the slides in synthetic resin such as DPX™.

3.8 Analysis of data

3.8.1 Microscopy

The tissue/cell autoradiographs are viewed using light microscopy. Bright field illumination is adequate for viewing medium to strong hybridization signals. Weak hybridization signals are 'apparently' clearer if viewed using dark field illumination. With dark field illumination, the tissue or cells appear dark whilst the grains appear white or refractile (*Figure 3*). For dark field optics, the microscope should be fitted with a dark field condenser. The best dark field image is obtained by putting immersion oil between the condenser and the glass slide.

3.8.2 Quantitation

One of the advantages of using radioisotopes in this application is the ability to quantitate signal by grain counting over particular cells. Absolute quantitation of signal from ISH is a contentious issue in several respects, since there are many variables both between different tissue sections and within the same section, e.g. uneven fixation and uneven probe penetration, autoradiographic efficiency, hybridization, efficiency, etc.

There is value in comparative grain counting; for example, to establish signal-to-noise ratios. This is particularly important when assigning genes to chromosomes by ISH or when detecting rare transcripts in cells. In these cases, statistical significance must be established (40).

Complex instrumentation is now available for image analysis so that comparative quantitation by statistical analysis is less tedious than performing manual grain counting with the light microscope.

3.9 Controls required for *in situ* hybridization

As with all histochemical methods, proper controls are essential for each ISH experiment to assess the reproducibility, specificity, and sensitivity of the reagents and procedures used.

For detection of mRNA, a comprehensive series of controls are detailed below:

- Northern blot analysis (see Section 3.9.1) should be carried out using *total* RNA extracted from the cells or tissues of interest. The blots are used to test the specificity of the specific antisense probe in question and any non-homologous or negative control probes used subsequently for *in situ* hybridization. It may be possible to eradicate spurious bands using altered stringency conditions or by recloning a portion of the probe.

  Northern blots or dot blots can also be used to establish the melting temperature (T_m) and therefore the optimal hybridization temperature for the probe of interest. Final stringency conditions are also determined in this way.
- Carry out ISH, using the probe of interest on a tissue or cell line known not to express the target transcript (checked by Northern or dot blotting total RNA).
- Many workers use sense transcripts of the sequence of interest as a negative control for ISH. (NB Check this on blots.)
- The specificity of the hybridization within the tissue section can be checked by competing out labelled antisense probe with unlabelled antisense probe.
- The overall hybridization procedure and solutions can be checked for each experiment by using a control probe, e.g. to β-actin which is highly conserved and expressed in all cell types.
- Some workers use RNase pre-treatment of tissues to show that signal generation is due to the presence of target RNA. This may be viewed as contentious when RNA probes are used (E. J. Gowans, pers. comm.).
- Omit specific probe to test for the presence of background grains following autoradiography.
- A blank 'subbed' slide should also be dipped to test the emulsion.

3.9.1 Northern blotting

A Northern blot provides information with respect to size and concentration of specific RNA transcripts in a mixture. (NB RNA molecules are extremely susceptible to degradation through cleavage with RNases at any stage during extraction, purification, separation, and blotting (see Section 3.4 for appropriate procedures to minimize RNase contamination.)

An average mammalian cell contains ~10–20 pg of RNA including 80–85% ribosomal (28S, 18S, and 5S), 10–15% low molecular weight RNA species (small nuclear transfer RNA) and 1.5% messenger RNA (31). Messenger RNA comprises a range of molecular sizes. It is possible to isolate mRNA from total RNA prior to Northern analysis by virtue of the 3′ poly A tail on these molecules. The method of choice is chromatography on oligo dT cellulose which can be obtained commercially.

When Northern analysis is used to assess probe specificity for *in situ* hybridization, it is important to use total RNA since this reflects the full range of target molecules *in situ*. There are several methods available to prepare total RNA (42, 43).

In summary, Northern analysis comprises the following basic steps:

(a) Size separate transcripts using denaturing gel electrophoresis.
(b) Blot the transcripts from the gel on to a solid support (e.g. nylon membrane).
(c) Fix the molecules to the support.
(d) Hybridize a specific radiolabelled probe to the blot.
(e) Wash off unhybridized probe.
(f) Visualize the hybrid using autoradiography.
(g) Interpret results.

i. Denaturing gel electrophoresis

The denaturing conditions disrupt secondary structure within the RNA molecules during electrophoresis enabling an accurate estimation of the length of the transcript. There are a number of denaturing gel systems currently in use including glyoxal gels, formaldehyde gels, and methyl mercuric hydroxide gels. The formaldehyde gel system will be detailed since in our experience, these systems are effective in producing good quality Northern blots which demonstrate high sensitivity.

Protocol 19. Renaturing gel electrophoresis of RNA

1. Incubate total RNA at 65 °C in the following solution:

Total RNA (~20 μg)	4.5 μl
Formaldehyde	3.5 μl

De-ionized formamide[a]	10.0 μl
5 × MOPS buffer pH 7.0[b]	2.0 μl
	20.0 μl

2. Chill on ice and add 2μl of sterile loading buffer to each sample.
3. Prepare a 1.5% agarose gel by melting 1.5 g of agarose in 62.1 ml of water. Cool the gel mixture to 60 °C. Add 20 ml 5 × MOPS buffer and 17.9 ml of formaldehyde. Mix and pour the gel immediately in a fume-hood. The gel running buffer is 1 × MOPS.
4. Load the total RNA samples on to the gel alongside suitable DNA or RNA molecular weight markers (treat the markers in exactly the same way as the RNA samples prior to loading). Electrophorese the samples until the Bromophenol Blue has migrated three-quarters of the way through the gel.

[a] See *Protocol 23*.
[b] MOPS buffer: 0.2 M 3-(*N*-morpholino) propane sulphonic acid, 50 mM sodium acetate pH 7.0, 5 mM ethylene diaminotetracetic acid (EDTA).

Protocol 20. Procedure for de-ionizing formamide

1. Mix 50 ml ANALAR formamide with 5 g mixed bel ion exchange resin (for example, Bio-Rad AG 501-X8) until the pH is neutral prior to use.
2. Filter the deionized formamide twice through Whatman's No. 1 filter paper.
3. Dispense the formamide into aliquots and store at −20 °C in tightly capped tubes ready for use.

ii. Transfer of RNA to a membrane support by capillary blotting (see Figure 4)

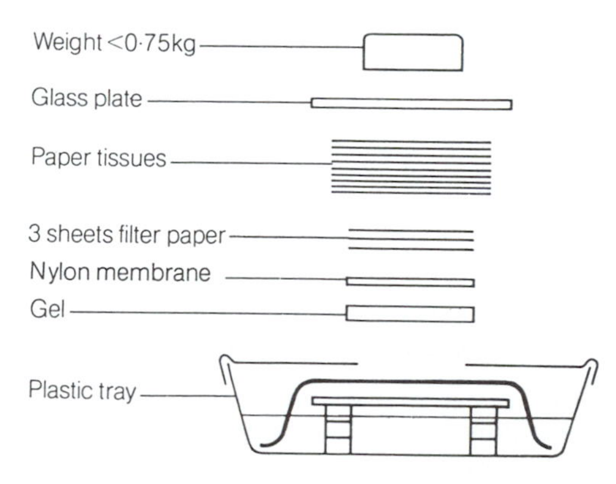

Figure 4. A typical blotting apparatus.

Protocol 21. Capillary transfer of RNA to a membrane support

1. Fill a tray or glass dish with blotting buffer (20 × SSPE). See *Protocol 14*.
2. Make a platform for the gel, e.g. an inverted gel former and cover it with a wick made from three sheets of Whatman 3 MM paper saturated with blotting buffer.
3. Place the gel upside down on the wick and avoid trapping air bubbles beneath it. Surround the gel with clingfilm to prevent blotting buffer from being absorbed directly in to the paper towels above the gel.
4. Cut a piece of nylon membrane to the exact size of the gel. Lower the membrane on to the gel taking care not to introduce air bubbles between the membrane and the gel. If bubbles do appear, squeeze them out by rolling a sterile pipette over the membrane.
5. Place three sheets of Whatman 3 MM filter paper cut to size and wetted with blotting buffer on top of the membrane.
6. Place a 5-cm-thick stack of absorbent paper towels on top of the 3 MM paper.
7. Place a glass plate on top of the paper to spread the pressure from a 0.75 kg weight.
8. Allow transfer to proceed for at least 12 h.

iii. Fixing RNA molecules to the membrane

Protocol 22. Procedure for fixing RNA to nylon membrane

1. After blotting carefully dismantle the apparatus. Before removing the membrane from the gel, mark the membrane with pencil to identify the tracks and loadings. NB Do not wash the filter.
2. Wrap the membrane in Saran Wrap and place the RNA side down on a standard UV transilluminator for 2–5 min.

NB Some brands of clingfilm are unsuitable for this application because they absorb UV light. We recommend the use of Saran Wrap. UV irradiation is the most effective means of fixing nucleic acids to nylon since a covalent linkage results. A wavelength of 312 nm is recommended. The power and wavelength of UV light from individual UV transilluminators may vary, so in order to achieve efficient UV crosslinking, it is essential to calibrate your transilluminator. Failure to do so can lead to very poor sensitivity of detection following hybridization. (NB Do not irradiate nitrocellulose membranes due to the risk of fire.)

3. Dry blots should be stored under vacuum in a desiccator. In this condition, they are stable for several months.

iv. Preparation of probe

RNA or DNA probes can be used; see Chapter 6. It is recommended that RNA probes are used in this instance since they are subsequently used for the *in situ* hybridization.

v. Hybridization of the Northern blot

Protocol 23. Procedure for hybridization of a Northern blot

1. Immerse the Northern blot in pre-hybridization solution containing 50% de-ionized formamide, 5 × SSPE, 5 × Denhardt's, 0.5% sodium pyrophosphate, 100 μg/ml denatured, sheared herring sperm DNA, 10% dextran sulphate, 2% SDS, incubate with shaking at an appropriate temperature for 1 h.

   Pre-hybridization can be carried out either in a heat-sealed plastic bag or in a suitable plastic box using 5 ml of solution per 100 cm^2 of membrane.
2. Add the RNA probe to freshly prepared hybridization buffer (as pre-hybridization solution detailed in step 1) at a concentration of 5 ng/ml. Mix the buffer thoroughly to achieve an even distribution of probe prior to adding the membrane to the pre-warmed buffer.
3. Incubate the membrane with radiolabelled probe in hybridization buffer at 42–65 °C for 12–16 h.
4. Wash the filter four times in 2 × SSPE, 0.1% SDS (w/v), 5 min each at room temperature.
5. Wash the filter twice in 0.1 × SSPE, 0.1% SDS (w/v) for 10 min each at 55–75 °C.
6. Treat the filter with 10 μg/ml RNase A in 2 × SSPE at 37 °C for 15 min.
7. Wrap the damp filter in clingfilm. (Do not allow the filter to dry completely if re-washing or re-probing is a possibility.)

---

vi. Autoradiography

Expose the filter to X-ray film. The appropriate length of time for autoradiography will depend on abundance of the target RNA molecule. For ^{32}P-labelled probes, autoradiograph at −70 °C using two intensifying screens and pre-flashed film for maximum sensitivity. For ^{35}S-labelled probes, autoradiograph *dried* filters at room temperature without clingfilm (see Chapter 4).

vii. Interpretation of results

In the simplest situation, a single band is observed when using a specific antisense RNA probe. However, this is not always the case when probing total RNA. Extra bands corresponding to spliced transcripts from different initiation sites or prematurely terminated transcripts are sometimes observed, depending on the probe of interest. It is important to have evidence that the extra bands are transcripts of the sequence of interest and not a closely related sequence from an entirely different gene. It may be possible to achieve this by varying the stringency of the hybridization and wash procedures, for example. It is useful to test a 'non-homologous' probe, e.g. a sense transcript of the sequence of interest for use as a negative control in subsequent ISH experiments using a Northern blot.

3.10 Simultaneous localization of proteins and nucleic acids

The development of the ISH technique has enabled researchers to question whether the localization of a protein at a specific site within a cell represents newly synthesized protein or stored sequestered protein. Immunocytochemistry has the power to detect subtle differences in molecules; for example, in different post-translational states assuming that selective antibodies are available.

There are different procedures in the literature for simultaneous localization of protein and RNA transcripts (44, 45). The major difference is whether the immunocytochemistry is performed before or after the *in situ* hybridization. Problems arise from RNase contamination of antisera and also from non specific binding of nucleic acid probe to diaminobenzidine (DAB) reaction products if ICC is performed prior to ISH.

In this application, radioisotopes are extremely useful as an alternative detection system distinguishing transcript detection from antigen detection in the same cell or tissue section.

4. Conclusion

We have focused on the use of specific and sensitive antibody and nucleic-acid probe for localization of biological molecules.

For subcellular localization the 'classical' techniques of organelle separation coupled with gel fractionation and blotting are particularly valuable since the size of molecules can be determined and subcellular preparations can be utilized for functional studies. Furthermore techniques are now being used to reconstitute *in vivo* systems based on subcellular fractions. Radioisotopes are widely used for high sensitivity (e.g. ^{125}I, ^{32}P) since resolution is not as critical on gels/blots as it is for microscopy. For the latter application ^{3}H-labelled probes are favoured for subcellular studies of mRNA since resolution is better, although development time is long. It seems clear that the future development of techniques for precise location of molecules will lie with non-radioactive techniques that are sensitive, fast and have high resolution, and lessons may be learnt from developments that have occurred in non-radioactive ICC. However, conventional ISH without the need for high resolution may be quite adequate using other labels including ^{32}P. Although EM techniques give precise localization of molecules to organelles they are time consuming and complex; in the future we may see the specificity of ICC/ISH coupled with more specific sub-cellular markers to identify the location of molecules using the light microscope. A parallel situation may occur in the field of cytogenetics where chromosome analysis may be speeded up using chromosome specific probes.

Acknowledgements

The authors are grateful to their colleagues at Amersham International for comments and data included in this chapter. Particular thanks are also extended to Chris Jones, Giorgio Terenghi, Judith Parke, and Robert Burgoyne for experimental data and protocols.

References

1. Cumming, R. and Burgoyne, R. D. (1983). *Bio. Sci. Rep.* **3,** 997.
2. Rickwood, D. (1989). *Centrifugation – A Practical Approach*. IRL Press, Oxford.
3. Cumming, R. and Burgoyne, R. D. (1985). In *Techniques in Immunocytochemistry* G. R. Bullock and P. Petrusz, Vol. 3, p. 55.
4. Hames, B. D. and Rickwood, D. (1981). *Gel Electrophoresis of Proteins*. IRL Press, Oxford.
5. Andrews, A. T. (1986). In *Electrophoresis. Theory Techniques and Biochemical and Clinical Applications*. Clarendon Press, Oxford.
6. Dunn, M. J. (1986). *Gel Electrophoresis of Proteins*. Wright, Bristol.
7. Dunbar, B. S. (1987). *Two-dimensional Electrophoresis and Immunological Techniques*. Plenum Press, New York.
8. Cumming, R. (1980). *J. Immun. Meth.* **37,** 301.
9. Hunt, S. P., Allanson, J., and Mantyh, P. W. (1986). In *Immunocytochemistry* (eds. J. M. Polak and S. Van Noorden), p. 99. Wright, Bristol.
10. Cumming, R., Burgoyne, R. D., Lytton, N. A., and Gray, E. G. (1983). *Neurosci. Lett.* **37,** 215.
11. Drury, R. A. B., and Wallington, E. A. (1980). *Carlton's Histological Technique*. Oxford University Press, Oxford.
12. Ish-Horowicz, D. and Burke, J. F. (1979). *Nucl. Acids Res.* **7,** 1541.
13. Westermark, B. (1974). *Int. J. Cancer* **12,** 438.
14. Hames, B. D. and Higgins, S. J. (1985). *Nucleic Acid Hybridization*. IRL Press, Oxford.
15. Sommerville, J. and Scheer, O. (1987). *Electron Microscopy in Molecular Biology – A Practical Approach*. IRL Press and Oxford University Press, Oxford.
16. Gall, J. G. and Pardue, M. L. (1969). *Proc. Natl. Acad. Sci. (USA)* **63,** 378.
17. Bentley Lawrence, J., Villnave, C. A., and Singer, R. H. (1988) *Cell* **52,** 51.
18. Rooney, D. E. and Czepulkowski, B. H. (1978). *Human Cytogenetics*. IRL Press, Oxford.

19. Davies, K. E. (1986). *Human Genetic Diseases*. IRL Press, Oxford.
20. Roberts, D. B. (1986). *Drosophila*. IRL Press, Oxford.
21. Manuelides, L. (1985). *Ann. N.Y. Acad. Sci.* **450,** 250.
22. Bentley Lawrence, J. and Singer, R. H. (1986). *Cell* **45**, 407.
23. Hentzen, D., Renucci, A., Le Guellec, D., Benchaibi, M., Jurdic, P., Gandrillon, O., and Samaret, J. (1987) *Molecular and Cellular Biology* **7**(7), 2416.
24. Bloch, B., Popovice, T., Choucham, S., and Kowalski, C. (1986). *Neuroscience Letters* **64,** 29.
25. Schuster, V., Mate, B., Wiegand, H., Traub, B., Kampa, D., and Neumann-Haefelin, D. (1986). *J. Infect. Dis.* **154** (2), 309.
26. Haase, A., Brahic, M., Stowning, L., and Blum, H. (1984). *Methods in Virology* **III** 189.
27. Maitland, N. J. Cox, M. F., Lynas, C., Prime, S., Crane, I., and Scully, C. (1987). *J. Oral Pathol.* **16,** 199.
28. Gowans, E. J., Burrell, C. J., Jilbert, A. R., and Marmion, B. P. (1983). *J. General Virol.* **64**, 1229.
29. Gupta, J., Gendelman, H. E., Naghashfar, Z., Gupta, P., Rosenhein, N., Sawada, E., Woodtaff, J. D., and Shah, K. (1985). *Int. J. Gynaecol. Pathol.* **4,** 211.
30. National Radiological Protection Board (1984). *Living With Radiation*. Her Majesty's Stationery Office Publ.
31. McAllister, H. and Rock, D. (1985). *J. Histochem. Cytochem.* **33,** 1026.
32. Bentley Lawrence, J., Singer, R. H. (1985). *Nucl. Acids Res.* **13,** (5).
33. Terenghi, G., Cresswell, L., and Fallon, R. (1988). *Proc. R. Microscop. Soc.* **23,** (1), 47.
34. Toutellotle, W.W., Verity, A. N., Schmid, P., Martinez, S., and Shapshak, P. (1987). *J. Virol. Meth.* **15,** 87.
35. Maddox, P. H. and Jenkins, D. (1987). *J. Clin. Pathol.* **40,** 1256.
36. Sambrook, J., Fritsch, E. F., and Maniatis, T. (1989). *Molecular Cloning: A Laboratory Manual*, 2nd edn. Cold Spring Harbor Laboratory Press, USA.
37. Cox, K., DeLeon, D. V., Angerer, L. M., and Angerer, R. C. (1984). *Dev. Biol.* **101,** 485.
38. Scott Young III, W., Mezey, E., and Siegel R. E. (1986). *Molecular Brain Research* **1,** 231.
39. Unger, E. R., Budgeon, L. R., Myerson, D., and Brigatti, D. J. (1986). *Am. J. Surg. Pathol.* **10,** 1.
40. Brahic, M. and Haase, A. J. (1978). *Proc. Natl. Acad. Sci. (USA)* **75** (12), 6125.
41. Rogers, A. W. (1979) *Techniques for Autoradiography*, 3rd edn. Elsevier/North Holland.
42. Messe, E. and Blin, N. (1987). *Gene Anal. Tech.* **4,** 45.
43. Chan, V. T.-W., Fleming, K. A., and McGee, J. O'D. (1988). Simultaneous extraction from clinical biopsies of high-molecular-weight DNA and RNA: Comparative characterization by biotinylated and 32-P labelled probes on Southern and Northern blots. *Anal. Biochem.* **168,** 16.
44. Shivers, B. D., Harlan, R. E., Pfaff, D. W. and Schachter, B. S. (1986). *J. Histochem. Cytochem.* **54**(1), 39.
45. Hoeffler, H., Childers, H., Montminy, M. R., Gechan, R. M., Goodman, R. H., and Wolfe, H. J. (1986). *Histochem. J.* **18,** 597.

8

Radioisotopes and immunoassay

ADRIAN F. BRISTOW,
ANDREW J. H. GEARING, and ROBIN THORPE

1. Introduction

Immunoassay is the quantitative measurement of a substance of interest ('analyte') using antibodies which bind specifically to that analyte. For the rest of this chapter, these two main components of the assay will be referred to as ligand and antibody respectively. In order to perform such an assay, it is necessary to measure the binding of the ligand to the antibody at very low concentrations and this has most commonly been achieved with the use of radiosotopes. The first practical application of the principle was the technique of radioimmunoassay, invented by Yalow and Berson in the late 1950s (1), for which the Nobel prize was subsequently awarded. This technique, and subsequently developed related techniques such as immunoradiometric assays and enzyme-linked immunosorbent assays, have revolutionized the biological sciences. Using these methods the minute quantities found in biological materials of a whole range of analytes such as protein hormones, steroids, viral antigens, cytokines, and clotting factors can be accurately measured without relying on difficult and sometimes imprecise measurements based on their biological activities. Indeed, it is hard to overestimate the impact that the development of immunoassays has had on the life sciences.

Although the basic principles of the radioimmunoassay and of the immunoradiometric assay have remained the same, both methods have been the subject of countless refinements and improvements, some of which have included the use of non-isotope labels (enzymes, luminescent labels), the development of novel methods of separating free from bound antigen such as magnetic particles, or of quantifying that ratio without separation (e.g. fluorescence depolarization), and the use of specific combinations of monoclonal antibodies. This chapter will attempt to cover the use of radioisotopes in immunoassays, with reference to a number of specific examples of assay methodology. This is a vast field, however, and the interested reader is referred to a number of excellent texts covering the various fields of immunoassay in the life sciences (2, 3).

2. Immunoassays: theoretical considerations

The theoretical basis of immunoassays employing isotopes has been reviewed by Ekins (4). Two basic types of immunoassay can be recognized; 'limited reagent' assays, and 'reagent excess' assays, the two basic methods being typified by radioimmunoassays (RIA) and immunoradiometric assay (IRMA) respectively.

Limited reagent assays have been described by a number of terms such as 'saturation assay', 'competitive protein binding assay' and 'displacement assay'. Limited reagent assays employing isotopes and antibodies are almost invariably referred to as 'radioimmunoassays', a somewhat unfortunate term since it does not in any way describe the theoretical basis of the assay method. The underlying principle in all 'limited reagent' assays is that by limiting the concentration of one of the reagents, the system is saturable. In the technique of radioimmunoassay, a radiolabelled ligand reacts with antibodies which bind specifically to that ligand. The amount of antibody available to react with the ligand is limited such that it is saturable, only a fraction of the total labelled ligand is bound to antibody. The reaction may be written as

$$\mathrm{Ab} + {}^{*}\mathrm{L} \rightleftharpoons \mathrm{Ab}\ {}^{*}\mathrm{L}$$

where Ab = antibody, and *L = radiolabelled ligand.

If the reaction is allowed to reach equilibrium in the presence of ligand that is not radiolabelled, two simultaneous reactions will take place;

$$\mathrm{Ab} + {}^{*}\mathrm{L} \rightleftharpoons \mathrm{Ab}\ {}^{*}\mathrm{L}$$

and

$$\mathrm{Ab} + \mathrm{L} \rightleftharpoons \mathrm{Ab\ L}$$

where L = unlabelled ligand.

Under limited antibody conditions therefore, L will compete with *L for available antibody, and the fraction of radiolabelled ligand that is antibody bound will fall as the amount of unlabelled ligand present increases. In practice, the fraction of antibody-bound label is measured using a system that separates bound and free ligand. Increasing concentrations of unlabelled ligand generate a displacement curve of the type shown in *Figure 1*.

The parameters of the assay are the percentage bound in the absence of unlabelled ligand (B_0) and percentage bound in the absence of antibody (non-specific binding, NSB) and the working range of the assay is the concentration range over which the displacement of radioactivity is linearly related to log L.

In 'reagent excess' immunoassays, the antibody is present in excess, such that at equilibrium all available ligand is antibody-bound, and the majority of antibody remains unused. In practice, the amount of antibody that has ligand bound to it is determined. Measurement of antibody–ligand complex

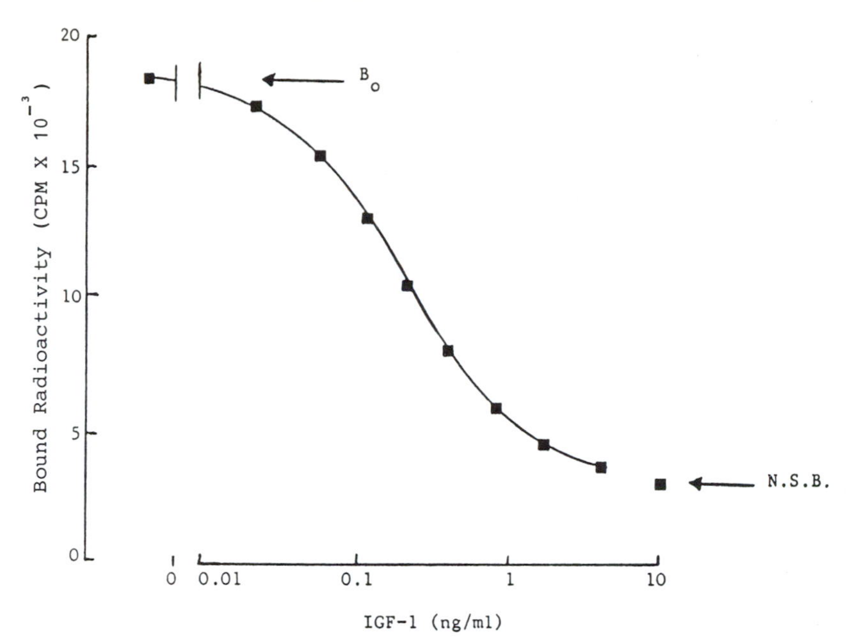

Figure 1. Radioimmunoassay of insulin-like growth factor-1 (IGF-1). Experimental procedure for this radioimmunoassay is given in *Protocol 4.* B_0 = per cent binding at zero IGF-1 concentration. NSB (non-specific-binding) = per cent binding at zero antibody concentration.

in this type of assay is most conveniently achieved by radiolabelling the antibody; the ligand-bound radiolabelled antibody is then separated from the free radiolabelled antibody by some convenient procedure. The most widely used immunoassay based on the excess reagent principle is the immunoradiometric assay (IRMA). In the first IRMA procedures to be described, the free labelled antibody was removed from the reaction by using a large excess of the ligand coupled to a solid phase such as dextrose beads, which can easily be removed by centrifugation (this procedure is shown diagrammatically in *Figure 2*). This approach suffers from the serious drawback that it is only applicable when very large quantities of the ligand are available. A much more useful, and widely used approach, is the two-site IRMA, in which two separate antibodies are used, that react with two independent sites on the ligand. One of the antibodies is used to radiolabel the complex, and the other to remove the complex from the radiolabelled pool. The latter is achieved most commonly by coupling the second antibody to a solid phase. The procedure is schematically illustrated in *Figure 3*.

In the two-site IRMA a linear response is obtained, in which bound radioactivity is directly proportional to the concentration of ligand in the reaction mixture. The two-site IRMA offers a number of advantages over the 'limited reagent' radioimmunoassay:

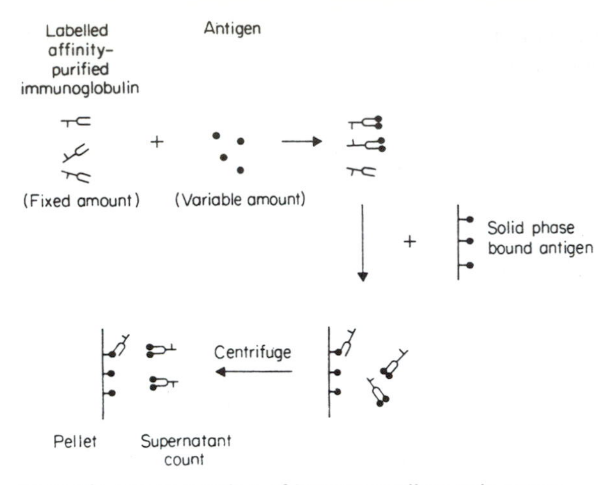

Figure 2. Diagrammatic representation of immunoradiometric assay.

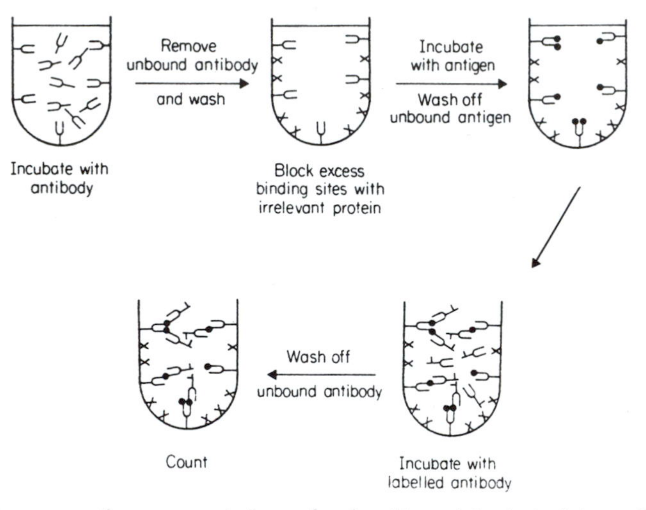

Figure 3. Diagrammatic representation of microtitre plate based two-site immunoradiometric assay (IRMA) for estimation of antigens.

- the IRMA is generally more sensitive than corresponding RIAs for the same ligand;
- in the RIA the ligand is radiolabelled by a process of chemical modification; the antibody may then not react with the labelled and unlabelled ligand to the same extent;
- the two-site IRMA requires a specific interaction between two different antibodies and their binding sites; the specificity of such methods is therefore greater than in the radioimmunoassay, in which immunologically similar but distinct molecules may interfere.

Despite these advantages, however, the usefulness of the two-site IRMA is often restricted by the requirement for two antibody-binding sites. Small molecules, or molecules that are poorly antigenic, may simply not produce the right combination of antibodies, and the radioimmunoassay remains a powerful and widely used analytical tool.

3. Radioimmunoassays

3.1 Reagents

3.1.1 Antibodies

Currently, two basic types of antibodies are available to the biologist, monoclonal antibodies and polyclonal antibodies. Monoclonal antibodies, although widely used in IRMAs, are very often of little use in radioimmunoassay. Instead, the analyst has to obtain polyclonal antisera, raised in such animals as the rabbit, guinea pig, sheep, or goat. A complete summary of methods of producing antisera for use in radioimmunoassay is beyond the scope of this book, and is available elsewhere (5).

(i) Immunogens

The immunogen is the preparation to be injected into the animal to elicit antibody formation. Generally speaking, proteins with molecular weight greater than 5000 will readily elicit antibody formation in experimental animals. Smaller molecules such as steroids or thyroid hormones are non-immunogenic without chemical modification and conjugation to carrier proteins, such as albumin or thyroglobulin, enabling the animal to see them as 'foreign', and so produce antibodies. It should also be noted that even for larger proteins, immunogenicity varies in different species of animal. Generally speaking, the more foreign a protein is the better will be the immune response. For instance, human insulin is a poor immunogen in the rabbit, whose own insulin is very similar in structure, but is a much better immunogen in the guinea-pig, which has insulin of rather different structure.

(ii) Evaluation of antisera

Antisera for use in radioimmunoassay are evaluated in terms of titre, avidity and specificity. Determination of the titre of an antibody is the most useful primary screening technique. The antibody titre, usually expressed as a dilution (e.g. 1/200 000) is the concentration of antibody that binds a certain percentage (usually 40–60%) of radiolabelled tracer under a given set of experimental conditions. Although such measurements are not absolute in that they vary with the experimental conditions used, they do represent the easiest way of monitoring the production of antibody during the immunization schedule.

The avidity, or association constant, is an absolute measurement of the extent to which an antibody binds to its ligand at equilibrium. In a radioimmunoassay, equilibrium is reached between labelled and unlabelled ligand binding to a limited concentration of antibody (Ab). The association constant (K_{ass}) for the reactions is given by

$$K_{ass} = \frac{AbL}{Ab \times L}.$$

It can be readily seen that the higher the association constant, or avidity, the lower the concentration of L required to form the Ab-1 complex. Hence there is a direct relationship between the sensitivity of a radioimmunoassay (the concentration of ligand that can be measured), and the avidity of the antibody. In practice, the most common method for determining the avidity of an antibody is Scatchard analysis, in which the percentage of radiolabelled ligand bound is measured as a function of antibody concentration. Data are then plotted on a graph where the *y*-axis (ordinate) is the ratio of bound to free ligand and the *x*-axis (abscissa) is the concentration of ligand in mol/litre. The slope of the line gives K_{ass} (litre-mol) (5).

Antisera for use in radioimmunoassay should have appropriate specificity. This means that while the substance to be measured should bind to the antibody, there should be no binding to any other substances present in the test sample. Many biochemical analytes such as steroids share common structural features, and even more complex structures such as proteins may exhibit immunological similarities. In practice few antibodies exhibit absolute specificity for one ligand, and most will show a decreased level of binding to structurally related ligands. Under these circumstances it needs to be shown that any other substances that bind to the antibody are not present in the test samples in sufficient concentrations to affect the assay significantly.

3.1.2 Tracers

Radiolabelled ligands used in radioimmunoassay are generally referred to as tracers. Preparation of the tracer is one of the most important aspects of a radioimmunoassay, and indeed the use of poor tracers is the most common cause of assay variability, or lack of sensitivity and/or precision. The most common isotope used in radioimmunoassay is ^{125}I, although other isotopes such as ^{3}H have been used, particularly for smaller molecules such as steroids. The major point of concern in considering tracers for use in radioimmunoassay is that it should be remembered that the mechanism of radioimmunoassay assumes that the antibody does not discriminate between the unlabelled ligand and the tracer (radiolabelled ligand). Although incorporation of tritium into steroids, or of ^{125}I into thyroxine, may be done synthetically without appreciably altering the chemical structure of the ligand, for the most part incorporation of radioiodine into biological macromolecules involves a process of chemical modification, and it should be recognized that the possibility always exists that this chemical modification alters the immunological properties of the ligand such that the antibody will discriminate between antibody and tracer.

Detailed procedures for radioiodination have been covered in Chapter 6. For the purposes of radioiodination, two basic methods have been used: those employing ^{125}I in the presence of oxidising agents, and those employing radioiodinated alkylating agents that do not require oxidizing conditions. Generally speaking, for robust ligands, oxidizing conditions such as the chloramine-T method are suitable. The side reactions that accompany such procedures, however, often modify amino-acid side-chains other than those being iodinated, and this can occasionally modify or destroy antibody-binding. The alternative, non-oxidizing procedures, such as the Bolton–Hunter method, are often suitable. Such methods, however, introduce quite bulky iodine-containing chemical groups, which may themselves modify the immunological properties of the tracer.

Although the radioiodination procedures described in Chapter 6 serve as a useful guide, detailed methods need to be developed for each ligand studied. Since the underlying principle of a radioimmunoassay requires displacement of tracer with equal concentrations of unlabelled ligand, it follows that the molar concentration of tracer should be as low as possible, and, therefore, that the specific activity of the tracer should be as high as possible. In practice, therefore, tracers for use in radioimmunoassay should be radiolabelled to as high a specific activity as possible without affecting the antibody binding. Evaluation of the suitability of a tracer for use in a radioimmunoassay is achieved using the parameters set out in Section 3.5.1.

3.1.3 Choice of working concentrations

In order to establish a radioimmunoassay, the working concentrations of antibody and of tracer must be optimized. Radioimmunoassay is a 'limiting reagent' technique, in which the sensitivity of the method increases as the concentration of reagents approaches zero. The precision of radioactivity measurements, however, decrease as the amount of radioactivity to be measured approaches zero. Optimization of working concentrations of antibody and tracer is therefore a balance between increasing the sensitivity and decreasing the precision by reducing the concentration of reagents.

Tracer

For a radioimmunoassay in which the tracer is labelled with ^{125}I, typical working concentrations would be 10 000–20 000 d.p.m./tube. Higher levels of radioactivity do not appreciably improve the precision, but reduce the sensitivity. At lower levels of radioactivity experimental error becomes unacceptably high.

Antibody

The antibody concentration should be chosen such that some fraction of the radioactivity is antibody bound at equilibrium. Using 20 000 d.p.m./tube of tracer, 20% binding would correspond to 4000 d.p.m./tube. This addition of increasing concentrations of unlabelled ligand would displace this bound radioactivity. In practice therefore, the assay would be measuring d.p.m./tube between 0 and 4000. Such a working range represents a reasonable compromise between sensitivity and precision, although with well-optimized conditions and experienced operators working ranges of 0–1000 d.p.m./tube or even lower can sometimes be used.

The concentration of antibody giving 20% binding needs to be determined under the conditions of the assay from an antibody dilution curve. A typical procedure is illustrated in *Protocol 1*.

Protocol 1. Determination of an antibody dilution curve for anti-interleukin-1β

Reagents

Human interleukin-1β, radiolabelled to a specific activity of 3.07×10^6 Bq/μg (83 μCi/μg) using the *N*-bromosuccinimide procedure.

Sheep anti-interleukin-1β antiserum

Bovine γ-globulins (Sigma)

Polyethylene glycol 6000 (BDH)

Diluent: 0.05 M sodium phosphate buffered isotonic saline (0.15 M NaCl) pH 7.4, containing 0.2% bovine serum albumin.

Method

1. Prepare twofold dilutions of antibody in diluent to cover the range 1/1000 to 1/1 024 000.
2. Prepare LP-3 (Luckham's) plastic tubes in triplicate containing:

   100 μl diluent
   100 μl tracer (10 000 c.p.m./100μl)
   100 μl antibody dilution

   Set up set of triplicates containing zero antibody.
3. Incubate at 4 °C for 16 h.
4. Add to each tube, 300 μl of PEG/γ globulin reagent.

   (5 g polyethylene glycol 6000/12.5 ml H_2O)
   (30 bovine γ globulin/2.5 ml H_2O)
   (5.0 ml 0.2 M Tris–HCl pH 8.5)
5. Vortex and incubate for 1 h at 4 °C.
6. Centrifuge (3000 r.p.m., 30′), carefully aspirate off the supernatants and determine the radioactivity in the pellets by γ counting.
7. Plot counts precipitated against $\log_{10}$ antibody concentration.

3.1.4 Standards

Radioimmunoassay is not an absolute technique. Concentrations of ligand in test samples are determined by reading the values off displacement curves obtained with standard preparations. Although for well-optimized radioimmunoassays the dose–response range will remain reasonably constant, there will inevitably be day to day drift, and a separate standard curve needs to be included in each assay. The assay standard needs to meet certain requirements. Most important, it must be immunologically identical with the ligand present in the test sample. Although this may seem a relatively easy criterion to meet it should be remembered that many molecules of interest to the biologist are complex structures that are only isolated from biological tissues after long and painstaking purification procedures. They may not even be present in homogeneous form in the body or cell to start with. As a result, few preparations of proteins or other biological macromolecules are actually homogeneous in biochemical terms. The different isoforms present may react differently with the antibody, especially when one considers that the antibody may have been raised to a different preparation of the ligand in the first place. The standard must also be stable in its stored form. An unstable standard will result in a gradual erroneous increase in radioimmunoassay values, as the standard degrades.

3.1.5 Samples

The component of a radioimmunoassay over which the operator often has the least control is the sample for analysis. The binding of antibodies to their specific ligands may be perturbed by a number of factors present in samples. These factors are generally known as matrix effects and are described below.

Solvent effects

Radioimmunoassay is frequently used to monitor fractionation procedures such as chromatography, or other biochemical procedures. Commonly, such procedures utilize conditions of high or low pH, chaotropic or organic solvents. Any of these may disrupt antibody–ligand binding, and samples would have to be appropriately treated, by neutralization, desalting, lyophilization, or simple dilutions before they could be examined by radioimmunoassay.

Proteases

Biological samples, including blood and tissue extracts frequently contain proteases that under the conditions of the radioimmunoassay, may degrade the tracer, resulting in artefactual results. The presence of such activity can be revealed by gel-filtration of the tracer following incubation with the sample. Proteases, if present can be blocked, with protease inhibitors, or the ligand of interest may be extracted from the sample before assay (see *Protocol 3*).

Binding proteins

Many ligands exist in biological samples coupled to specific binding proteins. Occasionally these present a problem in that they will mask the antibody-binding site, and prevent the assay from working. The example given in *Protocol 2* utilizes low pH to separate the ligand (insulin-like growth factor-1) from its binding protein.

Sample extraction

The most common cause of sample unsuitability is simply that the ligand of interest is not present in high enough concentration in the sample. Many procedures have been developed for extracting and concentrating the ligand from large volumes of sample. The method given in *Protocol 3* describes extraction and concentration of growth-hormone releasing factor from plasma.

Protocol 2. Sample pre-treatment for the removal of insulin-like growth factor-1 (IGF-1) binding proteins

Reagents

Clinical plasma samples
Formic acid
Ethanol

Method

1. To 100 μl plasma, add 25 μl of 2.4 M formic acid and 500 μl ethanol.
2. Mix and incubate for 30 min at room temperature.
3. Centrifuge (3000 r.p.m., 30 min) and collect the supernatants.
4. Neutralize 100 μl of supernatant with 600 μl radioimmunoassay buffer.

Protocol 3. Extraction of growth-hormone releasing factor from serum

Reagents

Serum or plasma samples
Vycor glass
Methanol
Hydrochloric acid
Assay diluent: 0.1 M phosphate, 0.025 M EDTA
0.15 M NaCl, 0.01% thiomersal
0.5% human serum albumin pH 7.4

Method

1. Dilute 200 μl plasma samples to 1 ml with assay diluent.

2. Add 1 ml of a 10 mg/ml suspension of activated Vycor glass in distilled water, and rotate tubes for 30 min at 4 °C.
3. Wash glass successively with 2 ml H_2O and 2 ml 1.0 M HCl.
4. Elute growth hormone from the glass with 1.0 ml 80% methanol, evaporate to dryness and reconstitute in assay diluent.

3.1.6 Assay matrix

The solvent, or matrix, in which the assay is performed, contains two main components; buffer salts, and an agent to prevent assay reagents from binding non-specifically to other proteins or the walls of the tube. The most common buffer used in immunoassay reagents is sodium phosphate-buffered isotonic saline (pH 7.4), although many other neutral buffers can be used. Two types of agents may be used to prevent non-specific binding: proteins, such as serum albumin, at 0.1% to 1% (w/v), and detergents such as Tween. Occasionally, immunoassay matrices may contain additional reagents such as protease inhibitors (for example, benzamidine, trasylol), chelating agents (for example, EDTA), or inhibitors of bacterial growth (for example, thiomersal). The detailed immunoassay procedures given in *Protocols 4* and *5* use typical immunoassay matrices exhibiting many of these features.

Clinical immunoassays, in which the ligand of interest is present in serum, may suffer quite markedly from matrix effects due to other substances present in the sample. A common way round this problem is to use as an assay diluent 'stripped' serum; that is, serum from which all the ligand of interest has been removed by some suitable procedure, such as charcoal or immunoaffinity absorption.

3.2 Assay conditions

A radioimmunoassay consists of three stages:

(a) Pre-incubation (sample/standard + antibody).
(b) Incubation (sample/standard + antibody + tracer).
(c) Separation [separation of bound from free tracer).

Increased sensitivity is usually achieved by increasing the preincubation and incubation times, bringing the reactants closer to equilibrium. Periods of three days each for the pre-incubation and incubation stages have been used. Where ultimate sensitivity is not a requirement, the pre-incubation may be reduced to a few hours, or even incorporated into the main incubation, which can itself be reduced to 16 h or shorter. Incubations are usually carried out at 4 °C, but higher temperatures may be used to reach equilibrium faster.

Most radioimmunoassay methodology uses plastic or glass test tubes, and incubation volumes of around 0.5 ml. There are however many alternative procedures. The assay may be performed in microtitre-well plates, or using antibody-coated tubes, or on filter-paper spots.

3.3 Separation techniques

The final stage of a radioimmunoassay is to separate the bound antibody from the free radiolabelled tracer. Many different procedures have been developed for different ligands. Some of the more commonly used methods are described below.

Adsorption

Low molecular weight ligands may be adsorbed on to charcoal, leaving antibody-bound ligand in solution.

Solvent precipitation

Where the ligand is soluble in organic solvent, the antibody-bound ligand may be removed by ethanol-precipitation.

Polyethyleneglycol

Polyethyleneglycol has the property of forming specific precipitating complexes with immunoglobulins (antibodies). It may be used therefore to separate free from bound tracer either alone, in combination with non-specific immunoglobulin to form a better precipitate (see *Protocol 1*), or in combination with second antibody (see below) to accelerate the separation.

Second antibodies

Immunoglobulins possess a constant region that, within a given species contains several common epitopes. Anti-antibodies (usually referred to as second antibodies) can therefore be prepared in a different animal which will precipitate the first antibody. Second antibodies are widely used in radioimmunoassay, either alone, or with the polyethyleneglycol (see *Protocol 5*), or coupled to a solid phase (see below).

Solid-phase second antibody

A particularly useful approach is to couple the second antibody to a solid-phase such as microcrystalline cellulose, allowing easy and rapid precipitation of the antibody-bound tracer fraction (*Protocol 6*).

Protocol 4. Radioimmunoassay for insulin-like growth factor-1 (IGF-1)

Reagents

Serum or plasma samples, pretreated to remove IGF-I binding proteins (*Protocol 2*).

^{125}I-labelled IGF-1. Radiolabelled using chloramine T to a specific activity of 7.4×10^6 Bq/μg (200 μCi/μg).

Rabbit anti-IGF-1.

Purified IGF-1 standard

Assay diluent: EDTA (0.1 M), NaCl (0.145 M), Tween 20 (0.1%), sodium azide (0.1%), sodium phosphate (0.2 M) pH 7.4.

Method

1. Carry out the assay in triplicate in LP3 (Luckham's) plastic tubes.
2. Prepare a standard curve containing IGF-1 standard at the following concentrations (ng/ml):
   40, 20, 10, 5, 2.5, 1.25, 0.625, 0.313, 0.156, 0.075, 0.039, 0.02, 0
3. Prepare tubes containing 100 μl sample or standard, plus 100 μl anti-IGF-1 at a dilution of 1/4000. An additional set in triplicate should be prepared containing 200 μl assay diluent only. These are used for measurement of non-specific binding (NSB).
4. Incubate at 4 °C for 6 h.
5. Add to each tube 100 μl ^{125}I-IGF-1 (10 000 c.p.m./100 μl). Include one set of triplicates which contains tracer only for determination of total counts.
6. Incubate for a further 16 h at 4 °C.
7. To each tube except those for the total counts add 100 μl stirred cellulose-linked anti-rabbit. Vortex and incubate at room temperature for 30 min. Add 1 ml water to each tube, centrifuge (1500 *g*, 10 min), aspirate off the supernatant, and determine the radioactivity in the pellets by gamma counting.
8. Typical results are given in *Figure 1*.

Solid-phase first antibody

The need for a separation stage is overcome if the first antibody is coupled to a solid support. A particularly attractive method is to couple the antibody to the walls of the test tube, separation being achieved simply by removing the incubation medium. Alternatively, cellulose-linked first antibody may be used and removed by centrifugation.

3.4 Specific protocols

Detailed protocols for specific radioimmunoassays are given in *Protocols 4, 5*, and 7. Although identical in fundamental principles, each of the assays uses different sample treatment, incubation conditions and separation procedures, illustrating many of the principles and procedures described earlier in this chapter.

Protocol 5. Radioimmunoassay for growth-hormone releasing factor (GRF)

Reagents

Serum samples, extracted as described in *Protocol 3*.

^{125}I GRF radiolabelled using chloramine-T to a specific activity of 1.295×10^7 Bq/μg (350 μCi/μg).

Rabbit anti-GRF.
Purified GRF standard.
Assay diluent: given in *Protocol 3*.
Donkey anti-rabbit antiserum.
Normal rabbit serum.
Polyethylene glycol 6000

Method

1. Prepare standard concentrations covering the range 10 ng/ml, as described in *Protocol 4*.
2. Pre-incubate standards or samples, with antibody (1/100 000) for 24 h at 4 °C. Include NSB tubes (see *Protocol 4*).
3. Add ^{125}I GRF (10 000 c.p.m. in 100 μl) and incubate for a further 48 h at 4 °C.
4. Add 100 μl of second antibody reagent :

donkey anti-rabbit	1/200
normal rabbit serum	1/2400
polyethylene glycol	4%

5. Incubate for a further 1 h, centrifuge (4000 r.p.m., 30 min) and determine the radioactivity in the pellets by gamma counting.

Protocol 6. Preparation of cellulose-linked second antibody

Reagents

Thin layer chromatography grade microcrystalline cellulose (Whatman)
Donkey anti-rabbit antiserum
0.1 M trisodium citrate pH 6.5
Cyanogen bromide
0.5 M sodium carbonate pH 10.5
2 M ethanolamine
4 M sodium hydroxide

Caution: Cyanogen bromide is extremely toxic by skin contact and inhalation. All procedures should be done in a fume-hood, and all equipment soaked in 4 M NaOH before washing.

Method

1. Dialyse 15 ml donkey anti-rabbit antiserum exhaustively against citrate buffer.
2. Suspend 250 g cellulose in 1 litre sodium carbonate buffer.

3. Dissolve 60 g cyanogen bromide in 250 ml distilled water (this may take 2–3 h).
4. Add the cyanogen bromide in 20 ml aliquots to the cellulose suspension. The mixture should be continuously stirred and the pH maintained at 10.5 with 4 M NaOH.
5. When the pH has ceased to fall, filter the cellulose through a Buchner funnel, and wash the cake with 2 litres of ice-cold citrate buffer.
6. Transfer the washed cake to a beaker, add the dialysed antiserum and 45 ml citrate buffer and incubate overnight at 4 °C.
7. Next morning, add 8 ml ethanolamine solution, and continue mixing at room temperature for 1 h.
8. Wash the cellulose in 0.05 M sodium phosphate buffered isotonic saline (0.15 m NaCl) pH 7.4 and resuspend in 600 ml of:

0.2 M	sodium phosphate	
0.1 mM	EDTA	pH 7.4
0.145 M	NaCl	
0.1%	sodium azide	
0.5%	BSA	
0.5%	Tween 20	

Protocol 7. Microtitre plate IRMA

Reagents

Phosphate buffered saline (PBS), 10 mM sodium phosphate, 130 mM sodium chloride pH 7.2
PBS containing 3% bovine haemoglobulin (Hb-PBS)
Purified antibody 1 (for capture)
^{125}I labelled purified antibody 2 (for detection)
Purified standard antigen
Samples containing antigen

Method

1. Prepare antibody 1 in PBS at 1–20 μg/ml (try 4 μg/ml initially). Add 50–200 μl to each well of a 96-well flexible microtitre plate and incubate overnight at 4 °C.
2. Discard antibody solution by holding plate over the sink and tapping briskly. Wash the plate three times with Hb-PBS by filling wells and discarding contents as above.
3. Block wells with haemoglobulin by filling wells with Hb-PBS and incubating for 30 min to 1 h at room temperature. Wash once with Hb-PBS.

4. Prepare dilutions of standard antigen in Hb-PBS (see *Figures 4–7* for typical ranges). Add 45–195 μl of dilutions to each well and include a blank (Hb-PBS); carry out triplicate assays for each point. Include unknown solutions in triplicate. Incubate overnight at 4 °C or for 2–4 h at room temperature or 37 °C.

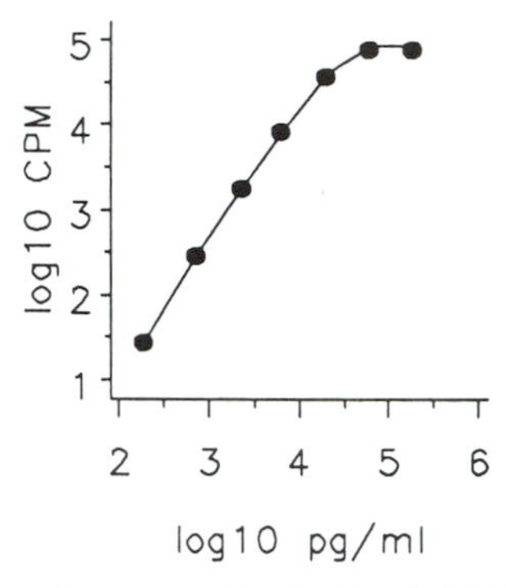

Figure 4. Dose–response curve for a two-site IRMA of rDNA derived human IL-1α using monoclonal antibodies which recognize different epitopes.

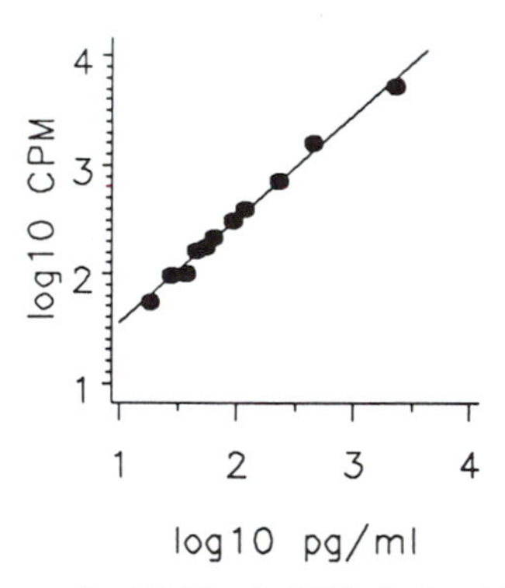

Figure 5. Dose–response curve for IRMA of rDNA-derived human IL-1α using a monoclonal antibody to capture antigen and ^{125}I-labelled sheep antibody to develop.

5. Discard well contents and wash three times with Hb-PBS. Add 45–195 μl ^{125}I labelled antibody 2 (3×10^5 c.p.m./well) diluted in Hb-PBS and incubate for 1–2 h at room temperature.
6. Discard well contents and wash three times with Hb-PBS. Cut out wells using a hot-wire plate cutter and determine the radioactivity bound to the wells by gamma counting.

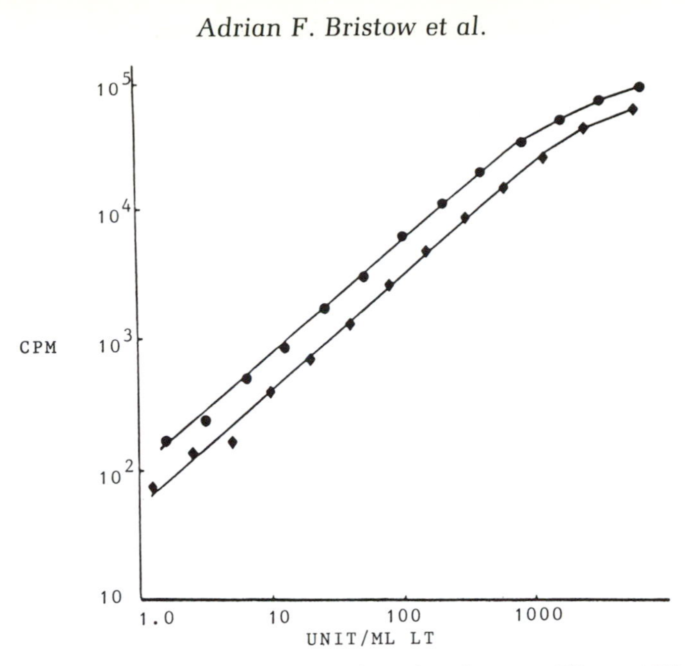

Figure 6. Bead-based IRMA for human lymphotoxin using two different MAbs. ◆ rDNA derived lymphotoxin (LT). ● natural LT. Courtesy of Dr Tony Meager.

3.5 Interpretation of results

3.5.1 Assay validation

A radioimmunoassay displacement curve, as illustrated in *Figure 1*, lies between two values, the B_0 value, or the fraction of tracer bound in the absence of unlabelled ligand, and the non-specific binding (NSB), or fraction of the tracer which is apparently bound in the absence of antibody. The log dose–displacement curve is sigmoidal, and the slope may be measured over the linear part of the curve. In practice, these three experimentally determined variables form useful diagnostic criteria for the performance of the radioimmunoassay in measuring analyte in biological samples.

Assay slope

Samples should always be measured at more than one dilution, in order that the slope of the displacement curve may be compared with that of the standard. Differences in slope indicate immunological differences between the assay standard and the ligand being measured in the samples. They also effectively invalidate the assay since different answers will be obtained, depending on which areas of the standard curve values are read off from.

Non-specific binding

The non-specific binding (NSB) value is the amount of tracer non-specifically

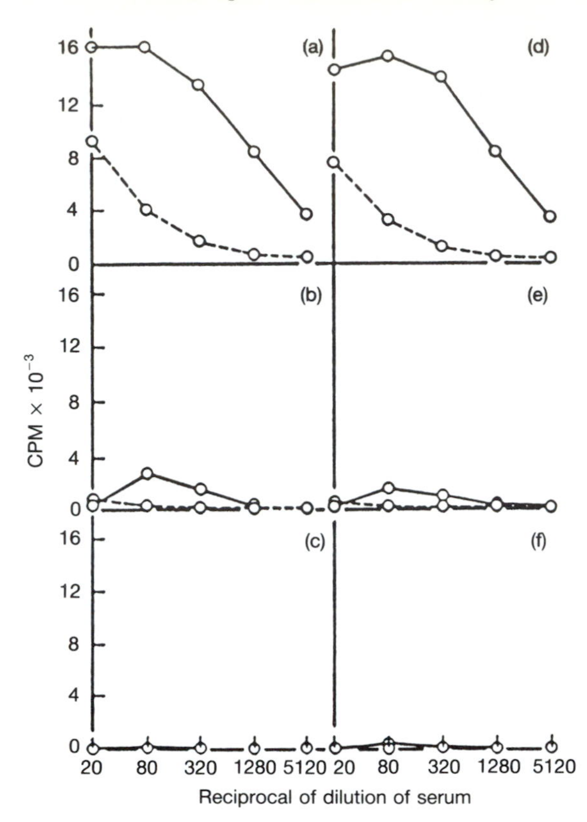

Figure 7. Two-site immunoradiometric assays for human IgE using MAbs and a polyclonal antiserum. (a), (b), and (c) using MAb 102 as capture antibody and (a) polyclonal anti-IgE serum. (b) MAb 102 and (c) MAb 117 as developing antibodies (d), (e), and (f) using MAb 117 as capture antibody and (d) polyclonal anti-IgE serum (e) Mab 102 and (f) Mab 117 as developing antibodies ○—○) using the international reference preparation for human serum IgE (85/502) (10 000 IU/ml): (○-----○) using pooled normal human serum.

precipitated or removed by the separation technique used. In practice, the displacement curve should, at high ligand concentration, reach the NSB value. If maximum displacement in the standard curve is higher than NSB it usually indicates that the tracer and the unlabelled ligand are not immunologically identical, this may arise from modification to the tracer during the iodination process, or, more commonly, may reflect ageing of the tracer, either by radioactive decay or radiolysis. If the maximum displacement obtained with the assay samples is lower than NSB, it indicates that the tracer is being degraded under the assay conditions, and that appropriate sample pretreatment is necessary.

Zero binding, B_0

If the bound fraction in the assay samples is higher than B_0, it indicates the presence in the samples of components that are affecting the antibody–ligand reaction. If the sample displacement curve is lower than B_0 at high dilution, it may also indicate tracer degradation.

3.5.2 Analysis of data

The simplest method of data analysis is to plot the standard curve as percentage binding against log ligand concentration, and to read the sample values off the curve. For some applications a more rigorous statistical analysis of the data is required. A number of approaches have been developed further, most of which use some form of data transformation to generate linear displacement curves, which are then analysed as parallel line assays. A complete description of the statistical analysis and validation of radioimmunoassays is beyond the scope of this book. The interested reader is referred to a number of excellent texts on this subject (for example, ref. 6).

4. Immunoradiometric assays

Originally, IRMAs were carried out using antibodies (or ligands) covalently coupled to a solid support in bead form for example Sepharose or cellulose. Currently, this approach is rarely used as it has been shown that antibodies and the majority of protein antigens stick particularly well by physical processes to most plastic and some types of glass. Not only is this easier and quicker than covalent coupling, it also does away with chemical procedures which may inactivate some antibodies, particularly the less stable murine immunoglobulins. The two-site IRMA technique is particularly well-suited for use with monoclonal antibodies (MAbs) or MAb/polyclonal antibody combinations. This overcomes the limitations of most MAbs when used in conventional RIAs and by careful selection of appropriate antibody combinations both extreme specificity and high sensitivity can often be achieved.

4.1 Choice of solid support

The most common solid supports currently used to immobilize the first ('capturing') antibody are the walls of plastic tubes and plastic microtitre plates or the surfaces of spherical plastic beads. Some manufacturers make plates specially for immunoassays, these plates are supposed to be particularly good for binding proteins to the well surfaces. The authors have found that the flexible type microtitre plates are perfectly adequate for IRMAs, and wells can be readily cut out from these using a hot wire, which is much less tedious than using the shears or band saw necessary for rigid plates.

The use of tubes for IRMAs is obviously more time consuming than using microtitre plates (particularly in the washing steps) but a larger volume can be accommodated, and this, in some cases can increase sensitivity. However, the use of large volumes obviously consumes larger volumes of reagents and requires a greater amount of sample and so this is often not a real advantage. Plastic tubes are available with moulded internal projections that increase the surface area available for antibody binding (for example, 'star' tubes) and these can increase sensitivity in some cases without excess consumption of reagents and requirement for a large sample volume.

Use of plastic beads for immobilizing antibody can also increase sensitivity as a fairly large volume of sample can be used. Beads are available with finely etched surfaces (for example, from Northumbria Biologicals) that increase the surface area available for protein binding.

4.2 Antibodies for IRMAs

Antibodies for IRMAs should be purified to as near homogeneity as possible. This is necessary as the sensitivity of the assay is directly related to the number of suitable binding sites on the solid support occupied by first (capturing) antibody. If the coating antibody is impure, contaminants may compete for the binding sites and hence reduce sensitivity. The second or 'developing' antibody must also be highly purified as it must be radiolabelled; the use of impure preparations will lead to these also becoming radioactive resulting in poor labelling efficiency of the antibody, decreased sensitivity and high assay backgrounds.

For IgG monoclonal antibodies it is usually sufficient to isolate the IgG fraction of ascitic fluid or culture fluid (if foetal calf serum or serum free medium is used during production). For this a combination of ammonium sulphate precipitation (40–45% saturation) and some form of ion-exchange chromatography is usually best and the authors have found the high performance liquid chromatography (HPLC) system described by Clezardin *et al.* (1985) most satisfactory. IgM MAbs can be problematical for use in IRMAs, but if used they can be purified by a variety of techniques including gel filtration on Sepharose 4B or equivalent.

Polyclonal antisera should also be highly purified, and in some cases affinity purification may be necessary. (See refs 7 and 8 for details of antibody purification.)

4.3 Radiolabelling antibodies

Radiolabelling antibodies with ^{125}I is almost always used for IRMAs. This is usually quite easily achieved using chloramine T as the catalyst (see Chapter 6), but some antibodies (particularly murine MAbs) radioiodinate poorly or are impaired in ability to bind antigen by the radioiodination procedure. If this is the case less harsh catalysts such as iodogen or lactoperoxidase may overcome the problem. The authors have found the procedure described in ref. 7 to be particularly good for use in IRMAs.

4.4 Assay design

4.4.1 Choice of conditions and non-antibody reagents

A general protocol for a two-site IRMA using microtitre plates is given in *Protocol 7* and a 'bead' version in *Protocol 8*. Tube-based versions are merely adaptations of plate assays. In general two-site IRMAs consist of five operations:

- coating solid phase with first (capturing) antibody;
- blocking unoccupied protein binding sites on solid phase;
- incubation with sample;
- incubation with radiolabelled second (developing) antibody;
- counting bound radioactivity.

Washing procedures are inserted between each operation to remove excess reagents, etc. In general it is best to coat plates/beads/tubes with antibody overnight at 4 °C and a coating concentration of 1–20 μg/ml IgG is optimal in most cases; higher concentrations often produce less sensitive assays. Blocking of unoccupied binding sites is best achieved by incubation for about 30 min with an irrelevant protein and the very cheap and effective bovine haemoglobin works well for this. Ovalbumin, bovine serum albumin, human serum albumin and many other proteins can be substituted but the authors caution against the use of non-protein blocking agents such as Tween that may block inefficiently in some cases.

Protocol 8. Bead IRMA

Reagents

- Phosphate buffered saline (PBS), see *Protocol 7*.
- PBS containing 2% bovine serum albumin (BSA-PBS)
- PBS containing 0.1% BSA (washing solution)
- Purified antibody 1 (for capture)
- ^{125}I-labelled purified antibody 2 (for detection)
- Purified standard antigen
- Samples containing antigen

Method

1. Prepare antibody 1 in PBS at 100–200 μg/ml. Coat beads in antibody by incubating in above solution overnight at 4 °C; 20 ml of this solution is enough for about 100 beads 6.5 mm in diameter.
2. Block plastic tubes e.g. Luckham LP4 with albumin by completely filling with BSA-PBS and incubating overnight at 4 °C.

3. Aspirate and discard the BSA-PBS from the tubes and prepare dilutions of the standard antigen in BSA-PBS (see *Figures 4–7* for typical ranges). Add 200 μl aliquots of dilution to the tubes and include triplicate assays for each point. Include a blank (BSA-PBS) and unknown solutions also in triplicate.
4. Remove beads from antibody solution and wash 5 times with washing solution. Blot to dryness on filter paper and add 1 bead to each tube. Incubate overnight at 4 °C or 4 h at 4 °C or room temperature.
5. Remove sample by aspiration and wash 5 times with washing solution. Add 200 μl of ^{125}I-labelled antibody 2 in BSA-PBS containing 1–3 $\times$ 10^5 c.p.m. to each tube and incubate overnight at 4 °C or 3–4 h at room temperature.
6. Remove labelled antibody and wash 5 times with washing solution. Determine the radioactivity bound to the beads by gamma counting.

Incubation with antigen containing sample is often most conveniently carried out overnight at 4 °C, but a shorter period, e.g. 3 h at room temperature or at 37 °C, usually produces no loss of sensitivity. This approach may be valuable if a single day assay is required.

Incubation time and amount of developing antibody required to produce the best IRMA must be determined empirically, but 1–2 h with about 300 000 c.p.m./well is optimal for most microtitre plate IRMAs.

4.4.2 Choice of antibodies

Careful evaluation of MAb or MAb/polyclonal antibody combinations for optimal performance in IRMAs is probably the most important factor in assay design (see *Figures 4–7*). In most cases it is necessary to select combinations that bind to different epitopes on the antigen and this is often difficult if polyclonal antisera are used. Immunochemical procedures can be used to establish that MAbs recognize different antigenic determinants and the reader is referred to a practical immunochemical text, for example (7), for such methodologies. If MAb/polyclonal combinations are used it is usually preferable to use the MAb to 'capture' antigen. If the antigen exists predominantly as a multimer for example immunoglobulins, then it may be possible to use the same antibody to capture antigen and develop the assay (see *Figure 7*).

5. Non-radioisotope immunoassays

There is an increasing trend to replace some radioisotopic radioimmunoassays with non-radioisotope equivalents. Although immunoassays based on chemiluminescent, fluorescent and electron spin labelling sytems have been described, the most common non-radioisotopic radioimmunoassays are enzyme-based systems usually described as enzyme-linked immunosorbent assays (ELISAs). Although a detailed discussion of ELISA and other non-radioactive type immunoassays is clearly beyond the scope of this book (for a general discussion of ELISA and detailed protocols, see ref. 9) a brief description of the commonest type of assay is given below.

5.1 Enzyme immunoassays

Any assay configuration, direct or competitive binding or two-site immunoradiometric can be modified to use enzymes. Enzymes are normally chosen which catalyse with simple kinetics the formation of stable, coloured or fluorescent products from colourless or non-fluorescent substrates. In many applications a positive/negative distinction can be read by eye; for example, in primary hybridoma screening or in commercial pregnancy tests. For quantitative work purpose-built spectrophotometers (ELISA readers) can read a 96-well microtitre plate in about one minute.

Table 1. Advantages and disadvantages of ELISA

Advantages:

- No radiation precautions needed.
- Can be read by eye or simple plate readers.
- Many samples can be processed in automated systems in a short time.
- Reagents are cheap and readily available.
- Enzyme conjugates are stable.

Disadvantages:

- Some reagents are toxic.
- Enzyme reactions prone to interference by biological samples.
- Conjugation of enzymes to proteins is not as easy or efficient as for radioisotopes.
- Enzyme groups are large and can interfere with antigen/antibody interactions.
- Requires determination of optimal conditions for enzyme reaction.

A wide range of commercial anti-species specific immunoglobulins or protein A products are available conjugated to the most commonly used enzymes, particularly horseradish peroxidase (HRPO), or alkaline phosphatase (AP). Conjugation of enzymes to antibodies can satisfactorily be performed in the laboratory using periodate- or glutaraldehyde-based methods (10, 7). Once formed these conjugates are very stable if kept at high concentration. For HRPO the most common substrate is orthophenylene diamine, and for AP *p*-nitrophenyl phosphate. Both systems use cheap reagents that generate a stable product and the choice of enzyme/substrate combination depends on the type of sample to be analysed: some samples (for example, bovine serum albumin) contain endogenous enzyme activity. Milk, or some cells such as macrophages contain peroxidase; and B lymphocytes contain alkaline phosphatase. Other samples can contain inhibitors of enzyme activity for example azide. Alternative enzymes such as β galactosidase, urease, or glucose oxidate can be used if HRPO or AP are unsuitable.

For all enzyme-based immunoassays it is essential to establish conditions for the enzyme catalysed reactions that give the optimal dynamic range for measurement of the analyte. The time and temperature of incubation for a given amount of enzyme-conjugate are usually the most significant parameters. In a multiplate assay it is also important to keep the temperature constant for every plate, and to either read the plates within a short time of each other, or to inactivate the enzyme with low (HRPO) or high pH (AP). The inclusion of negative controls and a standard titration of analyte on each plate should reduce problems of plate variation.

References

1. Yalow, R. S. and Berson, S. A. (1959). *Nature, Lond.* **184,** 1648.
2. Hunter, W. M. and Corrie, J. E. T. (eds.) (1983). *Immunoassays for Clinical Chemistry*, 2nd edn. Churchill Livingstone, Edinburgh.
3. Thorell, J. I. and Larson, S. M. (1978). *Radioimmunoassay and Related Techniques*. C. V. Mosby, St Louis, Mo.
4. Ekins, R. P. (1976). General principles of hormone assay. In *Hormone Assays and their Clinical Application* (eds. J. A. Loraine and E. T. Bell), pp. 1–72. Churchill Livingstone, Edinburgh.
5. Munro, A. C., Chapman, R. S., Templeton, J. G., and Fatori, D. (1983). Production of primary sera for radioimmunoassay. In *Immunoassays for Clinical Chemistry* (eds. W. M. Hunter and J. E. T. Corrie), pp. 447–456. Churchill Livingstone, Edinburgh.
6. Finney, D. J. (1978). *Statistical Methods in Biological Assay*, 3rd edn. Griffin, London.
7. Johnstone, A. and Thorpe, R. (1987). *Immunochemistry in Practice*, 2nd edn. Blackwell Scientific Publications, Oxford.
8. Baines, M. G., Gearing, A. J. H., and Thorpe, R. Purification of murine monoclonal antibodies. In *Methods in Molecular Biology*, Vol. 5, Chapter 54. Humana Press (in press).
9. Nakamura, R. M., Voller, A. & Bidwell, D. E. (1986). 27.1 in *Handbook of Experimental Immunology*, Vol. 1, *Immunochemistry* (ed. D. M. Weir). Blackwell Scientific Publications, Oxford.
10. Tijsenn, P. and Kurstak, E. (1984). *Ann. Biochem.* **136,** 451.

9

Summary of legislation in radiological protection in the United Kingdom

DAVID PRIME

1. Introduction

The use of equipment or materials which emit radiations is governed by two pieces of legislation: the Radioactive Substances Act (1960) and the Ionizing Radiations Regulations (1985). This legislation imposes a system of radiological protection which is more or less the same in all premises using radioactive materials or radiation emitting equipment. This system will be outlined in two sections dealing with the respective items of legislation.

2. The Radioactive Substances Act (1960)

This act controls the acquisition, storage and disposal of radioactive materials.

2.1 Acquisition

Before obtaining radioactive materials it is necessary to obtain a registration under this act. This registration will specify both the radionuclides or class of radionuclides and the total activities allowed.

2.2 Storage

Radioactive materials must be stored under certain conditions as outlined in the act. This will always mean a lockable store sufficiently shielded; normally to reduce the radiation dose to below 7.5 μSv/h on the outside of the store.

2.3 Radioactive waste

The disposal of waste is often particularly difficult. There are usually several different categories of waste of all phases generated. Every one of these disposal pathways is subject to an authorization under the Act.

Aqueous waste can often be disposed of down special sinks directly to the sewers. Utilizing this method of disposal requires knowledge of the total flow rate in the sewer in order to confirm that dilution is adequate. Organic solvents and other organic wastes are normally incinerated, although animal carcasses can sometimes be macerated and then disposed of as aqueous waste.

Since premises are often situated in centres of population, the disposal of gaseous radioactive waste is rarely permitted directly from the site.

Solid waste disposal routes are a function of activity and nuclide. Low levels can usually go, under special conditions, to local waste dumps where the radioactive material is immediately buried under normal refuse. Higher levels must be taken to the National Disposal Service, where special arrangements can be made for temporary or permanent storage.

Similarly, all other categories of waste not suitable for disposal by the above methods must be taken to the National Disposal Service.

3. The Ionizing Radiations Regulations (1985)

This set of regulations determine the way work with materials or equipment emitting ionizing radiations is carried out. It imposes working practices and a management structure. The regulations are divided into nine separate parts each of which will be briefly discussed below.

3.1 Interpretation and general

The first five regulations are grouped together in this section. The main features of these particular regulations are that all work involving radiation must be notified to the Health and Safety Executive so that the premises can be inspected appropriately and that there should be close co-operation between different employers. This last point is of particular importance in premises where a worker may spend time in more than one establishment and it is important that all radiation doses he receives are recorded.

3.2 Dose limitation

Many design features must be incorporated to reduce radiation exposure. Shielding and containment are of special importance and the general philosophy is that all radiation doses should be kept as low as is reasonably practical. This means that it is not merely necessary to keep within the dose limits, but to keep as far below these limits as is practicable. Furthermore, priority must be given to keeping individual doses to as low as is reasonably practicable. This means that in some cases a sharing out of radiation exposure must be carried out by increasing the numbers of persons carrying out a particular practise.

3.3 Regulation of work with ionizing radiation

This part is probably the most important section of the regulations. It commences by dealing with the definition of types of radiation areas. Wherever a considerable risk of exposure exists, a 'controlled area' must be declared. Only a special category of worker is allowed into these areas other than by a written and approved 'system of work'. Such special workers are designated as 'classified' and special records of radiation exposure and health must be kept. This section also calls for the appointment of Radiation Protection Advisers and Radiation Protection Supervisors. The former are full-time professionals in the subject who should be consulted on all matters concerned with radiation protection. The latter are the people who are in control of protection within their department, and act for the head of the department. It is these supervisors who should draw up local rules for work with ionizing radiations within their particular department consulting the adviser where necessary. Finally, this section deals with the training, instruction and information which should be given to all workers before they commence work with ionizing radiations.

3.4 Dosimetry and medical surveillance

This part of the regulations deals with individual and collective surveillance of radiation workers. Any approved and appropriate means of dose assessment is acceptable, but such assessment must be closely monitored to confirm the accuracy of the results obtained.

Medical examinations of all classified radiation workers should be carried out on an annual basis. Such examinations are not used to confirm dosimetric results since medical effects would be undetectable at normal dose levels, but to check that the health of the worker is compatible with the working conditions.

Records must be kept of all dosimetric results and medical examinations.

3.5 Arrangements for the control of radioactive substances

This section deals first with the keeping and moving of sources. Basically, the whereabouts of all sources should be known at all times and accounting methods should be employed to keep careful records of receipts and disposals. Special equipment and methods may be required for the movement of sources and it may even be necessary to declare special radiation areas during transport operations.

This part also calls for the provision of adequate washing and changing facilities as well as protective clothing for radiation workers.

3.6 Monitoring of ionizing radiation

This part of the regulations deals with monitoring of radiation in controlled and supervised areas. Equipment must be of the correct type and be regularly checked by special approved laboratories. Records have to be kept of all monitoring results.

3.7 Assessments and notifications

For sources of radiation of particular hazard or if an over-exposure of a worker is suspected or confirmed, special investigations and reports require to be submitted to the Health and Safety Executive. Major spills or losses of sources should also be notified to the Health and Safety Executive. Contingency plans should also be drawn up in order to cover likely incidents that could occur with the particular sources that one possesses.

3.8 Safety of articles and equipment

This part deals with responsibility of manufacturers and also the responsibility of individuals using equipment emitting ionizing radiation. All this equipment should be checked to ensure that users and, in the case of medical applications, patients are not receiving excessive doses of radiation.

3.9 Miscellaneous and general

This final part deals mainly with a timetable of implementation of the regulations.

4. EEC legislation

The Ionizing Radiations Regulations were formulated partially in order to bring the United Kingdom legislation into line with an EEC directive. Although interpretation of this directive is likely to vary from country to country, much of UK legislation will be very similar to that in other EEC countries. In particular, dose equivalent limits are the same in all EEC countries.

10

Federal regulations on use of radionuclides in the United States of America

NEAL S. NELSON

1. Introduction

In 1967, the World Health Organization reported that regulation of devices and materials emitting ionizing radiation was divided among fifty state governments, several city governments, six Federal agencies, a Federal Council and a Federal Commission (1). In a 1978 review, five Federal agencies and twenty-five states were reported to be regulating some aspects of radiation protection and control (2). Unfortunately the situation in the 1980s is not much improved.

Generally speaking, regulation is by laws, statutes, standards, regulations, and ordinances. Laws are the rules of conduct applied and enforced under the authority of established government, state or federal. They establish proper, permitted, denied, and penalized behaviour and statements of goals or intent (3). Statutes are acts of legislature, administrative regulations, or any enactment given the force of law by the state (3).

While laws and statutes state legislative requirements, interpretation may be necessary. Regulations are issued by the government agencies administering the laws. Such regulations are to define what is required by the law and how the legal requirements of the law may be met (3). Standards, like regulations, explain compliance with the law but provide a means of comparison or evaluation to show compliance with the law (3). Ordinances are legal requirements issued by a city or municipality.

Since itemizing State and Local regulations on control of ionizing radiation would be voluminous, only Federal regulations will be addressed here. However, individuals wishing to work with radionuclides or other sources of ionizing radiation must comply with all regulations, Federal, State, and Local.

The material presented in this review was prepared by the author and there is no express or implied liability or responsibility of the Environmental Protection Agency for accuracy or completeness of information.

When a federal agency wishes to propose a regulation or standard or change one, there is an extended public process. There may be: an advanced Notice of Proposed Regulation (ANPR) with a public comment period; a Draft Regulation or Standard, with a public comment period and then the Final Regulation or Standard. All of these advanced notice, draft and final published in the government newspaper the *Federal Record* (*FR*). Hence, citations x *FR* y are to volume and page in the *Federal Record*. The current regulations and standards are published each year in the *Code of Federal Regulations* with ancillary publications during the year. The *Code of Federal Regulations* (*CFR*) is published by the Office of the Federal Register. It is usually cited a *CFR* b referring to the title and section of the *CFR* where the regulation or standard is published.

Since the regulations and standards are under constant review and revision, those working with radionuclides should be familiar with the *Federal Record* and *Code of Federal Regulations*. In the United States, the Health Physics Society, through their newsletter, tries to keep their membership advised of activities of Federal agencies.

The United States Government established a broad authority for radiation control with the Atomic Energy Act of 1946 [Public Law 585]. This act established the Atomic Energy Commission which had authority to control source and byproduct materials and issue lisenses (4). This act was amended frequently and then rewritten in 1954 as the Atomic Energy Act of 1954 [Public Law 83-702] (5). Chapter 10 of this act dealt with 'Atomic Energy Licenses' and expanded the 1946 Act. In 1959, the Federal Radiation Council (FRC) was established to '... advise the President with respect to radiation matters directly or indirectly affecting health, including matters pertinent to the general guidance of executive agencies by the President ...' (6). Guidance promulgated by the FRC and signed by the President was the basis for Federal regulations on radiation exposure.

Presidential Reorganization Plan No. 3 of 1970 established the Environmental Protection Agency (EPA). All functions of the FRC and of the Atomic Energy Commission regulating environmental limits for radiation exposure were transferred to EPA at this time. The Energy Organization Act of 1974 [Public Law 93-438] abolished the Atomic Energy Commission and established the Energy Research and Development Administration with the responsibility of studying and promoting nuclear science and the Nuclear Regulatory Commission (NRC) with the responsibility of regulating source and byproduct materials (8).

The NRC has authority to enter agreement with the governor of any State, permitting that State to assume regulatory control over byproduct material, source material and special nuclear material (in quantities less than a critical mass). The State must be willing to assume regulatory control and the Governor of the State must certify the State has a regulatory control program for radiation which is adequate to protect public health and safety. The NRC must determine that the State programme is compatible with the Commission's regulatory programme and adequate to protect public health and safety (6). The 29 states mentioned above are Agreement States and regulative byproduct materials.

2. Licensing requirements

As noted earlier, the regulations promulgated by US Government agencies are published in the *Code of Federal Regulations* (*CFR*) which is updated on an annual basis. The regulations of the Nuclear Regulatory Commission (NCR) which has the responsibility of regulating source, special nuclear and byproduct material; can be found in 10 *CFR*, Title 10 of the *Code of Federal Regulations* (9). Changes between annual publication dates are also published as necessary to keep the regulations current. At the same time, the applicant for a license must also conform to any state regulations. It should be noted that it is illegal to receive, acquire, own, possess, use, or transfer any radionuclides if you do not have the appropriate license.

Part 30 of 10 *CFR* [10 *CFR* 30] lists types of licenses, general and specific, and requirements for licensing to obtain, keep and use radioactive materials. The general license is . . . 'effective without the filing of applications with the Commission or issuance of licensing documents to particular persons:' (9). Specific licenses for byproduct material are '. . . issued to named persons upon applications filed pursuant to the regulations of this part and Parts 32 through 35 and 39 of this Chapter (9)'.

It should be noted that byproduct material means any radioactive material (except special nuclear material) made radioactive by exposure to radiation incident to the process of producing or utilizing special nuclear material. Usually, this refers to reactor produced radionuclides.

Special nuclear material means plutonium, uranium-233, uranium enriched in isotope 233 or 235, and other material designated by the Commission but not including source material. Usually this refers to weapons grade material. Source material means any uranium or thorium or combination thereof in any physical or chemical form. It also includes ores which contain by weight, $^{1}/_{20}$ of 1% or more of uranium, thorium, or any combination thereof not including special nuclear material. Usually this refers to uranium or thorium ores or reactor fuel grade materials.

3. General domestic license

General licenses for possession and use of byproduct material contained in certain items and a general license for ownership of byproduct material is given in 10 *CFR* 31.

Items included for possession under a general license, if the manufacturer has complied with appropriate NRC regulations, are:

- not more than 18.5 MBq of polonium-210 in a static elimination device;
- not more than 1.85 GBq of tritium or 18.5 MBq of polonium-210 in an ion generating tube;
- measuring, gauging, or controlling devices containing tritium or not more than 3.7 MBq of beta-and/or gamma-emitting material or 370 kBq of alpha-emitting material;
- aircraft luminous safety devices containing not more than 370 GBq of tritium or 1.11 GBq of promethium-147;
- reference source of not more than 185 kBq of americium-241;
- not more than 1.85 MBq of strontium-90 in an ice-detection device; and
- pre-packaged clinical and laboratory test units containing not more than 370 kBq of iodine-125, iodine-131, selenium-75, or carbon 14; 1.85 MBq of tritium; 740 kBq of iron-59; or a reference source of 1.85 kBq of iodine-129 and 0.185 kBq of americium-241.

However, for clinical and laboratory use the recipient must also file an NRC form 'Registration Certificate—In Vitro Testing with Byproduct Material under a General License' and show adequate capability to monitor radiation in the laboratory.

Further provisions and restrictions can be determined by consulting 10 *CFR* 31 in the publication for the current years.

4. Specific licenses of broad scope

The provisions of the NRC in 10 *CFR* 3.33 for granting an individual a specific license are: (1) the application is for a purpose authorized by the Act, (2) the applicants proposed equipment and facilities are adequate to protect health and minimize danger to life or property, (3) the applicant is qualified by training and experiences to use the material for the purpose requested in a safe manner, (4) any special requirements in Parts 31 through 35 and 39 are satisfied, and (5) if the application involves conduct of any activity the Commission determines will significantly affect the quality of the environment, a special determination must be made.

The byproduct material that a licensed individual may hold is that quantity specified in the license. Three types of license are available, as specified in 10 *CFR* 33.11; Types A, B, and C. All applicants must comply with provision of 10 *CFR* 30.3.

Type A licenses are usually for quantities in the multicurie range. Requirements are covered in 10 *CFR* 33.13 which should be consulted for detailed requirements. In general, extensive administrative and accounting controls with managerial review, a radiation safety committee, and a qualified radiation safety officer.

Type B licenses are for curie (37 GBq) amounts of most byproduct isotopes with an overall limit on possession. Requirements are covered in 10 *CFR* 33.14 which should be consulted for detailed requirements. In general there is an overall limit defined in 10 *CFR* 33.100, administrative control and accounting with managerial review and appointment of a qualified radiation safety officer and required.

Type C licenses are for millicurie (37 MBq) amounts with an overall limit on possession. Requirements are covered in 10 *CFR* 33.15 which should be consulted for detailed requirements. In general, the applicant must state the material will be used by or under direct supervision of individuals who have received (1) a college degree at the bachelor level or equivalent training and experience in physical or biological sciences or in engineering, or (2) at least 40 h of training and experience in safe handling of radioactive materials, characteristics of ionizing radiation, units of radiation dose and quantities, radiation detection instruments, and biological hazards of exposure to radiation of the appropriate type and form of byproduct material used. In addition, the applicant must have established administrative controls for procurement, procedures record-keeping, managerial control and accounting, and review, to assure safe operations. Much of the non-medical research with radionuclides is conducted under a Type C license.

Additional types of specific licenses are issued by the NRC for other purposes. Specific parts of 10 *CFR* should be consulted for the requirements for these licenses, e.g.

10 *CFR* 32 Specific Domestic Licenses to Manufacture or Transfer Certain Items Containing Byproduct Material—covers manufacture of devices containing byproduct material.

10 *CFR* 34 Licenses for Radiography and Radiation Safety Requirements for Radiographic Operations—covers industrial radiography and sources.

10 *CFR* 39 Licenses and Radiation Safety Requirements for Well Logging.

5. Medical uses

A special type of specific license is covered by 10 *CFR* 35, Medical Uses of Byproduct Material. The requirements for licensing under Part 35 are rather extensive and cover licenses for use in institutions and by individual physicians, dentists, or podiatrists; use of sealed sources for certain types of medical uses; for teletherapy sources, and general licenses for certain quantities of byproduct material. The applicant should consult appropriate portions of 10 *CFR* 35 for specific requirements.

The license for institutions will require: adequate facilities for clinical care and radiation safety, a Radiation Safety Officer, and a Radiation Safety Committee; also that the physician designated as an authorized user has appropriate experience working with radioisotopes, and that general technical requirements of 10 *CFR* 35 have been met (9).

There are extensive training and experience requirements listed in 10 *CFR* 35.900. Requirements for the Radiation Safety Officer and physicians, dentists, or podiatrists who are identified as authorized users. The appropriate sections of 10 *CFR* 35 should be consulted if any use of radioisotopes for medical uses is considered (6). The subparts listed under Part 35 are:

Subpart A—General Information
Subpart B—General Administrative Requirements
Subpart C—General Technical Requirements
Subpart D—Uptake, Dilution, and Excretion
Subpart E—Imaging and Localization
Subpart F—Radiopharmaceuticals for Therapy
Subpart G—Sources for Brachytherapy
Subpart H—Sealed Sources for Diagnosis
Subpart I—Teletherapy
Subpart J—Training and Experience Requirements
Subpart K—Enforcement

The NRC also requires immediate telephone notification of therapy misadministration and written notification of therapy and diagnostic misadministration and maintenance of records of misadministration (9).

6. Exemptions

The NRC has exempted the Department of Energy in the case of high-level waste repositories in geological formations (9) and those licensed for land disposal of radioactive waste (9) from the provisions of 10 *CFR* 30 (9). Also use of byproduct material under certain Department of Energy and Nuclear Regulatory contracts has certain exemptions (9).

Common and contract carriers, freight forwarders, warehousemen, and the US Postal Service in the case of normal transport are exempt (9). Exempt concentrations of byproduct material a person may hold are listed in 10 *CFR* 30.70 Schedule A. Of course, these materials may not be introduced into any commodity designed for ingestion, inhalation by or application to a human being, nor may they be transferred to persons exempt from regulations (9).

Items using byproduct material such as self-luminous watch dials or hands, gas or smoke detectors, etc., have maximum quantities or levels of radiation as listed in 10 *CFR* 30.15, 10 *CFR* 30.16, 10 *CFR* 30.19 and 10 *CFR* 30.20.

Exempt quantities of radioisotopes a person may use are listed in 10 *CFR* 30.71, Schedule B. These byproduct materials may not be used in production of products intended for commercial distribution nor may they be transferred to persons exempt from regulations (9). These exempt quantities are in the microcurie range.

7. Radiation protection

Radiation protection requirements are listed in Part 20 of 10 *CFR* and include permissible doses, levels, and concentrations; precautionary procedures; waste disposal; records, reports and notification; exemptions and additional requirements, and enforcement. Since the provisions in this part are extensive, 10 *CFR* 20 should be consulted for specific requirements (9).

Individual doses of persons in restricted areas should not exceed.

- 2.5 mSv per calendar quarter to whole body; head and trunk; active blood forming organs; lens of eyes; or gonads
- 187.5 mSv per calendar quarter to hands and forearms; feet and ankles
- 75 mSv per calendar quarter to skin of whole body

However, a licensee may permit an individual in a restricted area to receive a total occupational dose to the whole body greater than those listed above, if:

- the whole body dose does not exceed 30 mSv in any calendar quarter, and
- the whole body dose when added to the accumulated occupational dose to the whole body does not exceed 50 (N–18) mSv when N equals the individuals age in years at his last birthday (9).

Limits of exposure in restricted areas ('Restricted Area' is any area where access is controlled by the licensee for purposes of protecting individuals from exposure to radiation and radioactive materials. 'Restricted Area ' shall not include any areas used as residential quarters) to airborne radioactive materials are given in 10 *CFR* 20 Appendix B, Table I, Column 1 (10 *CFR* 20.103). Exposure of minors (under 18 years of age) in a restricted area should not exceed 10% of the limits of individual doses per calendar quarter listed above. There is also a special limit for exposure to airborne radioactive material in 10 *CFR* 20, Appendix B, Table II, Column 1 (9).

Levels of radiation in unrestricted areas ('Unrestricted Area' is any area where access is not controlled by the licensee for the purpose of protecting individuals from radiation or radioactive material) should not give whole body doses in excess of 5 mSv per year or deliver doses in excess of 20 mSv per hour of 1 mSv in any 7 consecutive days (9). Radioactivity in effluents to unrestricted areas should not exceed limits specified in 10 *CFR* 20, Appendix B, Table II (9).

Licensees are required to make appropriate surveys, monitor personnel and provide caution signs, labels, signals, and controls (9). Magenta and yellow radiation trefoil emblems shall be posted with the warning—Caution

Radiation Area, Caution High Radiation area; Caution Airborne Radioactivity Area; or Caution Radioactive Materials as appropriate for the hazard (9). High radiation areas must be equipped with locks and devices warning of entry or with interlocks to prevent entry if a dose of 1mSv per hour could be exceeded. Exceptions to these requirements are given in 10 *CFR* 20.204.

Special procedures are required for picking up, receiving and opening packages. Packages should be received or collected expeditiously. With exceptions noted in 10 *CFR* 20.205, packages should be monitored within 3 h during the working day or within 18 h if received after normal hours. If removable radioactive contamination is in excess of 0.37 kBq (22 000 disintegrations per minute) per 0.01 m^2 of package surface or dose rates in excess of 2 mSv per hour at the surface or 0.1 mSv per hour at a distance of 1 m from the surface of the package, the licensee must immediately notify the NRC and the final delivering carrier (9).

Licensed materials stored in an unrestricted area must be secured or under the constant surveillance and immediate control of the licensee (9).

Waste disposal can be by transfer to an authorized recipient or by method approved by the NRC as required in 10 *CFR* 20.302 (9). Waste disposal into sanitary sewerage systems is possible but cannot exceed the limits of 10 *CFR* 20, Appendix B, Table I, Column 2, after dilution, so amounts releasable are related to sewerage flow and volume. 10 *CFR* 20.303 should be consulted for details. In any case, there are annual limits, not to exceed ten times the quantity of each isotope listed in 10 *CFR* 20, Appendix C, and the gross quantity released does not exceed 37 GBq (1 Ci) per year. Up to 185 GBq of tritium and 37 GBq of carbon-14 may be released each year. Excreta from persons undergoing medical diagnosis or therapy are exempt from this section (9).

No licensed material can be disposed of by incineration except material with 1.85 kBq or less of tritium or carbon-14 per gram of scintillation counting fluid or per gram of animal tissue averaged over the weight of the entire animal. Animal tissue not exceeding these concentrations can be disposed of by any means. However, tissues may not be used as human food or animal feed, records of receipt, transfer, and disposal must still be kept and all regulations governing toxic or hazardous material must be followed (9).

Requirements for disposal in a licensed land disposal facility and the manifest tracking system for control can be found in 10 *CFR* 20.311 which should be consulted for the detailed requirements. Sections 10 *CFR* 20.401 should be consulted for requirements on records of surveys, radiation monitoring, and disposal and 10 *CFR* 20.402 on reports of theft or loss of licensed material.

The NRC must be notified immediately if there is:

- whole body exposure of 0.25 Sv or more; skin exposure of 1.5 Sv or more; or exposure of feet, ankles, hands, or forearms of 3.75 Sv or more;

- release of radioactive material in a 24-h period exceeding 5000 times the limits in 10 *CFR* 20, Appendix B, Table II;
- loss of one week or more of facility operation; or
- property damage of $200 000 or more.

Notification must be made in 24 h if there is:

- whole body exposure of 50 mSv or more; skin exposure of 0.35 Sv or more; or exposure of feet, ankles, hands, or forearms of 0.75 Sv or more;
- release of radioactive material in a 24 h period; exceeding 500 times the limits in 10 *CFR* 20, Appendix B, Table II.
- loss of one day or more of facility operation; or
- property damage in excess of $2000 or more (9).

Detailed written reports of overexposure and excessive levels or concentrations must be made within 30 days. Detailed requirements are given in 10 *CFR* 20.405.

Requirements for personnel monitoring reports are given in 10 *CFR* 20.407, those for monitoring at termination of employment work in 10 *CFR* 20.408 and those for notification and reports to individuals in 10 *CFR* 20.409.

8. Other sections of Title 10, *Code of Federal Regulations*

The regulations of the NRC are extensive. In addition to the parts listed above, other parts should be consulted for specific requirements. For example:

10 *CFR* 9 Public records
10 *CFR* 10 Clearances
10 *CFR* 19 Notices, instructions, and reports to workers and inspections
10 *CFR* 21 Reporting of defects and noncompliance
10 *CFR* 71 Packaging and transportation of radioactive material

9. Radiation protection—Department of Labor

The Department of Labor, through its Occupational Safety and Health Administration (OSHA), also regulates and enforces radiation protection in the occupational setting under Title 29 of the Code of Federal Regulations in 29 *CFR* (10).

The occupational exposure limits enforced by OSHA are the same as those listed earlier for the NRC in 10 *CFR* 20 as are requirements for posting of signs and warnings and employee-records keeping. In addition the regulations specify evacuation warning signal characteristics a sound wave between 450 and 500 hertz modulated between 4 and 5 hertz and of at least 75 decibels at all locations where individuals who must evacuate are located. Design objective tests and checks are also specified (10).

Radioactive material packaged and labelled according to Department of Transportation regulations in 49 *CFR*, Chapter I, are exempt from labelling and posting requirements during shipment of inside containers are labelled properly (10). Notification requirements listed in 10 *CFR* 20 in case of an overexposure are extended to include the Assistant Secretary of Labor, or his authorized recipient (10).

10. Transportation—Department of Transportation

The Department of Transportation (DOT) regulates all commercial transportation of radioactive material, and Title 49 of the *Code of Federal Regulations* should be consulted for detailed requirements; the appropriate sections are 49 *CFR*, Parts 100 to 199 (11). Since the regulations are quite extensive covering labelling, packaging, warning signs and symbols, 49 *CFR* should be consulted by persons wishing to ship radioactive material, particularly Parts 171 to 177 (11). The Department of Transportation has prepared a review of the regulations for transportation of radioactive materials which states materials containing less than 74 Bq per gram are not considered radioactive for purposes of transportation and are not regulated by DOT (12).

11. Transportation—US Postal Service

The US Postal Service regulates under Title 39 of the *Code of Federal Regulations*. The Postal Service Regulations in the *Domestic Mail Manual*, Part 124.37 state that any package of radioactive materials required to bear the DOT Radioactive White-I, Radioactive Yellow-II, or Radioactive Yellow-III or which contain radioactive material in quantities in excess of those authorized in *US Postal Service Publication 6* are non-mailable (13). Likewise, advertising, promotional, or sales material which solicits or induces mailing of any article described in Part 124.37 as unmailable, is itself unmailable (13).

The *Domestic Mail Manual* should be available for review at any US Post Office, *US Postal Service Publication 6* (14) would have to be obtained from a library or from the US Government Printing Office.

12. Environment—Environmental Protection Agency

The US Environmental Protection Agency (EPA) issues regulations in Title 40 of the *Code of Federal Regulations* (15). EPA has issued final rules for radionuclides as hazardous air pollutants. Under this rule NRC-licensed facilities and facilities owned or operated by any Federal agency other than the Department of Energy must keep emissions of radioactive materials to amounts that will not cause a dose equivalent in excess of 0.25 mSv per year to the whole body or 0.75 mSv to the critical organ or any member of the public. Doses due to radon-220, radon-222, and their respective decay products are excluded from these limits (15). Facilities regulated under 40 *CFR*, Parts 190, 191, or 192; low-energy accelerators and users of sealed radiation sources are excluded from these limits (15).

Compliance is based on measured radionuclide emissions and date equivalents calculated based on EPC approved sampling procedures, EPC Code (AIRDOS-EPA and RADRISK), or other approved methods (15). Alternative emission standards can be approved by the Administrator of EPA after review of an application for such by the facility operator. Such alternative standards cannot allow exposure of a member of the public to over 1 mSv per year on a continuous basis or 5 mSv per year on a noncontinuous basis from all sources excluding natural background and medical procedures (15).

Facility operators can also request a waiver of compliance (15).

Department of Energy facilities not regulated under 40 *CFR*, Parts 190, 191, or 192 are subject to the same limits (15).

Appropriate sections of 40 *CFR* should be consulted for detailed requirements.

13. Human uses—Department of Health and Human Services

The Food and Drug Administration (FDA), part of the Department of Health and Human Services, regulated medical uses of radioactive materials in Title 21 of the *Code of Federal Regulations* (16). Under Part 310–New Drugs, the FDA listed a number of radioactive drugs, for which safety and effectiveness could be demonstrated by new-drug applications or by licensing by the Public Health Service. These drugs were required to have appropriate labelling by isotope, chemical form and use, and adequate evidence of safety and effectiveness had to be furnished. The NRC and the FDA then concluded these isotopes should not be distributed under investigational-use labelling if they were actually intended for use in medical practice (16). Additional provisions for manufacture or distribution of radioactive drugs and requirements for a new-drug application or application for approval to market a new drug are also detailed in this regulation (16).

The FDA has regulated research use of radioactive drugs (as defined in Section 310) in humans (16). In addition to being safe and effective, such drugs may not be intended for immediate therapeutic, diagnostic, or similar purposes, nor to determine the safety and effectiveness of the drug in humans (clinical trial). They may be used in a research project to obtain basic information on metabolism, physiology, pathophysiology, or biochemistry (16).

Conditions under which radioactive drugs for research are considered safe and effective are:

- the project is approved by a Radioactive Drug Research Committee
- the pharmacological dose is limited to cause no clinically detectable pharmacological effects
- the limits on annual radiation dose in adults is
  (a) Whole body active blood forming organs, lens of eye, and gonads

Single dose	30 mSv
Annual and total dose commitment	50 mSv

  (b) Other organs

Single dose	50 mSv
Annual and total dose commitment	150 mSv

  Persons under 18 years of age at their last birthday shall receive a radiation dose not exceeding 10% of that listed above.
- any radiation doses from X-ray procedures that are part of the research study must be included when determining total doses and commitments (16).

Detailed requirements on: composition of the radioactive Drug Research Committee, what reports are to be made, approval of the Radioactive Drug Research Committee, qualifications of investigators, requirement for NRC licensing, selection of human subjects, research protocol, treatment of adverse reactions, and institutional review board approval are also included in this section (16).

14. Future developments

In 1987 the EPA as authorized by Reorganization Plan No. 3 of 1970 exercised its Federal Guidance Function. The *Federal Register* of 27 January 1987 contained approval by President Reagan of 'Radiation Protection Guidance to Federal Agencies for Occupational Exposure (17). The 'Memorandum for the President' which was approved is 'Federal Radiation Protection Guidance for Occupational Exposure' recommended by the EPA Administrator (18). The Administrator recommended:

(a) there should be no exposure without expectation of an overall benefit from the activity;
(b) ALARA (as low as reasonably achievable), practices are applicable, economic and social factors being taken into account;
(c) radiation doses received as a result of occupational exposure should not exceed the limiting values for assessed dose to individual workers specified below:
  (1) *for career and genetic effects* the effective dose equivalent, H_E, received in any year by adult worker should not exceed 50 mSv and

(2) *for other health effects* in addition to the limitation on effective dose equivalent, the dose equivalent, H_T, received in any year by an adult worker should not exceed 150 mSv to the lens of the eye, and 500 mSv to any other organ, tissue (including skin) or extremity of the body.

The effective dose equivalent is defined as:

$$H_E = W_T H_T$$

where W_T is a weighting factor and H_T is the annual dose equivalent averaged over organ or tissue T.

Values of W_T and their corresponding organs and tissues are:

Gonads	0.25
Breasts	0.15
Red bone marrow	0.12
Lungs	0.12
Thyroid	0.03
Bone surfaces	0.03
Remainder	0.30

(the five other organs with the highest doses: other than lens of eye, skin or extremities; each organ given a weighting factor of 0.06)

(d) In the case of committed doses from internal sources of radiation, intake of radionuclides by an adult worker will be controlled so that:

(1) the anticipated committed effective dose equivalent from the annual intake plus any annual effective dose equivalent from external sources will not exceed 50 mSv and

(2) the anticipated magnitude of the committed dose equivalent to any organ or tissue from such intake plus any annual dose equivalent from external exposure will not exceed 500 mSv;
the committed effective dose equivalent from internal sources of ratiation, $H_{E,50}$ is

$$H_{E,50} = W_T H_{T,50}$$

where W_T is defined as in Recommendation (c) and $H_{T,50}$ is the sum of all dose equivalents to an organ or tissue T that accumulate in an individuals' life-time (taken as 50 years).

(e) occupational dose equivalents to individuals under 18 years of age are limited to one-tenth the values specified in Recommendations (c) and (d) for adult workers;

(f) the dose equivalent to an unborn as a result of occupational exposure of a woman who has declared she is pregnant should be kept as low as reasonably achievable and in any case should not exceed 5 mSv; however, no discrimination in employment will be allowed;

(g) individuals occupationally exposed to radiation should be instructed on basic health risk from radiation and basic radiation protection principles;

the degree and type of instruction should depend on the potential radiation exposures involved;

(h) there should be appropriate personnel and work-place monitoring and records kept to ensure conformance to these recommendations;

(i) radiation exposure control measures should be designed, selected, utilized, and maintained to ensure anticipated and actual doses meet the objectives of the guidance;

(j) the numerical values recommended should not be exceeded lightly, only in emergencies or after review of reactions by the Federal agency having jurisdiction (Th87). The recommendations also permit use of the International Commission on Radiological Protection (ICRP) quality factors, dosimetric conventions, models for reference persons, metabolic models, and dosimetric methods in determining conformance with the recommendations (Th87).

As this new Federal Guidance is incorporated in the regulations of the various Federal agencies, the requirements listed in the *Code of Federal Regulations* will change to reflect the new Guidance.

15. Conclusion

The regulations governing possession and use of radioactive materials and radiation protection are extensive and complex involving many Federal agencies. Selected major regulations on radiation of the Department of Labor, the Nuclear Regulatory Commission, the Department of Health and Human Services, the Environmental Protection Agency and the Department of Transportation have been extracted here. This has not been a detailed explanation of the regulations and the original references must be consulted for detail. In addition, most Federal agencies have a public information service or professional information service to answer questions or provide copies of regulations and other materials. These sources should also be employed to obtain correct and current information.

No attempt was made to try to review State or local regulations on radioactive materials or radiation protection. These will usually be site specific and so the appropriate authorities should be consulted. Likewise, no attempt was made to include every Federal agency. Regulations of the military, the Department of Defense, the Department of Agriculture, The Department of Commerce, the Department of State, etc., were considered sufficiently removed from the normal regulation of users of radioactive material, that they were not of general interest. For these agencies, the *Code of Federal Regulations* for that area of government should be consulted.

Finally, it should be remembered that the *Code of Federal Regulations* is republished every year. As new regulations are developed they are incorporated in the next issue of the *Code of Federal Regulations*. Therefore continual review of the regulations is necessary to maintain a current knowledge of the requirements of the regulations.

References

1. World Health Organization (1972). *Protection Against Ionizing Radiations*. World Health Organization, Geneva.
2. Peterson, H. T., Jr., and Mattson, R. J. (1978). Laws, Regulations and Standards. In *Nuclear Power Waste Technology* (eds. A. A. Moghissi, H. W. Godbee, and M. W. Carter), Chapter 3, pp. 99–166. American Society of Mechanical Engineers, New York.
3. Ballentine, J. A. (1969). *Ballentines Law Dictionary*, 3rd edn. Bancroft Whitney Co., San Francisco, California.
4. United States Code Annotated (1978). *Title 42: The Public Health and Welfare* Parts 1400 to 1890. West Publishing Co., St. Paul, Minn. See Parts 1801–1819.5.
5. United States Code Annotated (1973). *Title 42: The Public Health and Welfare*, Parts 2011 to 3100. West Publishing Co., St. Paul, Minn. See Parts 2011–2296.
6. Eisenhower, Dwight D. 2.1 E.O. 10831 (14 August 1959). *Establishment of the Federal Radiation Council. Federal Register* **24,** 6669.
7. *Code of Federal Regulations* (1971). *Title 3: The President, 1966–1970 Compilation*, pp. 1072–5. US Government Printing Office, Washington, DC.
8. United States Code Annotated (1983). *Title 42: The Public Health and Welfare*, Parts 4541 to 6500. West Publishing Co., St. Paul, Minn. See Parts 5801–5891.
9. Code of Federal Regulations (1 January, 1988) *Title 10: Energy*, Chapter I – Nuclear Regulatory Commission. Parts 0 to 50, and 51 to 199. US Government Printing Office, Washington, DC.
10. Code of Federal Regulations (1 July 1987). *Title 29: Labor*, Chapter XVII – Occupational Safety and Health Administration, Department of Labor, Parts 1900 to 1999. US Government Printing Office, Washington, DC.
11. Code of Federal Regulations (1 January 1987). *Title 49: Transportation*. Subtitle B – Other Regulations Relating to Transportation. Chapter I – Research and Special Programs Administration, Department of Transportation, Parts 100 to 199. US Government Printing Office, Washington, DC.
12. Department of Transportation (1983). *Review of the Department of Transportation (DOT) Regulations for Transportation of Radioactive Materials*. US. Printing Office, Washington, DC.
13. US Postal Service, *Domestic Mail Manual*. Issue 26, 3 April 1988. US Government Printing Office, Washington, DC.
14. US Postal Service (September 1983). *US Postal Service Publication 6 – Radioactive Materials*. US Government Printing Office, Washington, DC.
15. *Code of Federal Regulations* (1 July 1987) *Title 40: Protection of the Environment*. Chapter I – Environmental Protection Agency, Parts 61 to 80 and Parts 190 to 399. US Government Printing Office, Washington, DC.
16. *Code of Federal Regulations* (1 April 1988). *Title 21: Food and Drugs*. Chapter I – Food and Drug Administration, Department of Health and Human Services, Parts 300 to 499. US Government Printing Office, Washington DC.

17. Reagan, Ronald (1987). *Title 3: The President Recommendations Approved by the President Radiation Protection Guidance to Federal Agencies for Occupational Exposure. Federal Register* **52** (17), 2822.
18. Thomas, Lee M. *Memorandum for the President (1987) Federal Radiation Protection Guidance for Occupational Exposure. Federal Register* **52** (17), 2822–34.

A1

Unpacking and dispensing radioactive solutions

Radioactive materials are generally delivered in a sealed can within a large polystyrene dry-ice container or simple cardboard box. Inside the can there is polystyrene packing and a plastic or metal container for the isotope supplied in a glass vial.

Radioactive solutions are generally packed in borosilicate glass vials, 7 ml capacity, with a screw cap and two seals. The inner seal comprises a Teflon-faced rubber disc secured by a metal ring. Aqueous or alcohol solutions can be dispensed directly without removing this seal using a syringe. If the teflon disc and metal ring are removed, the vial can be resealed with the plastic screw cap that incorporates a second Teflon-faced rubber seal. If the vial contains benzene or toluene-based solutions the inner seal and ring must be removed and the vial resealed, if necessary, with the screw cap.

For small volumes the isotope may be supplied in a glass vial with a plastic insert. In this case the manufacturers recommend that the vial is centrifuged briefly before opening to ensure that the radioactive solution is at the base of the conical insert.

The glass vial containing the radioactive solution is generally contained within an outer plastic container (for ^{3}H or ^{14}C solutions, for example) that may incorporate a steel shield (for ^{125}I and ^{131}I) or a lead pot (for ^{32}P). The isotope should be stored within this outer container at all times and only removed from it if, or when, absolutely necessary.

Below is a suggested protocol for dispensing an aliquot of a strong β-emitter such as ^{32}P. You will require the following items described below:

- A spill tray lined with absorbent paper.
- A suitable bench radiation monitor.
- A Perspex body shield.
- Sterile syringe and needles.
- Paper tissues.
- A retort stand and clamps.
- A suitably labelled flask to receive the isotope solution, containing non-radioactive carrier, if appropriate. The label should record your name, the

date, the radionuclide, name of compound, the radioactive and molar concentration, and stock number

The receptacle should preferably be non-breakable with a screw cap; it should be placed within an outer container, such as a beaker, as a precaution against spillage and to act as a radiation shield.

Volumetric flasks are not ideal containers for radioisotope solutions; they are too easy to knock over and awkward for pipettes.

- A can opener of the butterfly type.
- Alcohol solution 70% by volume.
- Two pairs of large forceps.

Protocol 1. Procedure for unpacking and dispensing radioactive solutions

Prior to dispensing the radioisotope, study the data sheet describing your purchase. Do all necessary calculations regarding radioactive concentration and specific activity (see Chapter 2, Section 8.1) and prepare non-radioactive carrier solutions as necessary.

Wear a fully-fastened laboratory coat, your personal dosimeter, safety spectacles, and one or two pairs of disposable gloves. Monitor your gloves regularly throughout the procedure and avoid distractions such as the telephone. Always remember to: MAXIMIZE THE DISTANCE between yourself and the source, MINIMIZE THE TIME of exposure and MAINTAIN SHIELDING at all times.

1. Lay out the materials described above in a spill tray. Secure the receptacle for the dispensed isotope with the retort stand so that it cannot be knocked over. Place the radiation monitor just outside the tray and switch it on. Place the safety screen between yourself and the tray.
2. In the spill tray, open the sealed can containing the radioisotope. Remove the contents, monitor the empty can and the opener. They should not be radioactive and can be put to one side.
3. Remove the top of the lead pot, but leave the vial in the lower section at all times. (If necessary, allow the sample to thaw in its lead pot and/or under an upturned beaker for shielding; be patient and do not be tempted to hold the vial in your hand, the dose would be significant—see Chapter 2.). Manipulate the vial and cap with forceps, not your fingers.
4. Swab the top of the Teflon seal with a tissue soaked in 70% alcohol. Monitor the tissue, it should not be contaminated, and discard. Insert a sterile syringe needle through the seal until it has just penetrated. This will act as an air bleed. Insert a second needle attached to a syringe through the seal and withdraw the plunger the required amount. Surround the syringe needle and teflon seal with a paper tissue and gently withdraw the syringe.

5. Dispense the solution into the prepared receptable. Replace the needle guard on the syringe. Remove the air-bleed and replace needle guard. Swab the top of the Teflon seal and replace the lead top.
6. Monitor your gloves. Discard all contaminated tissues, etc., in the appropriate waste bag. Rinse the syringe in water prior to safe disposal.

   Note: Contaminated syringe needles should never be discarded in waste bags. Designated disposal receptables for sharp objects are available commercially.
7. Monitor the spill tray and contents and dispose of all liquid and solid radioactive waste via the routes designated in your Local Rules. Return radioactive sources to locked storage area.

A2

Properties and radiation protection data of isotopes commonly used in the biological sciences

Radioisotopes vary considerably with respect to their properties and potential hazard. The Table 1 below summarizes some generally useful information about isotopes used in biological experiments.

Table 1.

Property/Radiation protection criteria	Isotope					
	^{3}H	^{14}C	^{35}S	^{32}P	^{125}I	^{131}I
$T_{1/2}$	12.3 years	5730 years	87.4 days	14.3 days	60.0 days	8.04 days
Mode of decay	β	β	β	β	γ(EC)	$\beta + \gamma$
β-energy (MeV) E_{max}	0.019	0.156	0.167	1.709	Auger electrons	0.806
β-energy (MeV) E_{mean}	0.006	0.049	0.049	0.69	0.035	0.108
Monitor	swabs counted by liquid scintillation	β-counter	β-counter	β-counter	γ probe	β-counter or γ probe

Biological half-life (in organic form)	12 days[a]	12 days	44 days	14 days	42 days	8 days
Critical organ	whole body	whole body/fat	whole body/testis	bone	thyroid	thyroid
Maximum range in air	6 mm	24 cm	30 cm	7.2 m	>10 m	>10 m
Shielding required	none	Perspex 1 cm[b]	Perspex 1 cm[b]	Perspex 1 cm	lead 0.25 mm	lead 13 mm
Annual limit of intake (ALI), MBq[c] oral	3000[a]	90	80	10	1	1
γ dose rate at 1 m ($\mu Gy h^{-1} GBq^{-1}$)	—	—	—	—	34	51
Quantity for notification of Health and Safety Executive (Bq)[d]	5×10^6	5×10^5	5×10^6	5×10^5	5×10^4	5×10^4
Designation of controlled areas[d]						
Air contamination (Bq m^{-3})	2×10^5	1×10^4	9×10^4	2×10^3	3×10^2	2×10^2
Surface contamination (Bq cm^{-2})	4×10^6	2×10^4	2×10^4	2×10^3	1×10^2	1×10^2
Total activity (Bq)	3×10^{10}	9×10^8	4×10^9	1×10^8	1×10^7	1×10^7
Special considerations	Monitoring difficulties lead to high potential internal hazard. High level of cleanliness necessary.	—	—	Potential high source of external radiation. Lead shielding for quantities of ⩾370 MBq (10 mCi) Finger dosimeter recommended.	Iodine sublimes, work in fume hood. Spills should be treated with alkaline sodium thiosulphate solution prior to decontamination. Many iodine compounds penetrate rubber gloves—wear two pairs.	

[a] Figure given is for ^3H_2O. For compounds such as ^{3}H-thymidine biological half-life is 140 days and the ALI should be reduced to 60 MBq.
[b] Thin Perspex is adequate to reduce dose but has poor mechanical properties.
[c] Figures based on a whole body dose equivalent limit of 50 mSv y^{-1}. ALIs may be reduced to 30% of the figures shown above if a limit of 15 mSv y^{-1} is adopted (see Chapter 2, Section 4).
[d] Figures taken from *The Ionizing Radiation Regulations* 1985, for supervised areas divide by three. Figures may be reduced to 30% as note *c*.

A3

Types of monitor

1. Personal dosimeters

1.1 Film badges and thermoluminescent dosimeters (TLDs)

Film badges consist of an X-ray film, resembling that used by a dentist, held in a plastic badge-holder pinned to the lab coat at waist or chest height. The holder incorporates a number of different thicknesses of shielding to facilitate assessment of the dose once the film is developed. Films give an estimate of dose up to 1 Gy.

TLDs contain a phosphor (such as LiF) that becomes and remains excited once exposed to radiation. Heat treatment results in light emission whose intensity is proportional to the radiation absorbed. TLDs can record a dose up to 20 Gy.

Both types of dosimeter are processed by an approved laboratory (e.g. NRPB in the UK) on a regular basis such as monthly or quarterly depending on the relative likelihood of exposure. TLDs are becoming increasingly popular but film badges remain superior in some respects as they provide more information about the type and energy of radiation absorbed. This is likely to be of little importance to a biology laboratory, however, as the worker may only be exposed to one or two relevant isotopes such as ^{32}P or ^{125}I. Neither type of dosimeter is appropriate for ^{3}H, and they are of little or no value for work with ^{14}C or ^{35}S, as the energy of radiation is so low.

It must be remembered that personal dosimeters only give information retrospectively. Their use means that a record is kept of doses received and is transferred with change in employment. If a dosimeter is lost, an appropriate proportion of the maximum permitted dose is assigned to that period.

1.2 Finger badges

These are small TLDs held in plastic holders worn under disposable gloves. They are processed by an approved laboratory as for above. They are particularly useful for work with ^{32}P where finger doses can be potentially high.

1.3 Instantaneous dosimeters

These are pocket ionization chambers or quartz fibre electroscopes. They generally have a metal case and are, therefore, only appropriate for gamma emitters. They are useful for measuring doses over short periods if radiation exposure is likely to be relatively high.

2. Laboratory monitoring

Most laboratory monitors are one of three basic types:

Geiger–Müller counters,
gas proportional counters, or
scintillation counters.

The first two types of instrument are based on gas ionization as the mode of detection. Briefly, they consist of a small gas chamber containing two electrodes, a voltage supply and a scaler. When ionizing radiation enters the chamber through the thin end window of the chamber, gas is ionized and a current pulse is recorded (c.p.m.). Proportional counters are so called because the degree of ionization in the chamber is proportional to the applied voltage. The relationship differs between alpha and beta particles as the former has greater ionizing power. Geiger–Müller counters operate at a higher voltage, but the signal is independent of voltage within the operating range and there is no discrimination between alpha and beta radiation. They operate at a higher gas pressure than proportional counters and thus have lower sensitivity because the end window is thicker. However, they are usually lighter and cheaper than proportional counters.

Both types of gas ionization counter are suitable for detecting ^{32}P and will pick up ^{14}C and ^{35}S if an instrument with a thin end window is used. Neither counter will detect ^{3}H. Monitoring of this low energy emitter is carried out by taking solvent or aqueous wipe tests (or 'swabs') followed by liquid scintillation counting of the effluent. Alternatively, a gas flow proportional counter can be used, but they are bulky and not particularly reliable or sensitive for ^{3}H monitoring.

Most of the so-called contamination monitors are proportional counters with a large end window on the lower face of the instrument. They give a reading in Bq cm^{-2}, counts sec^{-1}, or Sv h^{-1}. All radioisotope laboratories should have at least one of these but the cheaper Geiger counters should also be available. Their lower cost means that more can be made available and they generally have a probe on a flexible cable for convenience.

Laboratories using gamma-emitters require a hand-held scintillation counter. These have a probe that is either integral or connected by a flexible cable. Depending on the instrument, readings are given in counts sec^{-1} or Sv h^{-1}.

Contamination monitors must be calibrated annually by an approved laboratory (e.g. NRPB in the UK).

A4

Curie–becquerel conversion chart

μCi/mCi/Ci	kBq/MBq/GBq	μCi/mCi/Ci	MBq/GBq/TBq
0.1	3.7	30	1.11
0.2	7.4	40	1.48
0.25	9.25	50	1.85
0.3	11.1	60	2.22
0.4	14.8	70	2.59
0.5	18.5	80	2.96
1.0	37	90	3.33
2.0	74	100	3.7
2.5	92.5	125	4.625
3.0	111	150	5.55
4.0	148	200	7.4
5.0	185	250	9.25
6.0	222	300	11.1
7.0	259	400	14.8
8.0	296	500	18.5
9.0	333	600	22.2
10.0	370	700	25.9
12.0	444	750	27.75
15.0	555	800	29.6
20.0	740	900	33.3
25.0	925	1000	37.0
		3000	111.0
		5000	185

Examples:
1 μCi = 37 kBq
50 μCi = 1.85 MBq
10 mCi = 370 MBq
3000Ci = 111 TBq

A5

Calculation of amount of radioactivity remaining (R_R) after a given half-life (H_L) has passed

H_L	R_R	H_L	R_R
0.00	1.00	1.55	0.342
0.02	0.986	1.60	0.330
0.04	0.973	1.65	0.319
0.06	0.959	1.70	0.308
0.08	0.946	1.75	0.297
0.10	0.933	1.80	0.287
0.12	0.920	1.85	0.277
0.14	0.903	1.90	0.268
0.16	0.895	1.95	0.259
0.18	0.883	2.00	0.250
0.20	0.871	2.10	0.233
0.25	0.851	2.20	0.218
0.30	0.812	2.30	0.203
0.35	0.785	2.40	0.189
0.40	0.758	2.50	0.177
0.45	0.732	2.60	0.165
0.50	0.707	2.70	0.154
0.55	0.683	2.80	0.144
0.60	0.660	2.90	0.134
0.65	0.638	3.00	0.125
0.70	0.616	3.10	0.117
0.75	0.595	3.20	0.109
0.80	0.574	3.30	0.102
0.85	0.555	3.40	0.095
0.90	0.535	3.50	0.088
0.95	0.518	3.60	0.083
1.00	0.500	3.70	0.077
1.05	0.483	3.80	0.072
1.10	0.467	3.90	0.067
1.15	0.451	4.00	0.063
1.20	0.435	4.10	0.058
1.25	0.421	4.20	0.054
1.30	0.406	4.30	0.051
1.35	0.393	4.40	0.047
1.40	0.379	4.50	0.044
1.45	0.367	4.60	0.041
1.50	0.354	4.70	0.039
		4.80	0.036
		4.90	0.034
		5.00	0.031

A6

Units commonly used to describe radioactivity

Unit	Abbreviation	Definition
Counts per minute or second	c.p.m. c.p.s.	The recorded rate of decay
Disintegration per minute or second	d.p.m. d.p.s.	The actual rate of decay
Curie	Ci	The number of d.p.m. equivalent to 1 g of radium (3.7×10^{10} d.p.s.)
Millicurie	mCi	$\text{Ci} \times 10^{-3}$ or 2.22×10^{9} d.p.m.
Microcurie	μCi	$\text{Ci} \times 10^{-6}$ or 2.22×10^{6} d.p.m.
Becquerel (SI unit)	Bq	1 d.p.s.
Gigabecquerel (SI unit)	GBq	10^{9} Bq or 27.027 mCi
Megabecquerel (SI unit)	MBq	10^{6} Bq or 27.027 μCi
Electron volt	eV	The energy attained by an electron accelerated through a potential difference of 1 volt. Equivalent to 1.6×10^{-19} joules.
Roentgen	R	The amount of radiation which produces 1.61×10^{15} ion pairs per kg of air (2.58×10^{-4} coulombs kg^{-1})
Rad	rad	That dose which gives an energy absorption of 0.01 joule kg^{-1} (J kg^{-1})
Gray (SI unit)	Gy	That dose which gives an energy absorption of 1 joule per kilogram. Thus, 1 Gy = 100 rad.
Rem	rem	That amount of radiation which gives a dose in man equivalent to 1 rad of X-rays.
Sievert (SI unit)	Sv	That amount of radiation which gives a dose in man equivalent to 1 gray of X-rays. Thus, 1 Sv = 100 rem.

Some useful addresses

Note: Many of the companies given have offices in more than one country, these addresses have not all been included but can be obtained from the offices given below.

1. Suppliers of radiochemicals

Amersham International plc, Lincoln Place, Green End, Aylesbury, Bucks HP20 2TP, UK
Tel: (0296) 395222 for technical enquiries; (0800) 515313 for ordering

Amersham Corporation, 2636 S. Clearbrook Drive, Arlington Heights, IL 60005, USA
Tel: (3123) 364 7100

C.E.A., CEN-Saclay, Gif-sur-Yvette, France, F-91191
Tel: (1) 69082860

ICN Biomedicals Inc., 3300 Hyland Avenue, Costa Mesa, CA 92626, USA
Tel: (714) 545-0113

ICN Biomedicals Ltd, Free Press House, Castle Street, High Wycombe, Bucks HP13 6RN, UK
Tel: (0494) 443826

ICN Radiochemicals, 2727 Campus Drive, Irvine, CA 92717, USA
Tel: (800) 854-0530

New England Nuclear, 549 Albany Street, Boston, MA 02118, USA
Tel: (800) 225 1572

New England Nuclear, Du Pont (UK) Ltd, Wedgewood Way, Stevenage, Herts SG1 4QN, UK
Tel: (0438) 734080

2. Suppliers of scintillation counters

Beckman Instruments Ltd, Scientific Instruments Division, Irvine, California, USA

Beckman Instruments Ltd, Cressex Industrial Estate, Turnpike Road, High Wycombe HP12 3NR, UK

ICN Biomedicals Inc., 3300 Hyland Avenue, Costa Mesa, CA 92626, USA

Kontron Ltd., Bernerstrasse–Sued 169, Zurich, Switzerland

Kontron Ltd, PO Box 88, St Albans, Herts AL1 5JG, UK

LKB–Producter AB, Fack Bromma, Sweden S-16125

LKB Instruments Inc., 12221 Parklawn Drive, Rockville, Maryland 20852, USA

LKB Instruments Ltd, LKB House, 232 Addington Road, Selsdon, S. Croydon, Surrey CR2 9PX, UK

Packard Instrument Company Inc., 2200 Warrenville Road, Downers Grove, Illinois 60515, USA

Packard Instruments Ltd, Caversham Bridge House, 13–17 Church Road, Caversham, Berks RG4 7AA, UK

Phillips Electronic Instruments Inc., 85 McKee Drive, Mahah, NJ 07430, USA

Pye-Unicam Ltd, York Street, Cambridge CB1 2PX, UK

Searle Analytical Inc., 2000 Nuclear Drive, Des Plaines, IL 60018, USA

3. Other instruments

Atomic Products Corporation, PO Box 657, Centre Moriches, NY 11934, USA

Baird Corporation Inc., 125 Middlesex Turnpike, Bedford, MA 01730, USA

Baird-Atomic Ltd, Warner Drive, Rayne Road, Braintree, Essex, UK

Berthold Instruments, D-7547 Wildbad 1, Calmbacher Strasse 22, PO Box 160, Federal Republic of Germany

Berthold Analytical Instruments Inc., 28 Charron Avenue, Nashua, NH 03063, USA

Berthold Instruments (UK) Ltd, 35 High Street, Sandridge, St. Albans, Herts AL4 9DD, UK

Canberra Industries, 45 Gracey Avenue, Meriden, CT 06450, USA

Canberra Instruments Ltd, 223 Kings Road, Reading, Berks RG1 4LS, UK

ESI Nuclear, Klempfern House, Holmesdale Road, Reigate, Surrey RH2 0BQ, UK

Health Physics Instruments Inc., 124 Santa Felicia Drive, Goleta, CA 93017, USA

ICN Biomedicals Inc., 3300 Hyland Avenue, Costa Mesa, CA 92626, USA

Mini Instruments Ltd, 8 Station Industrial Estate, Burnham-on-Crouch, Essex LM0 8RN, UK

Mullard Ltd, Mullard House, Torrington Place, London WC1E 7HO, UK

Nuclear Data Inc., Golf and Meacham Roads, Schaumberg, IL 60196, USA

Nuclear Data (UK) Ltd, Rose Industrial Estate, Cores End Road, Bourne End, Bucks, UK

Nuclear Enterprises Ltd, Sighthill, Edinburgh EH11 4EY, UK

Nuclear Equipment Chemical Corp., 165 Marine Street, Farmingdale, NY 11735, USA

Nuclear Measurements Corp., 2460 North Arlington Avenue, Indianapolis, Ind 46218, USA

Panax Nucleonics Ltd, Trowers Way, Holmethorpe Industrial Estate, Redhill, Surrey RH1 2PP, UK

Radiation Monitoring Devices Ltd, 44 Hunt Street, Watertown, MA 02172, USA

4. Some licensing and advisory authorities

4.1 UK registration

England

H.M. Inspectorate of Pollution, Romney House, 43 Marsham Street, London SW1P 3PY

Scotland

H.M. Industrial Pollution Inspectorate, Pentland House, 47 Robb's Loan, Edinburgh EH14 1TY

Wales

Welsh Office, Environment Protection Unit, New Crown Building, Cathays Park, Cardiff CF1 3NQ

Northern Ireland

Department of the Environment for N. Ireland, Environment Protection Branch, Stormont, Belfast BT4 3SS, Northern Ireland

For administration of radioactive products to humans

The Secretary of the Administration of Radioactive Substances Committee, DHSS (HS2A3), Hannibal House, Elephant and Castle, London SE1 6TE

4.2 USA

Nuclear Regulatory Commission, Division of Licensing and Regulations, Washington, DC 20545

Environmental Protection Agency, ANR 460 41 M. St. SW, Washington, DC 20460

4.3 Other relevant addresses

Health and Safety Executive, 25 Chapel Street, London NW1 5DT

International Atomic Energy Agency, PO BOX 100, Wagrammerstrasse 5 A-1400, Vienna, Austria

International Commission on Radiological Protection, Clifton Avenue, Sutton, Surrey SM2 5PU, UK

National Radiological Protection Board, Chilton, Didcot, Oxon OX11 0RQ, UK

World Health Organization, CH-1211, Geneva 27, Switzerland

5. Other miscellaneous addresses

British Nuclear Fuels plc [for Drigg disposals], UK Reprocessing Business Centre, Risley, Warrington, Cheshire WA3 6AJ, UK

Eastman Kodak Co., 343 State Street, Rochester, NY 14650, USA

Harwell Analytical Services, Harwell, Didcot, Oxon OX11 0RA, UK

Kodak Ltd, PO Box 66, Hemel Hempstead, Herts HP1 1JU, UK

National Disposal Service, Industrial Chemistry Group, Building 175, UKAEA, Harwell, Didcot, Oxon, UK

D.A. Pitman Ltd (for TLDs), Jessamy Road, Weybridge, Surrey, UK

Index